New Polymeric Products

New Polymeric Products

Fundamentals, Forming Methods and Applications

YONG LIU

College of Materials Science and Engineering, Beijing University of Chemical Technology, Beijing, China

JING GE

College of Materials Science and Engineering, Beijing University of Chemical Technology, Beijing, China

CE WANG

Alan G. MacDiarmid Institute, College of Chemistry, Jilin University, Jilin, China

PING HU

Department of Chemical Engineering, Tsinghua University, Beijing, China

ELSEVIER CIP

Elsevier
Radarweg 29, PO Box 211, 1000 AE Amsterdam, Netherlands
The Boulevard, Langford Lane, Kidlington, Oxford OX5 1GB, United Kingdom
50 Hampshire Street, 5th Floor, Cambridge, MA 02139, United States

ISBN: 978-0-443-19407-8

For Information on all Elsevier publications
visit our website at https://www.elsevier.com/books-and-journals

Publisher: Matthew Deans
Acquisitions Editor: Glyn Jones
Editorial Project Manager: Naomi Robertson
Production Project Manager: Sujithkumar Chandran
Cover Designer: Greg Harris

Typeset by MPS Limited, Chennai, India

Contents

Preface

Polymer science and engineering is a rapidly evolving and multidisciplinary field critical in developing advanced materials with unique properties and processing techniques. The design, synthesis, and processing of polymers involve challenges related to material properties, processing methods, characterization, and application-specific requirements. Despite these challenges, researchers and engineers have made tremendous progress in this field, quickly developing new polymer products and their processing methods.

This book aims to provide an overview of the latest developments in new polymer products and their processing methods. It covers a broad range of topics, such as the basics of polymer science, the design and synthesis of new polymers, processing techniques, characterization methods, and applications. The book focuses on cutting-edge research and industrial applications of new polymer processing, case studies, and real-world examples.

The book is divided into six chapters that cover different aspects of polymer processing. Chapter 1 briefly introduces the fundamentals of polymer science, including their development, characteristics, and molding methods. It covers the history and application prospects of polymer materials and introduces new and functional materials, such as polymer separation membranes, magnetic and operatic polymers. The chapter also explores the molding and processing methods used for different types of rubber, plastics, and fibers. Chapter 2 explores the application of polymer materials in the field of rubber, which is divided into three parts. The first part covers the basic chemical structure and background of traditional rubber varieties and the latest research progress. The second part focuses on the development status and prospects of a new type of bio-based rubber. The third part discusses the production and processing equipment of tire rubber products. Additionally, the chapter concludes with a discussion of the future direction of tires, including the spherical tire.

Chapter 3 covers the application of new polymer materials in the field of the fuel cell. It is mainly divided into fuel cell proton exchange membrane material and fuel cell fiber catalyst layers. The content includes the working principle of the above energy devices, the types of applied materials and preparation methods, the latest research progress, and future development prospects. Chapter 4 focuses on several high-performance fiber materials, including carbon fibers, aramid fibers, ultrahigh molecular weight polyethylene fibers, and nanofibers. The main properties, advantages, and preparation techniques of these high-performance fiber materials, as well as their applications and prospects in the aerospace, industry, and daily life, are introduced. In particular, the detailed preparation process and various performance characterizations of a high-performance shock-resistant vesicle are presented, and the excellent performance characteristics of high-performance fibers are expressed.

Chapter 5 focuses on the most widely used lens resin materials, including CR-39, polymethylmethacrylate, polycarbonate, and contact lens material. These lens parameters and characteristics, such as light transmittance, wear resistance, strength, and ultraviolet blocking, are mainly described. Several new contact lens materials are also introduced, their performance and processing characteristics are compared, and the future development trend of contact lenses is proposed. Chapter 6 presents the application of polymer materials in biomedicine. It is divided into six parts: artificial blood vessels, antituberculosis drug carrying fibers, wound dressings containing dragon blood, fiber plasters, controlled-release fibers, and antibacterial masks. This chapter mainly introduces the characteristics of polymer materials used in varied fields and the characteristics of biomedical polymers, such as biocompatibility, mechanical strength, application principle, preparation, and performance characterization of polymer materials. The application prospects and the latest research results are introduced. The last section introduces the antibacterial mask-copper oxide mask, which can effectively prevent novel coronavirus infection.

Overall, this book is intended for researchers, engineers, students, and professionals interested in polymer processing and its applications. It is written by experienced experts in the field who have academic and industrial expertise in the subject. The book aims to provide a comprehensive overview of the latest developments in polymer science, design, processing, and applications. We hope it will serve as a valuable resource for the advanced study of polymer processing. We want to thank all the authors, especially for Chapter 1, Miss Han Guo; Chapter 2, Mr. Tiancheng Ge; Chapter 3, Miss Jingyi Sun; Chapters 4 and 5, Mr. Xin Qu; Chapter 6, Miss Jing Ge. All the previous students who contributed to this book are gratefully thanked. The editors and staff at the publisher who helped bring this project to fruition are sincerely appreciated.

Yong Liu

CHAPTER 1

Introduction

Contents

1.1 Brief history of polymer materials

Materials are crucial for scientific and industrial development, and the emergence of new materials can bring revolutionary advances to technology and society [1,2]. Despite its relatively short history, polymer materials science has become an essential field of new materials alongside metals and inorganic nonmetals. Polymer materials have had a significant impact on human civilization and are now one of the three pillars of contemporary science and technology.

1.1.1 Status and function of polymer materials

In contemporary society, the role of materials in human life, survival, and development is undeniable. One essential material in modern times is polymer, which is increasingly used in various industrial fields and is becoming comparable with metals in its widespread use. To ensure the continued development of polymer materials in the future, it is necessary to analyze and study their current usage status. This study can play a crucial role in promoting the overall development of polymer materials [3–5].

New Polymeric Products
DOI: https://doi.org/10.1016/B978-0-443-19407-8.00001-4

Polymer materials possess unique and excellent properties that make them indispensable basic materials in various industrial fields, such as machinery, chemical industry, transportation, aerospace, and civil life. Moreover, polymer materials significantly influence the development of related fields. For instance, the use of fluorinated polymer materials with good corrosion resistance has effectively solved the storage problem of enriched uranium for the industry. Another example is the use of high-temperature and corrosion-resistant phenolic resin materials in the production of spacecraft, artificial satellites, intercontinental missiles, and other cutting-edge international products. Presently, various high-tech fields increasingly depend on polymer materials. However, with the continuous harsh application conditions, the properties of polymer materials must meet higher requirements [6].

Polymer materials have a widespread natural existence, and before the emergence of humans, various animals and plants in nature were primarily composed of polymers such as protein, nucleic acid, and polysaccharides (starch and cellulose) [7]. Since the emergence of humans, they have been utilizing these natural polymers, such as thatch, wood, and bamboo for housing construction, paint and natural rubber for vehicles, and cotton, hemp, silk, wool, leather, and horn, among other materials, with a long history of human use. Polymer materials play a special role in human survival and development.

1.1.2 Development of polymer materials

The utilization of natural polymer materials by humans has been a long-standing practice. However, due to the limitations of material science and technology, the polymer industry and science began to develop relatively late [8—10]. The chemical modification of natural polymers did not commence until the mid-19th century with the advent of rubber vulcanization, nitrocellulose, and other products. Synthetic polymer products did not emerge until the 20th century [11]. The modern concept of polymers was established and recognized in the 1930s. Subsequently, with the rapid and abnormal development of the petrochemical industry, the synthetic polymer industry grew rapidly, and the application of polymer materials became increasingly widespread, particularly since the 1950s. By the early 1980s, the annual production of synthetic polymer materials, such as plastics, synthetic fibers, and synthetic rubber, had exceeded 100 million tons worldwide, surpassing the total volume of all metals. Today, polymer materials are indispensable, ranging from everyday necessities to cutting-edge high-tech products, and polymer science is the most rapidly developing field of materials science [12].

To provide readers with a comprehensive understanding of the development history of polymer materials, we have summarized the key developmental events of polymer materials in Table 1.1.

Prior to 1949, China had no synthetic fiber or synthetic rubber industry, with only a few small plastic factories processing Bakelite, producing a cumulative output of less than 400 t. However, after the founding of the People's Republic of China, the country began to research polymer science gradually, established national academic organizations [13,14], published academic journals, formulated development plans, conducted professional education and academic exchanges both domestically and abroad, introduced large- and medium-sized technical equipment, and established the polymer material industry.

In the 1950s, research work in China was innovative, focusing on synthesis based on domestic resources and establishing test and characterization modes. This process led to the cultivation of a large number of technicians for production and research, which laid a foundation for in-depth research. In the 1960s, many special plastics, including fluorosilicone polymer, heat-resistant polymer, polycarbonate, polyformaldehyde, polyacrylamide, polypropylene, and other general engineering plastics, were developed to meet the needs of new technology. Rubber chemistry, physics, and other major varieties also rapidly developed during this time [15—17]. In recent years, research has been deepened, focusing on the synthesis method and mechanism of general polymers and the synthesis and application of functional polymers. By exploring the relationship between structure, performance, and processing with advanced technology and testing means, new varieties and theories with Chinese characteristics have been formed [18,19].

1.1.3 Development prospect of polymer materials

Polymer materials have established their versatility and potential in numerous fields and are expected to continue evolving in the future. These materials possess unique application properties, making them ideal for use in everyday life, military, medicine, construction, among others [13,20—22]. Polymer materials are not only products of technological advancements but also novel materials for polymer synthesis that hold

Table 1.1 Development history of polymer materials.

Age and development characteristics	Polymer industry		Polymer science
Processing and utilization of natural polymers before the 19th century	Food protein, starch, cotton, hemp, wood, bamboo, paper, paint, natural rubber, and other natural polymers		In 1833, Berzelius proposed the term "Polymer" (including aggregates linked with covalent, noncovalent bonds)
Chemical modification of natural polymers in the middle of the 20th century	Vulcanization of natural rubber	1838 year	In 1870, it began to realize that cellulose, starch, and proteins were macromolecules In 1892, the structure of isoprene was determined
	Nitrocellulose	1845 year	
	Nitrocellulose plastics	1868 year	
	Rayon factory	1889 year	
The preparatory period for the establishment of Polymer Materials Science in the early 20th century	Phenolic resin Sodium butadiene rubber	1907 year 1911 year	In 1902, it was recognized that proteins are polypeptide structures composed of amino acid residues In 1904, it was confirmed that cellulose and starch were composed of glucose residues In 1907, the concept of molecular colloids was presented In 1920, the study of cellulose crystallization began In 1920, the modern polymer concept of covalent bond-linked macromolecules was proposed
	Ester acid fibers and plastics	1914 year	
	Polyvinyl polyester, alkyd resin, PVA, PMMA, UF	1929 year	

(Continued)

Table 1.1 (Continued)

Age and development characteristics	Polymer industry		Polymer science
The founding period of polymer materials science from the 1930s to the 1940s	Plastic		In 1930, after the cellulose relative molecular mass determination study, the modern polymer concept was recognized In 1932, "The Polymer Organic Compound" was published In 1929–40, the polycondensation reaction theory was put forward In 1932–38, the rubber elasticity theory was verified In 1935–48, the chain polymerization reaction and copolymerization theory were proved From 1942 to 1949, the polymer solution theory and various solution methods for measuring the relative molecular mass were established In 1945, the primary structure of insulin was established Era 40, to determine the theory of emulsion polymerization
	PVC, PS, PCTFE, PVB, LDPE, PVDC	1931–40	
	UP, EP, PTFE, ABS, HDPE	1941–50	
	Fiber		
	PVC, PA66, PU	1931–39	
	PA6, pet, vinyl on, pan	1941–50	
	Rubber		
	Neoprene	1931year	
	Butyl rubber	1940–42	
	Butadiene styrene rubber	1940–42	

(*Continued*)

Table 1.1 (Continued)

Age and development characteristics	Polymer industry		Polymer science
The 1950s witnessed the establishment of the modern polymer industry and the great development of polymer synthesis	HDPE	1953–55	In 1953, the Ziegler–Natta catalyst and coordination anion polymerization were funded In the 1950s, the development of anionic active polymerization, cation ionic polymerization, and crystalline polymer research was open In 1957, the acquisition of a polyethylene single crystal was realized In 1958, the mesosphere prion structure was determined In 1951, the protein A helix structure was proposed In 1953, H. Staudinger received the Nobel Prize in Chemistry
	PP	1955–57	
	POM	1956 year	
	PC	1957 year	
	CIS polybutadiene rubber	1959 year	
	Many new products are emerging		
In the 1960s, polymer physics developed greatly	Emergence and development of engineering plastics		From 1960 to 1969, the further development of crystalline polymer, polymer viscoelasticity, rheology research, and the application and development of various modern research methods in the study of polymer structure, such as NMR, GPC, IR, thermal spectroscopy, force spectroscopy, and electron microscope, and the piezoelectric study of PVDF were realized In 1963, Ziegler and Natta won the Nobel Prize in Chemistry
	PI	1962 year	
	PPO	1964 year	
	Polysulfone	1965 year	
	PBT	1970 year	
	Development of high temperature resistant polymers		
	PBI	1961 year	
	Nomex fiber	1967–1972	
	Polyacrylamide		
	Isoprene rubber	1962 year	
	Ethylene propylene rubber	1961 year	
	SBS	50 years	

(*Continued*)

Table 1.1 (Continued)

Age and development characteristics	Polymer industry	Polymer science
The great development of Polymer Engineering Science in the 1970s (high efficiency, automation, and large-scale production)	Polymer blends (ABS, MBS, hips, NORYL, etc.) Polymer composites (such as large-scale production of glass fiber-reinforced resin matrix composites)	In the 1970s, PE, PP highly efficient catalyst was developed In 1971, a polyacetylene thin film was developed In 1972, neutron small-angle scattering was applied In 1973, Kevlar fibers and polymer blending theory were developed In 1974, P.J. Flory received the Nobel Prize in Chemistry In 1977, the metal conductivity was doped with polyacetylene
	3,010,000 ton PE and PP plants, bulk polymerization of PVC, use of large polymerization reaction equipment and new processes, the emergence of large processing equipment, and establishment of new synthesis methods	
In the 1980s, research on high-performance materials developed fine polymers, functional polymers, and biomedical polymers		The molecular design was proposed In 1983, the group transferred polymerization was developed

substantial significance for the future of the material industry. In the future, polymer materials are expected to develop toward high performance by reinforcing their temperature and corrosion-resistant characteristics, which will undoubtedly become the focus of research and development. In various industries, polymer materials play an important role and have a significant impact.

Environmental pollution is currently a major issue, causing disruptions to people's daily lives. Green polymer materials can effectively prevent further environmental pollution by utilizing modern science and technology to study their eco-friendly characteristics, significantly improving their application and promoting their multiple uses, thereby reducing the waste of resources [23–25]. The development of green polymer materials is essential for future research.

Polymer materials are extensively used in various fields [26], and their advantages and characteristics make them the focus of future development in the material industry, creating more value. In the medical field, polymer materials have a significant advantage over cermet materials, as they can be matched with natural teeth in appearance and enable esthetic repair [27,28]. The most commonly used dental polymer materials include denture-making materials with methyl methacrylate as the main component and composites with polymerizable resin as the matrix and inorganic filler or fiber as the reinforcing material. Zirconia is often used as a porcelain veneer for teeth, as shown in Fig. 1.1 [29].

Anticaries materials are dental products that help prevent tooth decay by blocking the pit and fissure spaces of teeth, blocking bacteria from entering or improving the acid resistance of enamel. These materials are categorized as resin-based pits, fissure sealers, and fluorine-containing materials according to their mechanism of action. Fluorine-containing materials contain natural resins, such as fluorocarbon paint, ethyl acetate coating material, and fluorinated foam [30]. Fluoride is an essential component in the prevention of dental caries as fluoride ions can inhibit the

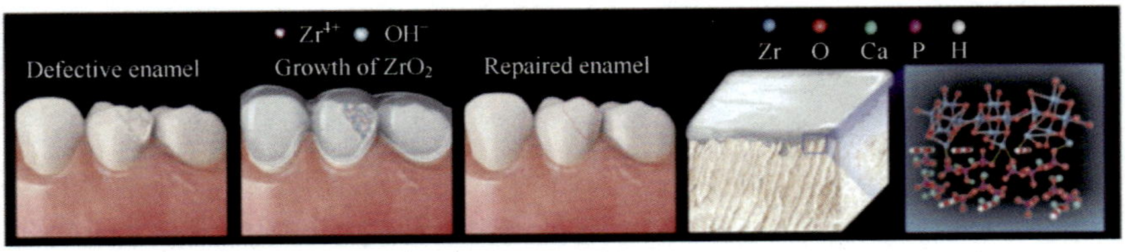

Figure 1.1 Schematic diagram of zirconia repairing enamel defect [29].

adhesion and aggregation of cariogenic bacteria, weaken the acidic environment of the tooth surface, and reduce the likelihood of demineralization on the tooth surface. Fluorapatite, formed by replacing hydroxyl ions with fluorine, is more stable than hydroxyapatite, making it more resistant to the erosion caused by acid-producing bacteria and thus effectively preventing dental caries.

1.2 New polymer materials

Polymer materials are composed of compound molecules with relatively large molecular masses and can be classified into natural, synthetic, and semisynthetic materials based on their origin. These materials include plastics, synthetic fibers, synthetic rubber, coatings, adhesives, and polymer matrix composites [31]. Despite their rapid development since the introduction of phenolic resin in 1907, conventional polymers have limitations in terms of mechanical strength, rigidity, and heat resistance, which restricts their use in modern engineering technology. Therefore the development and application of new polymer materials with high performance, such as conductive, biomedical, biodegradable, high temperature resistance, high strength, high modulus, high impact, and extreme conditions, are crucial to advance the field of polymer materials toward functionalization, intelligence, and refinement [32–34]. These advanced materials can address the challenges faced by conventional polymers and actively promote their development in new directions.

The demand for high-performance polymer materials has been increasing in various industries. For instance, the automotive industry requires polymer materials with excellent mechanical properties and heat resistance, and the aerospace industry demands materials that can withstand extreme temperatures and harsh conditions. The use of polymer materials in electronics and energy applications requires them to be conductive and have high dielectric properties. Moreover, biomedical and biotechnological applications require materials that are biocompatible, biodegradable and possess desirable physical and chemical properties. Therefore developing and applying high-performance polymer materials in these industries are a critical step toward improving product performance, reducing costs, and increasing sustainability. Furthermore, the use of recycled and sustainable polymer materials, such as bioplastics and biodegradable polymers, can mitigate environmental issues and promote circular economy principles.

1.2.1 Classification of polymer materials

Polymers, much like the letters of the English alphabet, can be combined in various ways to form a multitude of different structures with different properties and functions. In biomolecules, different monomers can be strung together by polymerase enzymes to form complex polymers with varying characteristics. The relative molecular mass of polymers can be much higher than that of low-molecular-weight compounds, with values ranging from 1.04 to 1.06 million. Due to their large size, polymers exhibit distinct physical, chemical, and mechanical properties that differentiate them from low-molecular-weight compounds.

At room temperature, most polymer compounds exist as nonvolatile solids or liquids. Depending on their structural forms, polymers can be classified as either crystalline or amorphous. Crystalline polymers have a regular molecular arrangement, while amorphous polymers have an irregular arrangement. It is worth noting that the same polymer compound can have both crystalline and amorphous structures. Synthetic resins, for instance, usually exhibit an amorphous structure [35–37].

Today, many polymer materials used globally are derived from natural gas, coal, and oil. The process of polymerization can be categorized as either addition or condensation polymerization. With the advancements in industrial technology, the production of polymer materials has grown tremendously over the years. Various plastics, synthetic rubber, synthetic fiber, and their related products are now widely used across various industries such as construction, transportation, machinery, and chemical industries [38,39]. The utilization of these polymer materials has extended beyond industrial settings and has become commonplace in many households.

According to the literature, polymers can be broadly classified into two categories: natural and synthetic. Natural polymers, in turn, can be categorized as inorganic and organic. Natural inorganic polymers include materials such as asbestos, graphite, diamond, and mica. On the other hand, natural organic polymers are produced by living organisms and comprise various structures that play vital roles in sustaining life. Examples of such structures include animal hair, tendon, skin, bone, claws, and plant cellulose. Additionally, natural organic polymers also include substances that store energy, such as liver sugar, starch, and proteins. Certain biological in vitro secretions, such as animal silk, spider silk, shellac, natural rubber, and fat from plants, are also categorized as natural organic polymers.

In contrast, synthetic polymers are produced through specific chemical reactions and polymerization methods using small molecular raw materials with known structures and relative molecular masses. Examples of synthetic polymers include polyethylene, polyvinyl chloride, polyacrylonitrile, and polyamide (nylon).

1.2.1.1 By performance

In polymer materials, three major categories exist based on their properties: plastic, rubber, and fiber. Plastics are divided into two categories based on their hot-melt properties: thermoplastic and thermosetting plastics, such as phenolic resin, epoxy resin, and unsaturated polyester resin. Thermoplastics are linear polymers that can be softened and molded when heated and can be repeatedly plasticized. This property allows for the recycling and processing of defective and waste products into new products. In contrast, thermosetting plastics have a body structure and solidify once formed. They cannot be reheated and softened again or processed and formed repeatedly, resulting in no recycling value for defective or waste products. Nevertheless, plastics are valued for their mechanical strength, especially those with a body structure, which makes them ideal for use as structural materials.

Fiber is another important category of polymer materials that can be divided into natural and chemical fibers. Chemical fibers such as nylon and polyester are classified as either manufactured or synthetic. Artificial fibers are created through chemical processing and spinning of natural polymers, such as short cotton wool, bamboo, wood, and hair [40,41]. Synthetic fibers, on the other hand, are synthesized from raw materials with low relative molecular mass. These fibers exhibit good strength and flexural properties and can be used as textile materials due to their ability to be drawn and formed.

Rubber is the third category of polymer materials and is characterized by its high elastic properties. Rubber can be classified as either natural rubber or synthetic rubber.

1.2.1.2 By application

In the field of polymer materials, a categorization can be made based on their properties and applications. This categorization includes general polymers, engineering polymers, functional polymers, biomimetic polymers, medical polymers, polymer drugs, polymer reagents, polymer catalysts, and biopolymers. General polymers are widely used due to their versatile applications and include the "Terrene" (polyethylene, polypropylene, polyvinyl chloride, and polystyrene) in plastics, the "four fibers" (nylon, polyester, acrylic, and vinyl) in fibers, and the "four glue" (styrene butadiene rubber, cis-polybutadiene rubber, isoprene rubber, and ethylene propylene rubber) in rubber.

On the other hand, engineering plastics refer to polymer materials with special properties such as high temperature and radiation resistance. These include polyoxymethylene, polycarbonate, polyinkstone, polyimide, polyarylene ether, polyacrylamide, fluorine-containing polymer, and boron-containing polymer. These materials have been widely used as engineering materials due to their unique properties [42—44]. Functional polymers, including ion exchange resin, photosensitive polymer, polymer reagent, and polymer catalyst, possess specific functionalities that enable them to be used in a wide range of applications. Medical polymers and medicinal polymers have specific requirements in medicine and physiological health and can also be considered functional polymers.

1.2.1.3 By main chain structure

Polymer materials can be classified based on their main chain structures into four categories: carbon chain polymer, heterochain polymer, elemental organic polymer, and inorganic polymer. Carbon chain polymers have a main chain made up of carbon atoms. Heterochain polymers contain oxygen, nitrogen, sulfur, and other elements in addition to carbon atoms in their main chain, such as polyester, polyamide, and cellulose. It should be noted that heterochain polymers are susceptible to hydrolysis. Elemental organic polymers are composed of atoms of elements other than carbon, oxygen, nitrogen, and sulfur, such as silicon, oxygen, aluminum, titanium, and boron. However, they have organic groups as side chains, such as polysiloxane. Inorganic polymers have both main chain and side chain groups composed of inorganic elements or groups. Examples of natural inorganic polymers include mica and crystal, while synthetic inorganic polymers include glass.

1.2.2 Types of new polymer materials

Traditional polymers are known for their poor mechanical strength, rigidity, and heat resistance in the context of the growing science and technology. However, to pursue higher performance, diversified functionality, and reduced rejection response, scientists have been working toward producing new and excellent polymer materials since the 1990s [45,46].

One such promising category is nano polymer materials, which utilize the small size effect and interface effect of nanomaterials to increase the material surface area and enhance the adsorption of target particles. This is a difficult feat to achieve with traditional materials, highlighting the considerable development prospects of new types of polymer materials.

1.2.2.1 Polymer separation membrane

A polymer separation membrane is a type of membrane that is made from polymer or composite materials and is designed to separate fluid mixtures based on selective permeability principles [47,48]. These membranes act as spacer layers that, under the influence of pressure, concentration, or potential differences, enable different components of the fluid mixture to be enriched on either side of the membrane, thus achieving separation, purification, concentration, and recycling objectives [49–51]. The critical indicators of membrane performance include the amount of fluid passing through the membrane per unit of time, the ratio of the permeability coefficient of different substances, and the rejection rate of certain substances. Here we demonstrate a facile supramolecular strategy that enables the highly selective capture of anionic PtII complexes in a simple manner (Fig. 1.2) [52].

Due to its high separation efficiency, selectivity, and low energy consumption, membrane separation technology has excellent application prospects in various fields, such as sewage treatment, seawater desalination, and chemical separation. One significant challenge is to achieve high selectivity for platinum recovery from wastewater because platinum has some chemical properties similar to other precious metals such as palladium, gold, and silver. However, the two-dimensional supramolecular polymer prepared by Chen has unique noncovalent bond interactions that enable the positively charged platinum complex of tri polypyridine to capture the tetrachloro platinate anion selectively. At the same time, other metal ions penetrate through the membrane pores nonselectively [53–58].

Polymer separation membranes are prepared using various methods, including tape casting, harmful solvent gelation, direct polymerization, surface coating, and hollow fiber spinning. Different polymer materials have been used as separation membranes, including cellulose esters, polysulfide, polyphenylene ether, aromatic polyamide, polytetrafluoroethylene, polypropylene, polyacrylonitrile, polyvinyl alcohol, polybenzimidazole, and polyimide [59]. Polymer blends and graft copolymers are also increasingly used to prepare separation membranes, and the methods for preparing these membranes include tape casting, flawed solvent gel method, micropowder

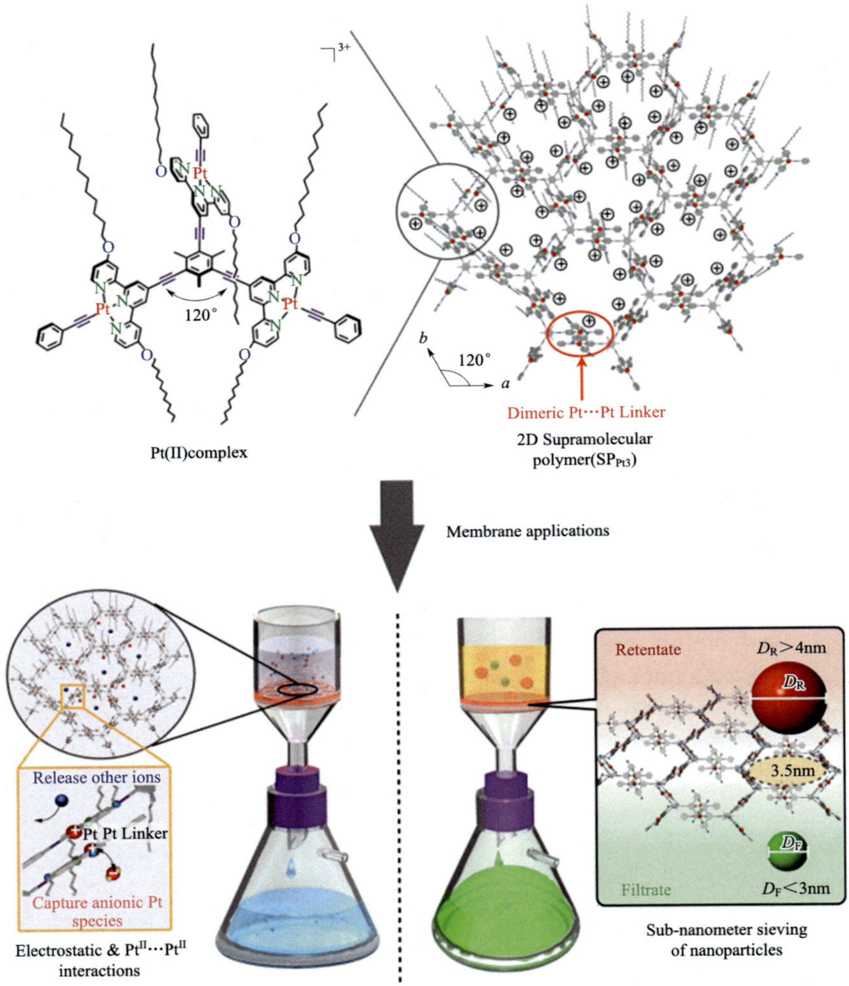

Figure 1.2 Application diagram of honeycomb two-dimensional supramolecular polymer separation membrane [52].

sintering method, direct polymerization method, surface coating method, controlled stretching method, radiation chemical etching method, and hollow fiber spinning method. Polymer separation membranes can separate liquid–solid, liquid–liquid, and gas mixtures and are composed of synthetic, semisynthetic, and natural polymers [60–63]. They are also called organic separation membranes to differentiate them from separation membranes made of inorganic substances.

The selective permeability properties of polymer separation membranes enable them to obstruct adjacent two phases' active or passive mass transfer, allowing fluid mixtures to be separated under the influence of pressure, concentration, or potential difference [64]. The separation process includes microporous filtration, reverse osmosis, gas permeation separation, evaporation, dialysis and electrodialysis, and liquid membrane. Membrane performance is determined by selectivity and permeability, with higher selectivity and lower permeability required for the substances that need to be intercepted. The performance is expressed as the ratio of the amount of fluid passing through the membrane to the material transmission coefficient per unit of time. These values must remain stable over an extended period before the membrane can be used in industry [65,66]. The shape of the polymer separation membrane can be divided into a hollow tube, hollow fiber, and flat plate. Different separation membranes can be obtained by changing the membrane preparation conditions.

Polymer separation membranes have wide-ranging applications in various fields, including seawater desalination, food and drug processing, wastewater treatment, and medical device manufacturing. In the nuclear fuel and metal refining industries, they are used for gas and hydrocarbon separation. In contrast, they are used for pure and ultrapure water preparation in environmental protection and sewage treatment. Polymer membranes are particularly useful for separating mixture systems that are difficult to separate, such as near-boiling-point mixtures, isotropic mixtures, isomer mixtures, and some thermosensitive substances. Reverse osmosis is a popular method of seawater desalination due to its low energy consumption and operational simplicity, which has replaced the multistage flash method. The use of polymer separation membranes in the food industry maintains the original flavor of food, while oxygen-enriched membranes are used in the medical industry for quick oxygen enrichment and ultrapure water preparation. Artificial kidneys and lungs assembled with separation membranes are used to purify the blood and treat renal insufficiency and as oxygenators in artificial heart and lung machines. In recent years, polymer separation membranes have undergone significant advancements in efficiency, selectivity, functional compounding, and diversification of forms [67,68].

1.2.2.2 Magnetic polymer materials

In early human society, naturally formed magnets were the primary source due to limited production and technical means. Despite technological advancements during the industrial revolution, magnets still suffered from defects such as brittleness, hardness, and poor processability [69,70]. To overcome these limitations, polymer magnetic materials were invented, which involved mixing the magnetic powder with plastic or rubber to prepare composite polymer magnetic materials that were lightweight, easy to process, and mold.

Polymer magnetic materials can be classified into structural magnetic polymer materials and composite polymer materials. Structural magnetic polymer materials include polymers such as oxyacetylene polymers, polyline, and nitrogen-containing groups instead of benzene derivatives. Meanwhile, composite magnetic polymer materials are magnetic bodies created by bonding polymer materials with various inorganic magnetic materials, filling and compounding, surface compounding, and laminated compounding [71]. Currently, ferrite polymer magnetic materials, which are composite magnetic polymer materials, are widely used. There are two main ways to design structural magnetic polymers: constructing a high-spin polymer with a significant magnetic moment based on the structure of a single-domain magnet and adjusting the low-spin polymer to obtain a high-performance magnetic polymer similar to ferrite with Fe and rutile structure [72,73].

Magnetic rubber, made from ferrite magnetic polymer, was initially used to manufacture door-sealing gaskets for refrigerated vehicles, refrigerators, and electric freezers. Its excellent machinability, easy forming, high-dimensional accuracy, good bremsstrahlung, lightweight, low cost, and easy mass production led to the development of magnetic rubber strips for fan motors and rotating tires, which are significant for small, lightweight, precise, and high-performance electromagnetic equipment. Magnetic polymers are versatile and can record acoustic, optical, and electrical information, making them widely used in many fields, such as electronics, electricity, instruments and meters, communication, and supplies [74]. For instance, they are used in the production of convergence components for color picture tubes, micro and exceptional motor magnetic steel, automotive instruments and meters, distributor gaskets, and magnetic rings of pneumatic components. As shown in Fig. 1.3, the comparison before and after the magnetization of the multipole magnetic rubber encoder can be seen.

Magnetic polymer microspheres possess the ability to respond to external magnetic fields rapidly. They can be modified with functional groups through copolymerization, making them highly versatile in many fields, including cell separation and analysis, radioimmunoassay, magnetic resonance imaging contrast agents, enzyme separation and immobilization,

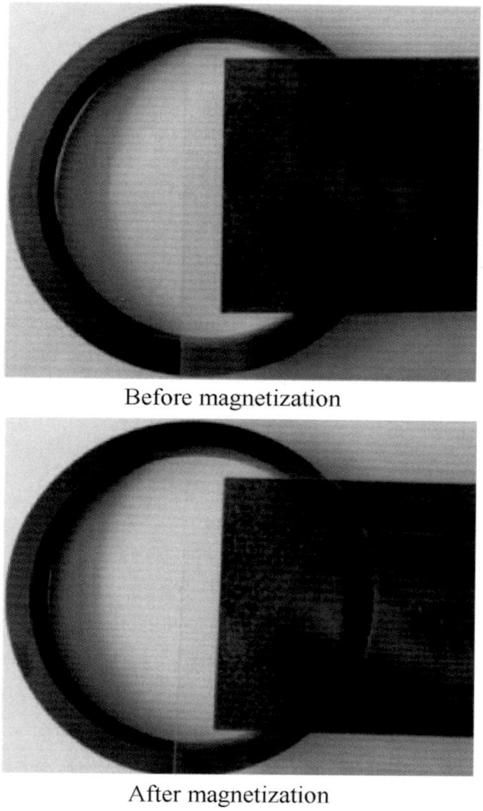

Figure 1.3 Comparison of multipole magnetic rubber encoder before and after magnetization [73].

DNA separation, targeted drug delivery, nucleic acid hybridization, clinical detection, and diagnosis [75]. For example, modified cellulose polysaccharide—polyphthalic anhydride copolymers can prepare magnetic cisplatin microspheres with a three-layer structure that possess good drug-controlled release characteristics and exhibit high application value in the treatment of malignant tumors [76]. Magnetic microspheres can also be used as drug carriers with functional groups on their surface to improve drug efficacy, reduce toxicity and side effects, and achieve targeted drug delivery via an external magnetic field.

In the field of stealth materials, the primary objective is to have strong absorption, while new materials must meet the requirements of being "thin, light, wide, and strong." Although most microwave absorbers are currently used to prevent radar detection, their high density makes them unsuitable for use in aircraft. Thus exploring new lightweight microwave absorbers with a wide frequency band and high absorptivity is crucial for the development of stealth materials [77]. According to electromagnetic wave theory, the frequency band can be broadened, and the absorptivity can be enhanced only by incorporating electrical and magnetic losses. Therefore the combination of magnetic polymer microspheres and conductive polymers can yield new microwave absorbers with promising applications in stealth technology and electromagnetic shielding.

1.2.2.3 Optical polymer materials

Optical functional polymer materials are a class of materials that possess the ability to absorb, transmit, convert, and store light. These materials include optical recording materials, photoconductive materials, optical processing materials, optical conversion system materials, optical plastics, photoresist materials, photosynthetic materials, and optical display materials. Linear optical materials, such as safety glass, prisms, and lenses, can be made using functional optical polymers. In contrast, nonlinear optical elements, such as plastic optical fibers, can be produced by utilizing the curve propagation characteristics of polymer materials. High-performance polycarbonate and plexiglass are the basic materials for advanced information storage elements [78–80].

Photosensitive resins are a type of polymer that undergo rapid photochemical reactions when exposed to light, leading to physical and chemical changes within or between molecules. The photosensitive group in the polymer completes the light absorption process, or a photosensitive compound added to the photosensitive material initiates the chemical reaction after absorbing light energy [81]. Photoresists are primarily used in the electronic industry, where they act as photoresists to promote high integration, miniaturization, and high reliability in the development of electronic devices [82]. Photoresists are produced by dissolving the photosensitive resin in an appropriate solvent and adding a sensitizer. In the manufacturing of semiconductor electronic devices or integrated circuits, photo etching technology involves the selective etching of the surface of silicon crystal or metal. A pattern is first exposed on the object's surface using a photoresist, and then it is developed with the appropriate solvent to obtain the pattern composed of the photoresist. The exposed part of the machined surface is then removed with a suitable corrosion solution to achieve the desired shape.

Photochromism or photochromic phenomenon occurs when the color of compounds containing photochromic groups changes under the influence of specific wavelengths of working light and returns to the original color due to other light and heat. Photochromic materials have various applications in information recording media and other areas. These materials offer several benefits, such as simple operation, high resolution, image cancellation after imaging, multiple reuses, fast response, low sensitivity, and sharp image retention time [83,84].

1.2.3 Functional polymer materials

Functional polymer materials have unique features, such as catalytic, conductive, photoresponsive, and biological activities. These materials can absorb, transform, or store materials, energy, and information. With their lightweight, diverse types, and remarkable specificity, functional polymer materials have wide-ranging applications in machinery, information technology, biomedicine, and other fields [85–87]. Current research focuses on a range of functional polymer materials, including optical functional polymers, liquid crystal polymers, electronic functional polymers, medical functional polymers, environmental degradation polymers, and adsorption and separation functional materials [11,88]. Compared with traditional materials, functional polymer materials have several advantages and distinctive properties. Beyond the typical properties of conventional polymer materials, functional polymer materials exhibit specialized functional groups. Moreover, functional polymer materials are advancing toward greater intelligence, with the development of materials such as self-repairing functional polymer materials and shape memory materials.

1.2.3.1 Reactive functional polymer materials

Reactive functional polymer materials refer to materials containing polymer reagents and catalysts, which link reactive active or catalytic centers to the polymer chain. These materials achieve macromolecular ionization of small molecular reagents or catalysts. Common polymer reagents include oxidation, reduction, alkylation, acylation, halogenation, and solid-phase synthesis reagents, depending on their chemical activity. On the other hand, polymeric catalysts include ion exchange, transition metal complex catalysts, phase transfer catalysts, and immobilized enzymes for acid–base catalysis [89,90]. Reactive polymer materials must possess high reactivity, selectivity, and specificity and are mainly used in chemical synthesis and reactions. Due to their unique properties, such as porosity, high selectivity, and chemical stability, polymeric reagents and catalysts expand the application range of chemical reagents and catalysts, offering potential in various fields.

1.2.3.2 Optical functional polymer materials

Optical functional polymer materials are a type of polymer that can absorb, store, transmit, and convert light energy. Photofunctional polymer materials, such as light stabilizers, photosensitive coatings, fluorescent agents, light conversion materials, photochromic materials, and photoconductive materials, have extensive applications in various areas, including optical fibers, solar energy, integrated circuits, and photocells [91,92]. For instance, Fig. 1.4 [93] shows that after irradiation with ultraviolet light, the diazole group in the photosensitive coating on a capillary undergoes a photochemical reaction to convert the ionic bond into a covalent bond. The resulting covalently connected coating inhibits protein adsorption on the capillary's inner surface, leading to improved electrophoretic separation performance, excellent repeatability, and stability compared with noncovalently coated or bare capillaries. This environmental and safe protein separation technology is a significant improvement [94].

Optical functional polymer materials also have a wide adjustable resistivity range, good processability, simple processing, corrosion resistance, and low cost, making them promising materials in various fields. Decades of research by researchers and industries have led to significant progress in

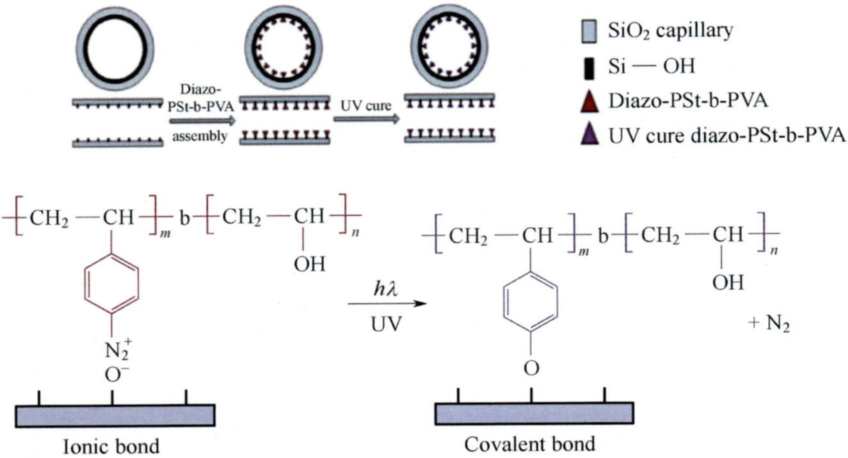

Figure 1.4 Preparation of photosensitive diazole polyvinyl alcohol/styrene covalent coating on capillary surface by ultraviolet irradiation [93].

this field. The range of functional polymer photoelectric materials is continuously expanding, which has significantly improved the performance of relevant devices. Currently, optical functional polymer materials can meet the application requirements of electronic paper, microprocessors, flexible displays, and chemical biosensors. This advancement is driving the rapid development of printable, flexible large-area electronics and wearable electronics in the new era [95–97].

1.2.3.3 Shape memory polymer

Shape memory materials are a class of materials that can recover their original shape when subjected to external stimuli after being deformed and fixed. Based on the nature of the stimulus required to trigger the shape memory effect, shape memory polymer materials can be classified into thermoinduced, electroinduced, photoinduced, and chemical-induced types, with thermoinduced shape memory polymer materials being the most widely used. Due to their lightweight, large deformation capabilities, and ease of processing, shape memory polymer materials have applications in various fields such as medicine, packaging, and construction. In one study, a polymer nanocomposite with a thermal response shape memory function was prepared using the melt blending method. As shown in Fig. 1.5, the material was composed of polycaprolactone (PCL), styrene-butadiene-

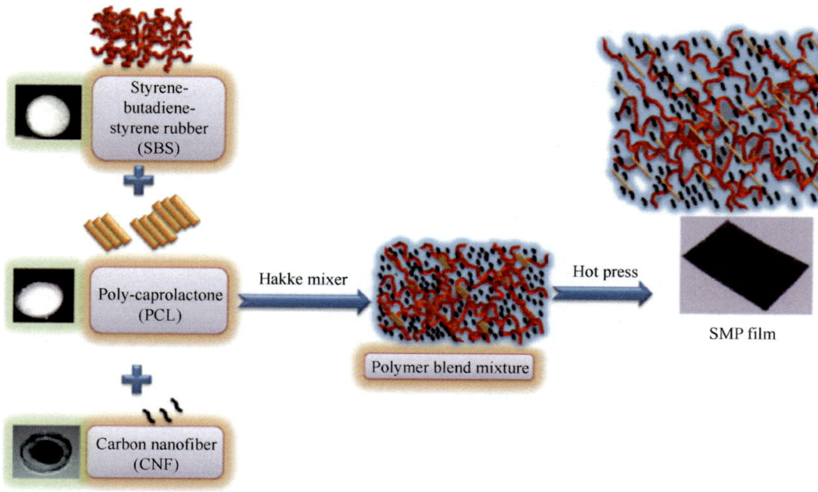

Figure 1.5 Schematic diagram of shape memory polymer nanocomposites synthesized by SBS, PCL, and CNF [98].

styrene block copolymer (SBS), and carbon nanofiber (CNF). PCL served as the switching polymer due to its low melting point, enabling thermal response shape memory performance. The combination of PCL and SBS provided better elasticity and flexibility, while CNF improved the thermal and electrical conductivity of the material. In another study, a shape memory material was synthesized using free radical polymerization from lauric acid, oleic acid, glycidyl methacrylate, and styrene. This highly translucent material was insoluble in common organic solvents and water and had a moderate cross-linking density. ESM copolymer also showed excellent shape memory ability and biodegradability [98].

1.2.3.4 Polymer hydrogel

Polymer hydrogel is a highly absorbent, hydrophilic material that forms a three-dimensional network structure with excellent biocompatibility and mechanical properties. It can respond to external stimuli, such as light, electricity, temperature, pH, and magnetic fields, which makes it highly suitable for various applications, including ion transport membranes, sensors, controlled release switches, drug carriers, drug delivery, cell packaging, and tissue engineering [99–104]. Hydrogels have been identified as potential artificial substitutes for biological tissues due to their three-dimensional cross-linked polymer networks and high water content. This endows them with good hydrophobicity and biocompatibility. These characteristics make them highly applicable in various fields, such as sensing, driving, drug and protein release, wound dressing, waste treatment, absorbents, and tissue engineering. Despite its broad potential uses, the practical application of hydrogels can be challenging due to their limited mechanical properties, such as the high water content that results in insufficient deformation and load-carrying capabilities compared with natural materials [105]. Moreover, the limitations of hydrogels, such as brittleness, complex preparation process, time consumption, and inability to self-repair, further restrict their application. Therefore developing hydrogels with high resilience and self-repairing capabilities is urgently needed to overcome these limitations [106,107].

To address the limitations of hydrogels, researchers have been exploring dynamic cross-linking to create a more efficient and recyclable energy dissipation system, which contributes to adequate energy consumption and self-healing ability [54]. By employing this approach, tough and self-healing poly ferric arylates can be produced using a simple method. The resulting Fe^{3+} hydrogels possess a unique dual dynamic cross-linked

MBN structure, which endows the hydrogel with robust mechanical properties due to the synergistic effect of different cross-links and hierarchical bonds. The physical cross-linking of Fe^{3+} and hydrogen ions forms a double-bond network, increasing the hydrogel's strength and toughness. Fig. 1.6 shows the mechanical properties of Fe^{3+}hydrogel. The diffusion of iron ions has a complex effect on the diffusion of polymer chains, which can accelerate. The synergistic effect of ionic and hydrogen bonds contributes to the excellent mechanical properties of PAA-Fe^{3+} hydrogel.

The mechanical properties of hydrogels can be evaluated through a series of tensile tests. Iron ions are found to improve the properties of hydrogels in two primary ways. First, the inclusion of iron ions results in an increased cross-linking density of the gel, which enhances its ability to bear stress. Second, the addition of iron ions leads to the reorganization of ionic bonds, thereby increasing the energy consumption rate of the gel. Table 1.2 shows the effect of different Fe^{3+} concentrations on the effective cross-linking density of hydrogels.

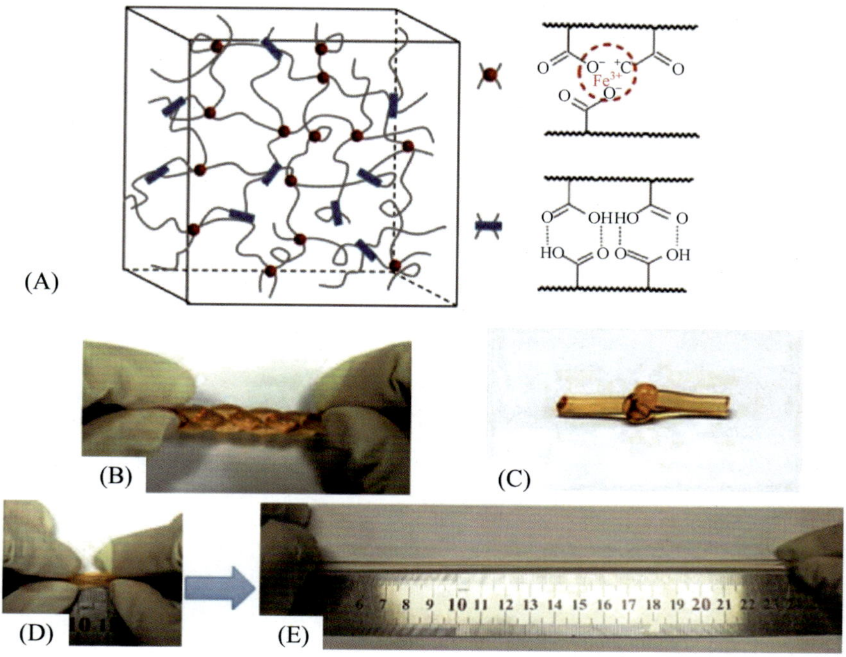

Figure 1.6 (A) PAA-Fe network structure description of Fe^{3+} hydrogels and their physical cross-links mechanical properties of PAA-Fe, (B) twist, (C) knot, (D, E) tension [108].

Table 1.2 Effective cross-linking density of different composition hydrogels [108].

Fe^{3+} concentration (mol.%)	PAA content (wt.%)	$\Sigma\lambda^a$ (kPa)	N^b ($mol \cdot m^{-3}$)
0.50	15	9.30	2.14
0.50	20	18.03	4.16
0.50	25	36.79	8.48
0.50	30	56.27	12.97
0.10	20	10.98	2.53
0.20	20	14.28	3.29
1.00	20	21.18	4.88
1.25	20	17.57	4.05

[a]Stress at 100%strain.
[b]The effective cross-linking density.

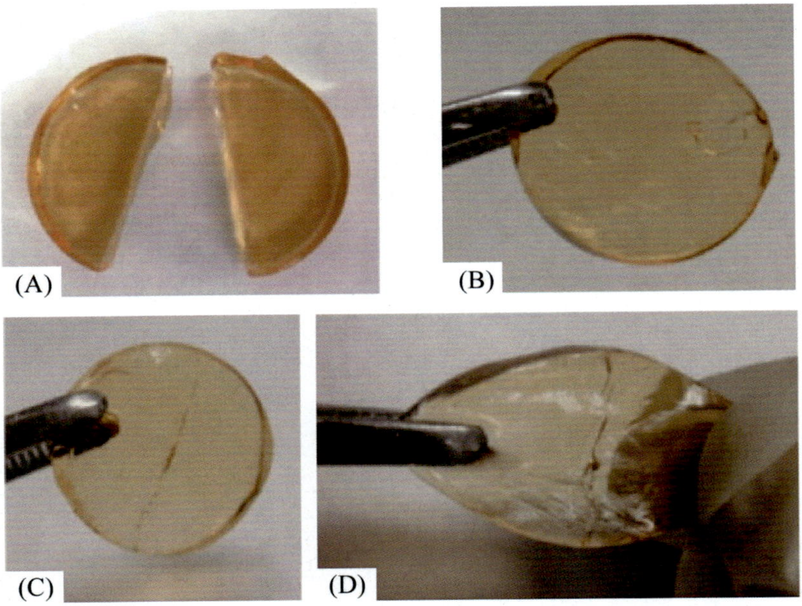

Figure 1.7 PAA-Fe^{3+} self-healing hydrogel. (A) the circular hydrogel was cut into two parts, (B) the hydrogel was self-healing at 50 degrees 2 H, (C) the hydrogel was self-healing at 25 degrees 2 h, and the interface could be seen healing, (D) the hydrogel continued to stretch [108].

Hydrogels can be endowed with improved mechanical properties and self-healing ability through supramolecular interaction. For instance, the PAA-Fe^{3+} hydrogel is capable of complete self-healing at 50°C without any visible healing interface, as shown in Fig. 1.7B. Two hours after healing, the gel was observed to be sufficiently strong to withstand stretching, as demonstrated in Fig. 1.7C and D, indicating a rapid self-healing rate. Additionally, Fe^{3+} hydrogels can achieve self-healing to a certain degree by simply contacting two cutoff hydrogel samples on the cutting surface without any healing agent.

Polymer hydrogels are materials with high water absorption and retention, making them useful in various applications. The use of hydrogels in sanitary materials has a long history. It is the most mature application, providing benefits to women, children, those with incontinence, and individuals with mental health issues [109–124]. Various joint diseases, such as rheumatoid arthritis, osteoarthritis, and natural cartilage injuries, may require the replacement of a joint with an artificial one. Friction and wear are significant factors that limit the life span of artificial joints. The use of lubricants can reduce the physical friction of inflammatory arthritis, further reducing synovial joint injury. As shown in Fig. 1.8, a novel hydrogel, polyvinyl alcohol/polyethylene glycol/graphene oxide (PVA/PEG/GO), was prepared by physical cross-linking and used as a lubricant to improve the lubricity and prolong the lubrication time of artificial joints. The PVA/PEG/GO hydrogel has excellent slow-release lubrication and self-healing properties, providing a new option for designing long-term lubricating joints. This development is expected to promote the progress of artificial joint lubrication applications.

In addition, due to their slow release and mild toxicity, polymer hydrogels have potential applications in medicine and medical treatment. However, the increasing prevalence of drug-resistant bacterial infections and slow

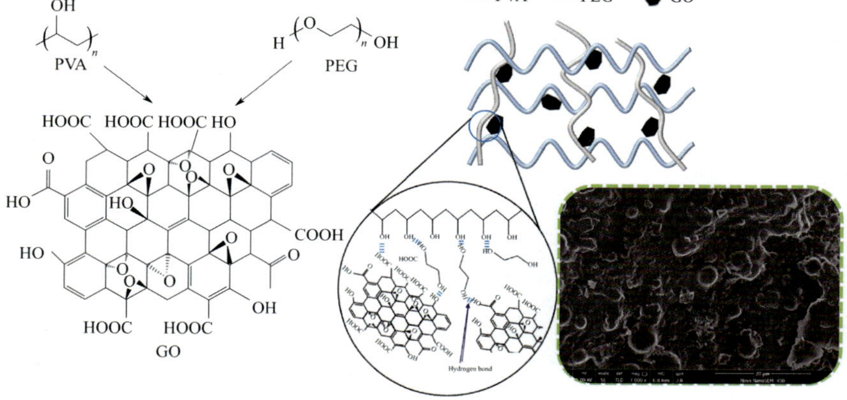

Figure 1.8 Molecular structure and binding mechanism of PVA, PEG, and GO [125].

wound healing of chronic infections have presented a significant challenge, necessitating the development of new antibiotics and wound-healing dressings [126–129]. Fig. 1.9 shows the image of the wound healing site of the 3rd, 7th, and 14th angel water gel dressing. Recent studies have found that modifying glycidyl methacrylate with polyethylene glycol monomethyl ether, methacrylamide dopamine, and zinc ions as substrates for photocross-linking can produce multifunctional antibacterial, antioxygen, hemostatic dressings that can disinfect drug-resistant bacteria and promote wound healing.

The hydrogel-based drug delivery system offers numerous benefits, including the ability to create a localized drug reservoir, prolong the drug release period, and provide a supportive three-dimensional structure similar to the synovial region. These features make hydrogel-based drug delivery systems an attractive option for drug delivery applications. Fig. 1.10 is the schematic diagram of the long-term treatment of osteoarthritis with a hydrogel system.

In the industry, polymer hydrogel materials have diverse applications, such as industrial dehydrating agents, coating thickeners, cable wrapping material, drilling lubricants, waterproof and leak-stopping agents for buildings, curing agents, anticondensation agents, and cement protective agents.

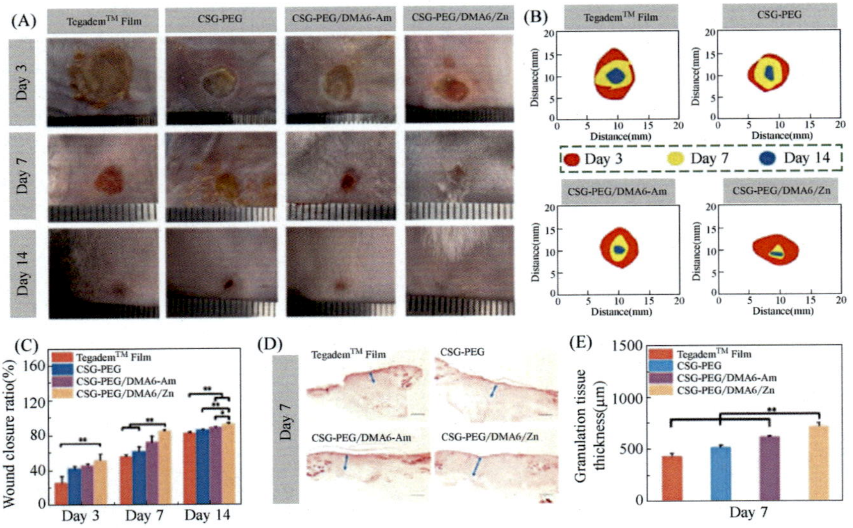

Figure 1.9 (A) Images of the wound healing site on the 3rd, 7th, and 14th days. (B) Schematic representation of the wound healing site on the 3rd, 7th, and 14th days. (C) Statistical data of wound closure ratio (n = 5). (D) Images of regenerating granulation tissue on the 7th day (granulation tissue: blue arrows), scale bar: 400 μm. (E) The results of the thickness of the regenerated granulation tissue on the 7th day (n = 5). *P < .05, **P < .01 [130].

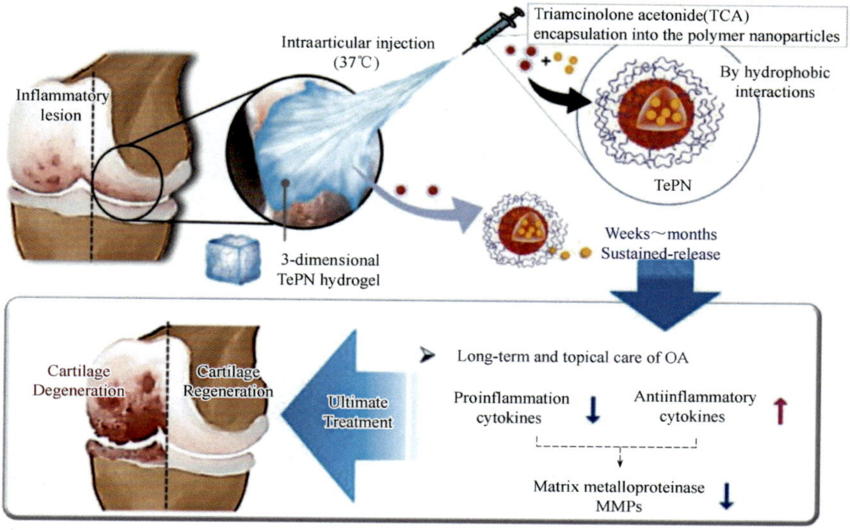

Figure 1.10 A schematic diagram of one-time injection Ten hydrogel system for long-term treatment of osteoarthritis [131].

They can also be used as fresh-keeping packaging material for food, fruit, and vegetables, a wetting agent for daily necessities and cosmetics, a hair styling agent, and a perfume slow-release agent. In environmental protection, polymer hydrogels can act as sewage treatment agents, gel flame retardants, and fireproof cloth in firefighting equipment, as well as humidity sensors, moisture measurement sensors, and water leakage detectors in the electronic industry. Additionally, they are widely used as immobilization carriers of artificial snow expansion toy enzymes, microorganisms, and oil—water separators. In seawater desalination processes, hydrogels are emerging as a promising new material [132,133]. They have promising applications in forward osmosis desalination, solar distillation, electrodialysis, and capacitive seawater desalination. Solar steam generators based on hydrogels can produce drinking water more affordably than other desalination methods.

1.3 Molding and processing of polymer materials

Polymer molding is an essential engineering technology that converts polymers into functional materials or products. However, these materials or products are typically not pure polymers but instead composed of resins supplemented with fillers, additives, colorants, and other materials [134]. The processing method is a crucial factor that ensures product quality and yield. The performance of the final product depends not only on the raw materials but also on the processing methods. Generally, polymer material processing includes four stages: preparation of raw materials, including pretreatment, batching, and mixing polymers and additives; deformation or flow of the raw materials into the desired shape [135]; curing of the materials or products; and postprocessing and treatment to improve the appearance, structure, and performance of materials and products.

Plastic products are usually made of polymers or polymer mixtures combined with other components. After heating, these materials are molded into a specific shape under certain conditions, cooled, shaped, and trimmed. This process is referred to as plastic molding and processing. The molding and processing methods for thermoplastic and thermosetting plastics differ due to their unique behavior after heating [136,137]. There are several plastic molding and processing methods, including extrusion, injection, calendaring, blow molding, and molding, with the first four methods accounting for more than 80% of all plastic products. The molding, casting, and transfer molding methods are mainly used for thermosetting plastics.

1.3.1 Molding and processing of plastic products

1.3.1.1 Extrusion molding

Extrusion molding is a crucial polymer processing method in which materials are heated, plasticized, and pushed forward by a screw to create cross section or semifinished products via the machine head. It is widely considered the most critical molding processing method in the field of polymer processing and the primary method for processing plastic materials [138,139], molding chemical fibers, thermoplastic elastomers, and rubber products. Extrusion molding, also known as extrusion molding or extrusion, is a molding method that involves the formation of heated molten polymer materials into continuous profiles with a constant cross section through the die, promoted by pressure with the help of the extrusion action of a screw or plunger. The extrusion process comprises feeding, melting and plasticizing, extrusion, shaping, and cooling. It can be divided into two stages: the first stage involves plasticizing the solid plastic and making it pass through the die with a unique shape under pressure to form a continuum with a similar section. The second stage involves making the extruded continuum lose its plastic state and become solid to obtain the required product [140].

Extrusion molding is a primary molding method for thermoplastics, accounting for about half of all plastic products [141]. Almost all thermoplastics can be formed using the extrusion method to produce continuous production, equal cut pipe, plate, film, wire and cable coating, and various special-shaped products. Additionally, extrusion molding can be used for plasticization, coloring, and blending of thermoplastics. In this process, the thermoplastic polymer is evenly mixed with various additives, plasticized and melted through mechanical shear force, friction heat, and external heat in the extruder barrel, and then pushed through the screw to be extruded into the forming die through the filter plate.

The characteristics of an extruder depend primarily on the number and structure of screws. The most commonly used extruder is the single-screw extruder, which has only one screw in the barrel. On the other hand, the twin-screw extruder, with two screws rotating in the same direction or reverse meshing in the barrel, has a better plasticizing capacity and quality than the single-screw extruder.

In traditional extrusion equipment, an external heating element and mechanical shear are used to create a steady-state plasticizing extrusion mechanism for processing polymers. This process, which employs a multi-system discrete structure, has several drawbacks, including low energy utilization, high energy consumption, high noise, large volume and weight, high manufacturing cost, and difficulty in improving the quality of extruded products. A new approach that introduces mechanical vibration generated by an electromagnetic field into the polymer plasticizing extrusion process has been proposed to address these issues. This new method is based on concepts such as polymer dynamic plasticizing extrusion, direct electromagnetic energy exchange, and integrating machinery, electronics, and electromagnetic technology. By incorporating these new principles, it is possible to fundamentally solve the problems associated with traditional extrusion equipment and improve the quality of extruded products while reducing energy consumption and production costs. Fig. 1.11 [142] shows the principle structure of the multisystem discrete structure used in traditional extrusion equipment.

In the traditional extrusion process, the motor and external heating element adopt the indirect energy exchange mode, which leads to low energy utilization, high energy consumption, and other defects. To solve these problems, a new electromagnetic dynamic plasticizing extrusion equipment has been developed that introduces mechanical vibration into the process of polymer plasticizing extrusion [143]. By placing the

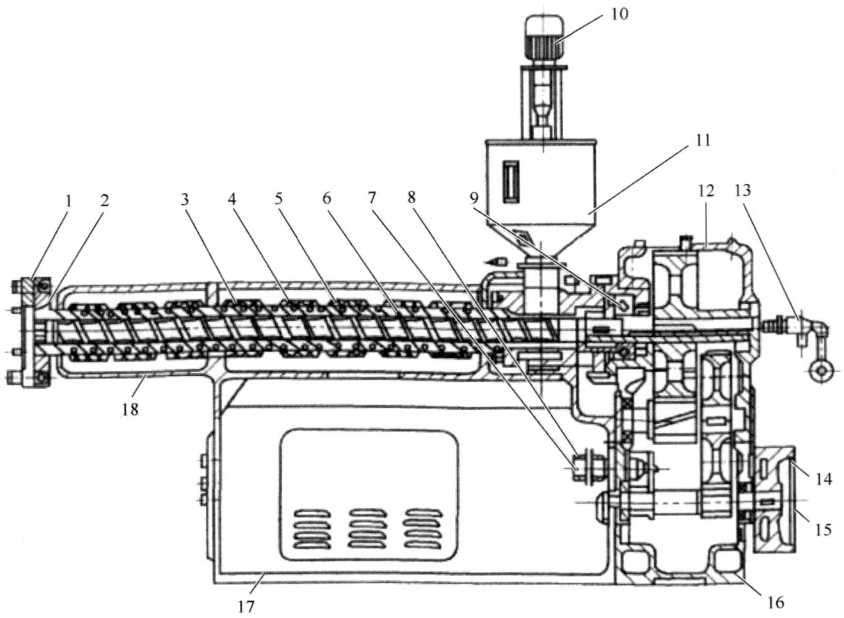

Figure 1.11 Schematic diagrams of the principle and structure of traditional screw extruder. 1—flange, 2—split plate, 3—screw, 4—cooling water pipe, 5—heater, 6—barrel, 7—gear pump, 8,10—motor, 9—bearing, 11—hopper, 12—Gear reducer, 13—rotary joint, 14—V pulley, 15—motor, 16—reduction box, 17—base, 18—hood [142].

plasticizing and extrusion equipment components in the inner cavity containing the rotor, the rotor can directly participate in the plasticizing and extrusion process. The vibration converts electromagnetic energy into heat, pressure, and kinetic energy to complete material transportation, plasticization, and extrusion. This equipment has many advantages, including reduced volume weight, manufacturing cost and energy consumption, low noise, good adaptability to materials, and good plasticizing and mixing effects.

The design and optimization of extrusion plasticization systems and single-screw precision extrusion technology are often based on experience and experiment. However, with the development of computer-aided engineering (CAE), simulation software has become a reliable and powerful tool for the design and optimization of modern extrusion equipment. To simulate various molding processes more accurately, scholars worldwide are continuously studying new models, algorithms, and simulation systems that combine simulation software with product design and manufacturing and develop integrated technology. It can be expected that computer simulation technology will be widely used and become a standard tool for solving various plastic molding and processing problems.

High-precision extrusion equipment in foreign countries is typically equipped with a melt gear pump, which is a positive displacement conveying equipment installed between the extruder and the die. The pump is operated by two precisely designed gears meshing with each other, and fluctuations generated by the extrusion system can be isolated from the head and downstream equipment, mainly because the gear pump has a separate drive motor. It is also a "pressure generator" that can always flow the melt to the head with stable pressure and flow, ensuring that the extrusion head produces dense and dimensionally accurate products [144].

1.3.1.2 Injection molding

The injection molding process involves melting plastic in the heating barrel of an injection molding machine and injecting it into a closed mold cavity using a plunger or reciprocating screw. This process is highly efficient, can produce products with complex shapes, accurate sizes, or inserts, and is suitable for most thermoplastics and some thermosetting plastics with good fluidity [145]. In recent years, reaction injection molding, which involves chemical reactions during the injection process, has developed rapidly. An injection molding machine comprises an injection device, a clamping device, and an injection mold, with two specifications for maximum injection volume or weight and maximum clamping force. Other important parameters include plasticizing capacity, injection rate, and injection pressure [146,147]. Fig. 1.12 shows a horizontal injection molding machine, and Fig. 1.13 shows a vertical injection molding machine.

The primary component of an injection molding machine is responsible for heating and plasticizing the plastic material into a flowing state, which is then injected into the mold cavity under pressure. Injection molding techniques typically involve plunger type, preplasticization type, and reciprocating screw type methods. The machine functions to close the fixed die and moving die of the mold, enabling the opening and closing action of the die and ejection of the finished product.

The injection molding process was developed based on the principles of metal die casting. Injection molding can produce products with intricate shapes, accurate sizes, or metal inserts in a single operation, thus making it widely used, accounting for less than 20% of the total molding processing. This process typically involves five stages: plasticization, mold

Figure 1.12 Horizontal injection molding machine [148].

Figure 1.13 Vertical injection molding machine [149].

filling, pressure maintenance, cooling, and demolding. The mechanical components of molten plastic in the injection barrel can either be a plunger or a screw. The former is classified as a plunger injection machine, while the latter is a screw injection machine [150–152]. Injection machines with an injection volume of more than 60 g each time are typically screwing injection machines. In contrast to extruders, the screw of an injection machine can move both forward and backward.

Commonly used injection molding products usually generate a gate, runner, and other waste materials, which require removal and repair. This procedure can be time-consuming and wasteful, thus limiting production efficiency. In recent years, gateless injection molding has been developed to address these issues. This innovative approach includes a manifold component (also known as the flow channel element) from the nozzle of the injection machine to the mold. The runner is required to be heated continuously for thermoplastics to keep the materials in the runner molten at all times, hence the term "hot runner" [153]. Conversely, for thermosetting plastics, the flow channel should be kept at a low temperature and is referred to as the "cold flow channel." Products obtained through Canongate injection molding typically do not require trimming.

Injection molding is a popular manufacturing process that is primarily used for thermoplastics but has recently been adopted for thermosetting plastics as well. In the case of thermosetting plastics, the material sheet needs to be kept in the thermoplastic stage to prevent hardening. This is achieved through a method called spray molding. In transfer molding or casting molding, the softened thermosetting plastic in the material sheet is pushed out with the help of a push rod, leaving no material in the barrel [154].

As the size and aspect ratio of injection parts increase, maintaining uniform polymer melt heating and sufficient clamping force during injection molding becomes challenging. Reactive injection molding has been developed to overcome this issue. This process essentially completes most polymerization reactions in the mold, reducing the viscosity of injection materials by more than two orders of magnitude. Reactive injection molding has been widely used to produce polyurethane foam and reinforced elastomer products [155].

Gas-assisted injection molding is a process that injects a polymer melt into the mold cavity and then injects compressed air into the melt through the nozzle and flow channel to form a gas sandwich product. This process offers advantages such as low consumption, high quality, and new structure, making it popular among plastic product manufacturers.

There is a growing demand for plastic factories to use this technology to provide high-grade and high-quality plastic parts. Strengthening and paying attention to the development of gas-assisted injection molding devices, molds, and process application technology are of great significance to promoting the application of gas-assisted injection molding technology and improving the grade of injection products in China [156,157].

In comparison to traditional injection molding, gas-assisted injection molding involves incorporating independent gas-assisted technology components that play a crucial role in the entire process. These components serve as a critical connection between product selection, structural design, mold design, and injection molding machines, ultimately producing qualified products. The high technical content of this new technology is due to its uniqueness and complexity, which brings significant advantages distinct from traditional injection molding. In the gas-assisted injection molding process, there is a gas—liquid two-phase flow of polymer melt and high-pressure gas. This highlights the importance of the components of gas-assisted technology, which act as a crucial link in the gas-assisted injection molding process.

1.3.1.3 Blow molding

Blow molding is a plastic molding technique used to produce hollow thermoplastic products. The process involves prefabricating plastic into pieces and forming them into a tubular blank by washing them into a simple shape. This blank is then heated and subjected to hot air or preheated and blown into the cold air, within the high elastic deformation temperature range and lower than its flow temperature. The result is a model-shaped hollow product. A blow molding die is installed at the front end of the extruder to blow the extruded tube blank into a film tube using compressed air, which is then folded and wound into a double-layer flat film after air cooling. This process is known as blow molding film formation [158—161]. The molding process for blow-molded hollow products involves extruding the embryo with an extruder or injection machine, placing it in the mold, cooling it, and setting it with compressed air to make it conform to the mold surface.

Blow molding, which refers mainly to hollow blow molding, is widely used for forming hollow products by blowing hot-melt parson closed in the mold with gas pressure. It is a standard plastic processing and rapidly developing plastic molding method. Unlike injection molding, blow molding equipment is less expensive, has strong adaptability and good formability, and can produce products with complex fluctuation curves. Blow molding originated in the 1830s and was not widely used until 1979. Currently, blow molding grade plastics, such as polyolefin, engineering plastics, and elastomers, are commonly used. Blow molding products find application in industries such as automobiles, office equipment, household appliances, and medical treatment. The technology can produce up to 60,000 bottles per hour and large blow-molded parts. Additionally, multilayer blow molding technology has been significantly developed [162]. Blow molding equipment has adopted closed-loop control systems using microcomputers and solid-state electronics. Computer CAE/CAM technology is becoming increasingly mature, making blow molding machinery more professional and characteristic.

Off-axis extrusion blow molding technology, also called 3D or 3D blow molding, has been developed to meet the demand for complex and tortuous conveying pipe products. The process involves extruding the parson, locally inflating it, and attaching it to one side of the die. The extruder head or die is then rotated. The multilayer blow molding process is often used to manufacture impermeable containers. The improved process involves adding a valve system to replace plastic raw materials in the continuous extrusion process, enabling the production of complex and soft products alternately. When producing large parts such as fuel tanks or automobile outer structural plates, it is necessary to reduce the pressure in the die cavity during cooling to adjust the processing cycle [163,164]. The solution is to store the melt in the melt groove at the front end of the extrusion screw and then extrude the parson at a reasonably high speed to minimize changes in the parson wall thickness and eliminate shrinkage and extrusion expansion.

The raw materials utilized for blow molding possess abundant properties that can enhance their processing performance. Blow molding grade pen materials exhibit exceptional features such as high strength, excellent heat resistance, robust gas barrier, transparency, ultraviolet radiation resistance, and suitability for producing various plastic bottles. These materials also offer remarkable attributes such as high filling temperature, excellent barrier performance to carbon dioxide gas and oxygen, and chemical resistance. Consequently, the use of such materials in product packaging containers and industrial products has increased significantly, leading to rapid developments in injection and multilayer blow molding.

In addition, the precision and efficiency of blow molding machinery and equipment have become crucial in manufacturing. The concept of "precision and high efficiency" encompasses not only high speed and pressure in the production and forming process but also the ability of the equipment to produce products with high stability in terms of appearance, size fluctuation, and piece weight fluctuation. Therefore the dimensions and shape geometry of each part of the produced products has high accuracy with minimal deformation and shrinkage. The appearance, internal quality, and production efficiency of the products should also meet high standards. Auxiliary operations such as debarring, cutting, weighing, drilling, and leak detection have also undergone automation to improve the production process, which is a developmental trend. As shown in Fig. 1.14 [165], it is a blow molding machine.

In the plastic forming stage of the product, the parison is inflated at high pressure to achieve a close fit with the mold cavity. At this stage, product form is influenced by the deformation of the parison due to high-pressure expansion and contact deformation between the parison and the die cavity. The primary process parameters that affect wall thickness distribution are material shrinkage, blowing pressure, and time. Other important factors include mold material, structure, exhaust system, and cooling system, including cooling water channel distribution and inlet temperature [166]. Although many factors impact the quality of blow molding products, optimizing the blow molding parameters can

Figure 1.14 Blow molding machine [165].

significantly enhance product quality, provided that production conditions and requirements are established. Optimization of process parameters can improve production efficiency, reduce raw material consumption, and optimize the comprehensive performance of products [167].

1.3.2 Rubber product forming and processing

Rubber products are commonly molded through various processes such as plastic refining, mixing, calendaring, extrusion, molding, and vulcanization. Among these processes, molding technology, transfer molding technology, wrapping molding technology, and injection molding technology are the leading methods used in rubber product manufacturing [168,169]. Injection molding technology, in particular, has gained significant attention and development in recent years. It has undergone three stages, namely plunger injection, screw reciprocating injection, and screw plunger injection, leading to the development of corresponding injection molding machines.

The plunger injection molding machine is shown in Fig. 1.15. Plunger injection molding machines were the earliest rubber injection molding equipment developed. On the other hand, screw injection molding machines utilize the extruder's screw, which rotates to melt the rubber and then moves axially to inject it into the mold cavity. The rubber is fed into the extruder from the feed port and undergoes strong shear under the screw's rotation, leading to a quick rise in its temperature. When the rubber reaches the front end of the screw, it is fully and evenly plasticized. As the screw moves backward while rotating, the axial power mechanism pushes it forward with a firm thrust to inject the rubber into the mold cavity.

The screw plunger injection molding machine combines the advantages of both plunger and screw injection molding machines and is widely used in rubber injection equipment. The machine's injection part comprises a screw plasticizing system and a plunger injection system. First, the hard rubber is fed to the plasticizing screw system, which plasticizes the rubber and extrudes it into the plunger injection system. Finally, the rubber is injected into the mold cavity by the plunger. A check valve installed at the end of the screw extruder ensures that the rubber flows in a particular order. After plasticization, the rubber enters the injection system through the check valve and pushes up the plunger. The narrow nozzle channel and the significant resistance prevent the rubber from flowing out

Figure 1.15 Plunger injection molding machine [170].

of the nozzle. Therefore, when the plunger injects the rubber from the nozzle into the die cavity at high pressure, the rubber does not flow back into the extruder due to the check valve's action [171].

1.3.3 Forming and processing of fiber products

In materials science, fiber is defined as a thin, flexible material with a length much greater than its diameter. For textile applications, the length ratio to the diameter of fibers is typically greater than 1000:1. Typical textile fibers possess a diameter ranging from several microns to tens of microns, with a length exceeding 25 mm. Fibers can be classified into two categories: natural fibers, including cotton, wool, silk, and hemp, and chemical fibers, produced through chemical processing using natural or synthetic compounds. The fiber processing cycle consists of three primary stages: spinning solution preparation, spinning, and postprocessing of primary fibers. Generally, the fiber-forming polymer is melted or dissolved into a viscous liquid, which is subsequently pressed out from a small hole in a spinneret with a spinning pump. The resulting stream of slime aggregates or condenses to form fibers, which are then subjected to postprocessing according to specific requirements. Melt spinning and solution spinning are the most commonly employed spinning methods in industry, with several novel approaches also being developed.

1.3.3.1 Melt spinning

Melt spinning is a technique utilized in producing fibers. Polymers are heated and melted into a viscous liquid, sprayed through small holes in a spinneret, and cooled in air or water to form solid fibers [172–174]. The process begins by feeding polymer raw materials into a screw extruder, where they are heated and melted before being sent to the metering pump to control and maintain the steady flow of the polymer melt. Subsequently, the melt is filtered and forced through the spinneret to form a fine liquid stream rapidly cooled by cold air from the temperature regulating bellows. The cooled filament is then subjected to prestretching by a wire guide roller, which further reduces its diameter. The resulting primary fiber is then wound onto a cylinder or collected into a barrel to form the final fiber product. Melt spinning is a high-speed process, with spinning speeds of thousands of meters per minute achievable.

In the melt-spinning process, the bulk polymer is melted in a screw extruder and subsequently sent to the spinning assembly through a spinning pump. After filtration, the melt is extruded from the spinneret pores and rapidly solidified by passing through a cooling medium before being stretched into a filament by a winding device below. The resulting filament is referred to as the primary fiber, and further processing is required to produce the final fiber product [175]. Fig. 1.16 shows a schematic diagram of the melt-spinning process of graphene/PA6. Sometimes, a greenhouse is added to the spinneret to prevent rapid cooling of the filament, which can interfere with fiber formation. The drawing speed of the winding device can be very high, up to 1500–3000 m/min, depending on the material type and rheological properties. Melt spinning has a vast drawing ratio and high yield. It can be adjusted in a wide range, making it suitable for producing fibers from materials such as polyester, nylon, and polypropylene.

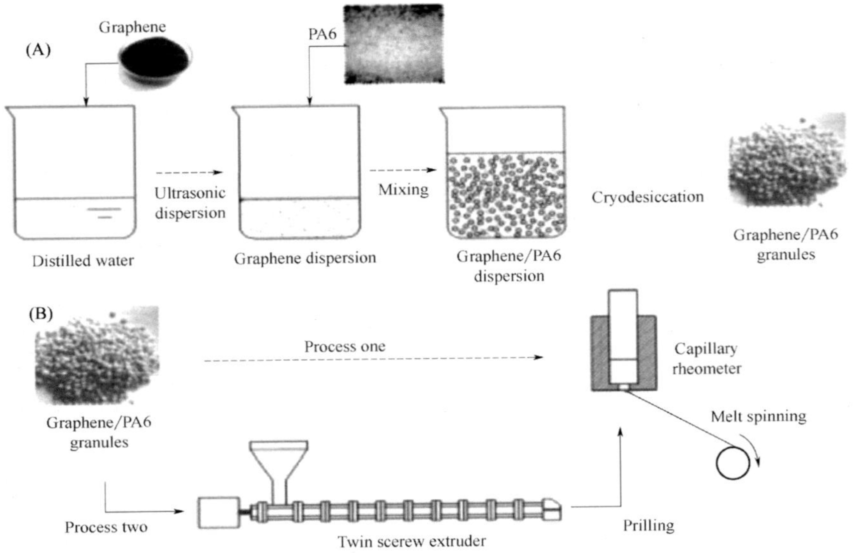

Figure 1.16 (A) The production process of graphene/PA6 particles. (B) Schematic representation of the melt spinning process for graphene/PA6 [176].

1.3.3.2 Solution spinning

Solution spinning is a process in which discoloring compounds and reagents are directly added to the spinning solution, making it a suitable spinning method. For instance, acrylonitrile/styrene/vinyl chloride copolymer solution and chromo tropic compounds were spun into a water bath to obtain photochromic fibers, which change their color from no color to dark green under sunlight or ultraviolet radiation. This spinning method applies to producing various items, including clothing, curtains, carpets, and toys. The solution-spinning process involves dissolving the polymer in a solvent to create a viscous spinning solution. The spinning solution is then sprayed into a stream using the spinneret and solidified through the solidification medium to form fibers. This method is referred to as the solution-spinning method [177]. It can be classified into two types based on the solidification medium.

In wet spinning, the solidification medium is a liquid. The thin stream of mucus pressed out from the spinneret's small hole passes through the liquid, solidifying the fibrous polymer in the thin stream into filaments. The spinning stock solution is obtained by dissolving the fiber-forming polymer in a suitable solvent to obtain a solution with specific composition, viscosity, and good spin ability [178]. This solution can also be obtained by homogeneous solution polymerization. Before spinning, the polymer solution must be prepared through mixing, filtering, and defoaming to remove gel blocks, impurities, and bubbles in the solution, thus making the properties of the spinning fluid uniform and homogeneous. In viscose fiber production, prespinning preparation includes a ripening process to make viscose have the required spin ability. Fig. 1.17 [179] shows the wet spinning equipment.

In dry spinning, the solidification medium is the dry gas-phase medium. The slime stream pressed out from the spinneret's small hole is introduced into the corridor with hot air flow, which quickly volatilizes the solvent in the slime stream. The hot air flow removes the volatilized solvent vapor, and the slime stream quickly changes into filaments after removing the solvent [180,181].

The schematic diagram of dry spinning is shown in Fig. 1.18. The performance of the dry spinning process is typically determined by the rate of solvent evaporation from the filament, as well as the curing rate of the polymer. The volatilization speed of solvent plays a crucial role in limiting the efficiency of this process. Additionally, the slow curing rate of the polymer has a significant impact on the properties of the resulting filament. These factors ultimately determine the characteristics of the dry spinning process.

Unlike wet spinning, dry spinning solutions have a higher concentration and viscosity, typically ranging from 18% to 45%. Dry spinning can withstand more stretching from the spinneret, generally up to two to seven times, resulting in finer fibers compared with wet spinning. The yarn's mechanical resistance during spinning is lower than that in wet spinning, allowing for faster spinning speeds of up to 300−600 m/min and even up to 1000 m/min. However, the dry spinning speed is lower than that of

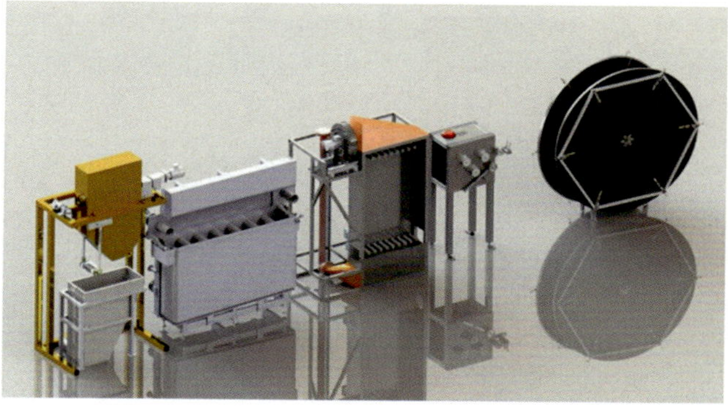

Figure 1.17 Wet spinning equipment [179].

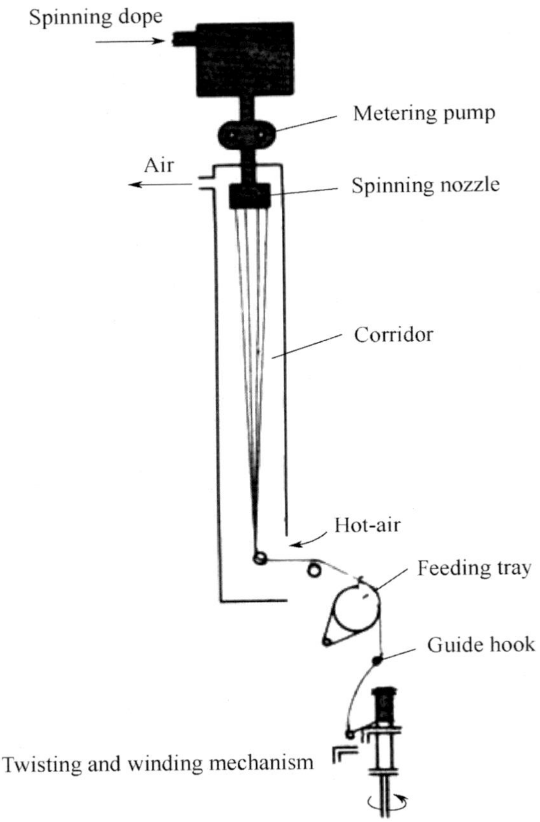

Figure 1.18 Schematic diagrams of dry spinning [182].

melt spinning due to the solvent's limited volatilization speed. As the curing of the filaments is slow, the number of spinneret holes is limited to no more than 1200 for dry spinning staple fibers, while wet spinning staple fibers can have tens of thousands to over 100,000 spinneret holes. Thus, the production capacity of a single spinning position is generally lower for dry spinning than for wet spinning. Dry spinning is ideal for producing long fibers with uniform structures and good quality. It is suitable for a variety of raw materials and has a wide range of applications. Dry spinning is commonly used for spinning chemical fibers and is preferred over wet spinning due to its higher quality output. Although dry spinning has a low output, it is used for producing acetate fiber and perchloroethylene fiber.

Dry wet spinning is a chemical fiber spinning method that combines the advantages of both dry and wet spinning techniques. Fig. 1.19 shows

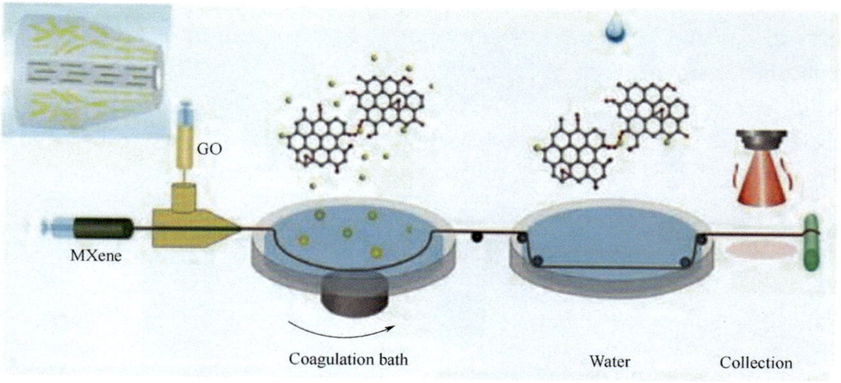

Figure 1.19 Schematic diagrams of dry and wet spinning [183].

a schematic diagram of dry—wet spinning. The spinning solution is extruded from the spinneret and travels through the air before entering the coagulation bath. The gas in the space can be air or other inert gases. Compared with wet spinning, the small flow of the spinning solution used in dry wet spinning can withstand greater spinneret stretching in the air, resulting in a longer stretching zone [184,185]. This longer stretching zone induces a slight velocity gradient in the axial deformation of the liquid flow without causing significant deformation in the swelling area. In contrast, during wet spinning, the stretching of the spinneret occurs at a shorter distance, and the velocity gradient is large, causing severe deformation in the swelling area. Consequently, dry wet spinning allows for higher draw ratios and spinning speeds when compared with wet spinning.

As industries such as aviation, space technology, and national defense develop, new requirements for the properties of synthetic fibers have arisen [186,187]. Many new fibrous polymers have been synthesized, which cannot be processed using conventional spinning methods such as melt spinning and solution spinning. Various new spinning methods have been developed, such as electrospinning, dry—wet spinning, liquid crystal spinning, gel spinning, phase separation spinning, emulsion or suspension spinning, and reactive spinning. These new methods have provided researchers with more options for processing new synthetic fibers.

1.3.3.3 Preparation by electrospinning

The following section briefly outlines the fiber-forming process prepared through electrospinning. Electrospinning involves the electrostatic atomization of polymer fluid to produce small polymer jets that solidify into fibers after traveling a long distance. Electrospinning is a unique fiber manufacturing process, where polymer solution or melt is jet spun in a solid electric field, causing the droplet at the needle tip to change from spherical to conical and then extend to obtain the fiber, as shown in Fig. 1.20 [188]. This process allows the production of nanodiameter polymer filaments [189].

Various factors impact the preparation of nanofibers by electrospinning, including solution properties such as viscosity, elasticity, conductivity, and surface tension, as well as control variables such as the static voltage in the capillary, potential at the capillary port, and interval between the capillary port and collector. Environmental parameters such as solution temperature, air humidity, temperature in the spinning environment, and airflow velocity also play a role. However, the fibers produced through this method have low strength, a rough and rugged texture, and can even be brittle. Consequently, postprocessing procedures are required to obtain fibers with stable structures, excellent performance, and textile processing [190,191]. Additionally, many chemical fibers are blended with natural fibers, requiring the cutting of continuous filaments

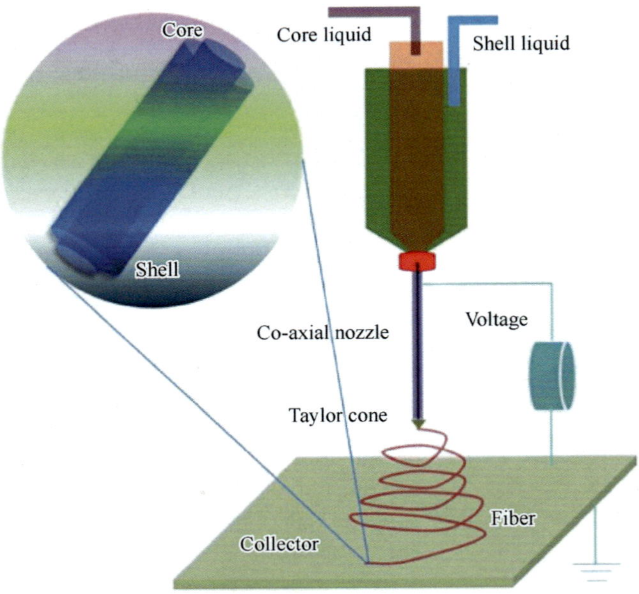

Figure 1.20 Electrospinning process of shell–core nanofibers [188].

during postprocessing to produce fibers with a certain length and curl, similar to natural fibers, such as cotton and wool, to meet the requirements of textile processing.

The postprocessing of fibers is a crucial step in the textile industry and depends on the type of spun fiber and the specific requirements of textile processing. Postprocessing can be categorized into short and long fibers and can produce fibers with unique properties, such as elastic and expanded yarns. For staple fibers, postprocessing is typically carried out on a long flow line, which involves several processes such as bunching, drafting, washing, oiling, drying, heat setting, crimping, cutting, and packaging. The content and sequence of postprocessing processes may vary based on the fiber variety.

Compared with short fibers, the processing technology and equipment structure for postprocessing long fibers are more complex. Long fibers require strands of wires to be treated separately instead of being bundled as short fibers. This ensures that each strand of silk undergoes treatment under the same conditions. The postprocessing process of filaments includes drawing, twisting, double twisting, heat setting, winding, grading, packaging, and other processes. The complexity of the process is attributed to the requirement of treating each strand separately and uniformly.

References

[1] Cai JY. Application status and development trend of polymer chemical materials. Chemical Management 2021;4:78–9.

[2] Chen P. Application status and development trend of polymer chemical materials. Contemporary Chemical Research 2021;17:15–16.

[3] Jiang Z. Difficulties and countermeasures in the cultivation of polymer materials and engineering professionals. Chemical Management 2020;33:28–9.

[4] Li M. Analysis on the development of integration of high mechanical performance polymer materials and competitive sports. Aging and Application of Synthetic Materials 2021;50(02):163–5 134.

[5] Liu S, Zhou H, Wang Y, et al. Research on processing and molding technology of polymer materials. Chemistry and Bonding 2021;43(3):228–30.

[6] Nie Y. Educational reform of the course of polymer materials related principles for the major of material chemistry—review of fundamentals of polymer materials. Chemical reagent 2021;43(05):710.

[7] Feng G. Development of polymer chemical materials in China. Chemical Design Communication 2020;46(10):23 93.

[8] Wang C, Ou C, Fan J. A brief talk on the current situation and development of polymer materials. Southern Agricultural Machinery 2017;48(07):108 110.

[9] Wang R. Development status and application analysis of functional polymer materials in China. Chemical Design Communication 2020;46(03):93 102.

[10] Zhao Y. Discussion on the development status and application trend of polymer materials. Modern Marketing (Late Edition) 2017;(07):273.

[11] Hao L, Li Y, Li J. Properties and applications of functional polymer materials. Chemical Design Communication 2021;47(05):67–8.

[12] Anna R, Rudolf JS. Development of a Lateral Flow Immunoassay (LFIA) to screen for the release of the endocrine disruptor bisphenol a from polymer materials and products. Biosensors 2021;11(7):231.

[13] Manga B, Liu J, Wang L. Recent development of *n*-type thermoelectric materials based on conjugated polymers. Nano Materials Science 2021;3(2):113–23.

[14] Chashchilov DV, Samoylenko V, Atyasova EV, et al. Development of a polymer composite material with an epoxy binder for producing a high-pressure cylinder. Chemical and Petroleum Engineering 2021;1–7.

[15] Dunshuai Q, Ting Q, Hue H. Acceptor–acceptor-type conjugated polymer semiconductors. Journal of Energy Chemistry 2021;59:364–87.

[16] Trushlyakov V.I., Russkikh G.S., Blesman A.I., et al. Development of polymer composite materials and structures with their subsequent utilization after the mission. Journal of Physics: Conference Series 2021;1901(1): 012099.

[17] Maria R, Jonna L, Leonardo P, et al. Tribological behaviour and transfer layer development of self-lubricating polymer composite bearing materials under long duration dry sliding against stainless steel. Wear 2021;484–5.

[18] Chen H. Application and development trend of polymer materials. Science and Technology Information 2018;16(27):96–7.

[19] Cheng Z, Zhao Q. Analysis of the development status and application trend of polymer materials. Science and Technology Innovation Guide 2017;14(32):69–70.

[20] Raphael C, Paula S, Marcia B, et al. Circular technology road mapping (TRM): fostering sustainable material development. Sustainability 2021;13(13):7036.

[21] Dubinskii SV, Kazmin EA, Kovalev IE, et al. Development of vibrothermography as a method for nondestructive testing of products made of polymer structural materials with the use of forced mechanical vibrations. Russian Journal of Nondestructive Testing 2021;57(6):465–75.

[22] Li X, Zhang Y. DNA methylation changes induced by BDE-209 are related to DNA damage response and germ cell development in GC-2spd. Journal of Environmental Sciences 2021;109:161–70.

[23] Jin Z. Research status and development prospect of functional polymer materials. Scientific and Technological Innovation and Application 2017;(04):106–7.

[24] Lu H. Application and development prospect of chemical polymer materials. Yunnan Chemical Industry 2018;45(11):19–20.

[25] Qiao J. Thoughts on the transformation and development of China's polymer material industry. Petrochemical Industry 2015;44(09):1033–7.

[26] Zhou P. Development and application of polymer materials. Science and Technology Innovation and Application 2015;(11):69.

[27] Li D. Preparation and application of biomedical polymer materials. Yunnan Chemical Industry 2021;48(06):76–8.

[28] Shen J. Development and basic research of biomedical polymer materials. Master's thesis, Jiangsu: Nanjing University of Science and Technology; 2004.

[29] Xu X, Sun H, Li J. Research progress and industrial transformation of dental medical polymer materials. Journal of Functional Polymers 2021;34(02):126–43.

[30] Wang J, Chen R, Fu J, et al. Research progress in synthesis and modification of biomedical polymer materials. Plastic 2021;50(03):83–7 92.

[31] Ophelia W. Integrating innovative polymer chemistry research into the introductory general chemistry two course sequence-fostering STEM interest and retention. Abstracts of Papers of American Chemical Society 2016;252.

[32] Swinarew AS, Swinarew B, Gabor J, et al. New kind of polymer materials based on selected complexing star-shaped polyethers. Polymers 2019;11(10):1554.
[33] Suleymenov IE, Sedlakova ZZ, Kopishev EE. New polymer materials for optical sensor systems. Journal of Inorganic and Organometallic Polymers and Materials 2019;29(3):758–64.
[34] Prima IO, Pamungkas B. Polyaniline as novel polymer materials for dry electrode based electrocardiography ECG. Jurnal Elektronika dan Telekomunikasi 2018;18(1):1–8.
[35] Wang S, Wu Y, Wang L. Teaching reform and practice of basic course of materials science and engineering. Shandong Chemical Industry 2018;47(12):176–7.
[36] Huang N. Research on the application of new polymer materials. Information recording materials 2019;20(04):19–20.
[37] Li Y, Zhao X, Wang H. Application status and development trend of polymer materials. Chemical Design Communication 2022;48(10):46–8.
[38] Wang Z, Liu J, Wang J, et al. Research on new functional polymer materials. Modern Chemical Industry 2007;(2):514–16 518.
[39] Xiao Z. New polymer materials and applications. Times Finance 2017;33:253.
[40] Research Xu J. and Application of new functional polymer materials. Journal of Basic Science of Textile University 2002;03:258–63.
[41] Hoods M, Skylar G, Polar M. Surface topography of a new polymer material for tooth implants. Journal of Biomedical Materials Research 1974;8(5):213–19.
[42] Francesco R, Simone B. A micron-scale surface topography design reducing reduces cell adhesion to implanted materials. Scientific Reports 2018;8(1):1–13.
[43] Muscat A, Beyersdorf J, Volos KD. Poly(carbamoyl sulphonate) hydrogel is a new polymer material for cell entrapment. Biosensors and Bioelectronics 1995;10(8).
[44] Sepe M. Tracing the history of polymeric materials. Plastics Technology 2021;67 (9):26–8.
[45] Li J, Liu H, Li Y, et al. Discussion on the ideological and political construction mode and evaluation system of the course Fundamentals of polymer material molding die. Polymer Bulletin 2021;(09):88–92.
[46] Liang W, Fu L, Chen L. Exploration on of teaching reform of fundamentals of material science for polymer materials specialty. Guangdong Chemical Industry 2020;47 (01):154–61.
[47] Hoagie L, Shan X, Banyu Z, et al. The phase structural evolution and gas separation performances of cellulose acetate/polyimide composite membrane from polymer to carbon stage. Membranes 2021;11(8):618.
[48] Huang S, Chen J, Zoo D. A preliminary study of polymer inclusion membrane for lutetium(III)separation and membrane regeneration. Journal of Rare Earths 2021;39 (10):1256–63.
[49] Du H. Polymer separation membrane and its application. East Sichuan Academic Journal 1997;(02):111–14.
[50] Hou S, Wang X, Dong X, et al. Research progress of anti-pollution polymer separation membranes. Applied Chemistry 2017;34(05):502–11.
[51] Li M, Chen G, Yao K. Intelligent polymer separation membrane. Chemical Progress 1997;(06):6–11.
[52] Zhen C, Kwun-Wa CA, Chun-Heir WV, et al. A supramolecular strategy toward an efficient and selective capture of Platinum(II) complexes. Journal of the American Chemical Society 2019;141(28):11204–11.
[53] Guseva MA, Alentiev DA, Bakhtin DS, et al. Polymers based on exo-silicon-substituted norbornenes for membrane gas separation. Journal of Membrane Science 2021;638:119656.
[54] Yong JK, Kim MS, Jeon HJ, et al. Mechanical performance of polymer materials for low-temperature applications. Applied Sciences 2022;12(23):12251.

[55] Thalami EM, Mahdi A, Habibollah Y, et al. Design and fabrication of high performance membrane for carbon dioxide separation via blending poly(ethylene oxide-b-amid 6) with dense, glassy, and highly CO_2-public approximated polymers. Reactive and Functional Polymers 2021;105014.
[56] Ma C, Huang H, Gu J, et al. Polymer separation membrane materials and their research progress. Material Guide 2016;30(09):144−50 157.
[57] Shi J. New trends in research and development of polymer separation membranes. Shanghai Chemical Industry 1984;02:28−32.
[58] Wang D. Development and application of polymer separation membrane. Filtration and Separation 1999;(01):38−40.
[59] Wang W, Xue Y, Bu M, et al. Research progress, development and utilization of polymer separation membrane materials. Guangzhou Chemical Industry 2016;44 (01):27−8 65.
[60] Jing D, Zhan H. State of the art and prospects of chemically and thermally aggressive membrane gas separations: Insights from polymer science. Polymer 2021;229:123988.
[61] Li C, Human D, Weiguo P, et al. Poly (vinyl alcohol-co-ethylene) (EVOH) modified polymer inclusion membrane in heavy rare earths separation with advanced hydrophilicity and separation property. Chemical Engineering Journal 2021;426:131305.
[62] Liang W, Xiang G, Fang Z, et al. Blending and *in situ* thermally crosslinking of dual rigid polymers for anti-plasticized gas separation membranes. Journal of Membrane Science 2021;638:119668.
[63] Port S, Hoping G, Krzysztof D, et al. New reactive ionic liquids as carriers in polymer inclusion membranes for transport and separation of Cd(II), Cu(II), Pub(II), and Zn(II) ions from chloride aqueous chloride solutions. Journal of Membrane Science 2021;638:119674.
[64] Wang B. Polymer separation membrane. Chemical World 1990;04:45−6.
[65] Wang X. Polymer separation membrane and its separation technology. Science and Technology Today 1988;(03):34.
[66] Wu X, Zhao Y, Wang X. Separation membrane polymer materials and progress. Plastics 2001;02:42−8.
[67] Chen C. Reasons and countermeasures for the decline of polymer separation membrane performance. Water Purification Technology 1989;04:5−9.
[68] Zhou G, Chen J, Zhang X, et al. Modification methods of polymer separation membrane materials. Journal of Xinyang Normal University (Natural Science Edition) 2003;(03):363−4 368.
[69] Cao M, Zhao Z, Zhang Y, et al. Recent research on polymer magnetic materials. Engineering Plastics Application 2005;33(7):64−6.
[70] Deng F, He W, Jiang Y, et al. Research progress in organic polymer magnetic materials. Polymer Materials Science and Engineering 2010;26(2):171−4.
[71] Pang B. On the properties and applications of organic polymer magnetic materials. Science and Technology Information 2016;14(35):117−19.
[72] Shao B. Research progress of organic polymer magnetic materials. Peer 2016;(6):23.
[73] Tao C, Wu L, Du J, et al. Research and application progress of magnetic polymer materials. Materials Guide 2003;17(4):50−3.
[74] Wang D. Polymer magnetic materials. Modern Chemical Industry 1991;(3):51−2.
[75] Wang L, Liu Z. Preparation of magnetic polymer microspheres and its application in analytical chemistry. Materials Guide 2006;20(6):36−40.
[76] Wen Y, Liu T. Research progress of magnetic polymer materials. Modern Plastic Processing and Application 2005;17(5):53−7.
[77] Yang W. A brief talk on magnetic polymer materials. Metallurgy and Materials 2020;40(2):106−8.

[78] Shao W, Teng Y, Zhu H. Research and application of polymer magnetic composite produced by injection method. Rubber and Plastic Technology and Equipment 2003;29(9):36–9.
[79] Zhang C. Preparation and characterization of magnetic particles and their application in induction heating of thermoplastic polymer materials. Doctoral thesis, Nanjing: Nanjing University of Technology; 2010.
[80] Cao Y. Polymer optoelectronic functional materials and their applications: flat panel display, white light illumination, solar cells. New Chemical Materials 2010;38:215.
[81] Geng Y, Gao H, Wu Y, et al. Patterned preparation of polymer functional materials and their applications in the field of optoelectronics. Journal of Polymer Science 2020;51(5):421–33.
[82] Han C, Hou B, Zheng Z, et al. Research progress of functional polymer materials. Materials Engineering 2021;49(6):55–65.
[83] Li W, Tang S. Biodegradable polymer materials (continued). Shanghai Plastics 2002;(2):10–13.
[84] Jiao J, Wu Y. Environmentally degradable polymer materials. Materials Guide 2004;18(8):29–31.
[85] Xu Y. Research progress of degradable polymer materials. Gold Fields 2012;5:307.
[86] Zhang C, Chen N, Li L. Advanced manufacturing of polymer micro nano functional composites and functional devices. Polymer Materials Science and Engineering 2021;37(1):209–17.
[87] Han Y. Discuss the application of functional polymer materials. Building Engineering Technology and Design 2014;(26):559.
[88] Li J, Jing T. Functional polymer materials. Intelligence 2011;(11):59.
[89] Lv C, Quan F, Liang F. New functional polymer materials. Guangzhou Chemical Industry 2013;41(20):7–8 11.
[90] Qian H. Analysis on of the current situation and future development of functional polymer materials. Chemical Management 2018;1:93.
[91] Song M. Application and development prospect of functional polymer materials. Liaoning Silk 2021;(2):34–50.
[92] Wang M, Wu W, Xiao L. Research and practical application of functional polymer materials. Chemical Management 2019;(14):121–2.
[93] Zhang Z, Li B. Research progress of functional polymer materials. Encyclopedia Forum Electronic Journal 2021;(18):654.
[94] Zhang G, Wang H, Ming Z. Functional polymer materials. Chemistry and bonding 2003;(6):307–8.
[95] Wang Z, Wang T, Zhan M, et al. Research progress of functional polymer materials in the field of flexible electronics. Journal of China University of Science and Technology 2019;49(11):878–91.
[96] Wu S, Lu Y. Development status and the prospect of functional polymer materials. Chemical Design Communication 2016;42(4):82.
[97] Zhang D, Shi Y, Li B. Functional polymer materials and their applications. New Chemical Materials 2004;32(12):5–8.
[98] Zheng J. Analysis of the development prospect of functional polymer materials. Southern Agricultural Machinery 2017;48(24):56.
[99] Li X, Li Y, Zhu X, et al. Research progress of polymer hydrogel materials. Functional Materials 2003;(04):382–5.
[100] Liu F, Zhuo R. Synthesis of temperature and pH-sensitive hydrogels and their applications in controlled release of biomacromolecules. Polymer Materials Science and Engineering 1998;(02):55–8 63.
[101] Liu F, Zhuo R. Preparation and application of hydrogel. Polymer Bulletin 1995;(04):205–16.

[102] Lu G, Yan Q, Su X, et al. Research progress of porous hydrogels. Chemical Progress 2007;(04):485−93.
[103] Zhai M, Ha H. Synthesis, properties and applications of hydrogels. College Chemistry 2001;(05):22−7.
[104] Yin D, Zhou Y, Liu Y, et al. Latest research progress of hydrogels. New Chemical Materials 2012;40(02):21−3 71.
[105] Cheng H, Wang J, Yan X. Research progress of hydrogel. File 2021;11(5):351−3.
[106] Chen P, Yang F, Gu Z, et al. Research progress of antioxidant hydrogels. Journal of Functional Polymers 2021;34(2):182−94.
[107] Lin M, Song M, Liang X, et al. Research progress of toughened hydrogel. Polymer Materials Science and Engineering 2019;35(11):174−80 190.
[108] Liu X, Hong M, Shi FK, et al. Multi-bond network hydrogels with robust mechanical and self-healable properties. Chinese Journal of Polymer Science 2017;35(10):1253−71.
[109] Beigang L, Haiyang Y. Excellent biosorption performance of novel alginate-based hydrogel beads crosslinked by lanthanum(III) for anionic azo-dyes from water. Journal of Dispersion Science and Technology 2021;42(12):1−13.
[110] Akbar SA, Mohammad GM, Meisam TM, et al. Hydrogel materials as an emerging platform for desalination and the production of purified water. Separation & Purification Reviews 2021;50(4):1−20.
[111] Kim SG, Nowicki KW, Gross BA, et al. Injectable hydrogels for vascular embolization and cell delivery: The potential for advances in cerebral aneurysm treatment. Biomaterials 2021;277.
[112] Zhang L, Chen L, Zhong M, et al. Phase transition temperature controllable poly (acrylamide-co-acrylic acid) nanocomposite physical hydrogels with high strength. Chinese Journal of Polymer Science 2016;34(10):1261−9.
[113] Yuhan S, Zhaoliang W, Yongchao W, et al. A self-healing carboxymethyl chitosan/oxidized carboxymethyl cellulose hydrogel with fluorescent bioprobes for glucose detection. Carbohydrate Polymers 2021;274.
[114] Dong T, Li S, Li X. Smart MXene/agarose hydrogel with photothermal property for controlled drug release. International Journal of Biological Macromolecules 2021;190:693−9.
[115] Hee KJ, Hasan TM, Sung LD, et al. Temperature and pH-responsive *in situ* hydrogels of gelatin derivatives to prevent the reoccurrence of brain tumor. Biomedicine & Pharmacotherapy 2021;143:112144.
[116] Shi F, Zhong M, Zhang L, et al. Toughening mechanism of nanocomposite physical hydrogels fabricated by a single gel network with dual crosslinking-the roles of the dual crosslinking points. Chinese Journal of Polymer Science 2017;35 (01):25−35.
[117] Yang Z, Chen X, Xu Z. Anti-freezing starch hydrogels with superior mechanical properties and water retention ability for 3D printing. International Journal of Biological Macromolecules 2021;(190):382−9.
[118] Zhao D, Shen Z. Dual-functional calcium alginate hydrogel beads for disinfection control and removal of dyes in water. International Journal of Biological Macromolecules 2021;(188):253−62.
[119] Feng S, Liu F. Exploring the role of chitosan in affecting the adhesive, rheological and antimicrobial properties of carboxymethyl cellulose composite hydrogels. International Journal of Biological Macromolecules 2021;(190):554−63.
[120] Kazemi HB, Shahrzad J, Reza MG, et al. Hydroxyapatite grafted chitosan/laponite RD hydrogel: Evaluation of the encapsulation capacity, pH-responsivity, and controlled release behavior. International Journal of Biological Macromolecules 2021;1 (190):351−9.

[121] Jijo T, Vianni C, Anjana S, et al. An injectable hydrogel having a proteoglycan-like hierarchical structure supports chondrocyte delivery and chondrogenesis. International Journal of Biological Macromolecules 2021;1(190):474–86.

[122] Mahalakshmi P, Vignesh S, Aathira P, et al. In-situ silver nanoparticles incorporated *N, O*-carboxymethyl chitosan-based adhesive, self-healing, conductive, antibacterial and anti-biofilm hydrogel. International Journal of Biological Macromolecules 2021;188:501–11.

[123] Sisi Y, Fan L, Yucheng L, et al. Quantum dots-based hydrogel microspheres for visual determination of lactate and simultaneous detection coupled with a microfluidic device. Microchemical Journal 2021;171:106801.

[124] Nascimento CD, Feitosa JP, Simmons R, et al. Durability indicatives of hydrogel for agricultural and forestry use in saline conditions. Journal of Arid Environments 2021;195.

[125] Yang Y, Liang Y, Chen J. Mussel-inspired adhesive antioxidant, antibacterial hemostatic composite hydrogel wound dressing via photo-polymerization for infected skin wound healing. Bioactive Materials 2022;8:341–54.

[126] Gong J, Katsuyama Y, Kurokawa T, et al. Double-network hydrogels with extremely high mechanical strength. Advanced Materials 2003;15(14):1155.

[127] Shi F., Xie X. Dual physical crosslinked hydrogel with high ductility and self-healing ability. National Polymer Academic Paper Conference; 2013.

[128] Liu M, Cheng R, Qian R. Study on swelling characteristics of polyvinyl alcohol hydrogel. Journal of Polymer Science 1996;(02):234–9.

[129] Cu Y, Sheng K, Li C, et al. Self-assembled graphene hydrogel via a one-step hydrothermal process. ACS Nano 2010;4(7):4324–30.

[130] Feng H, Hailin L, Zishuo Y, et al. Slow-release lubrication of artificial joints using self-healing polyvinyl alcohol/polyethylene glycol/graphene oxide hydrogel. Journal of the Mechanical Behavior of Biomedical Materials 2021;124:104807.

[131] Seo BB, Kwon Y, Kim J, et al. Injectable polymeric nanoparticle hydrogel system for a long-term anti-inflammatory effect to treat osteoarthritis. Bioactive Materials 2022;7:14–25.

[132] Jin S, Liu M, Chen S, et al. Response and application of intelligent polymers and hydrogels. Journal of Physical Chemistry 2007;(03):438–46.

[133] Zhu W, Duan S, Ding J. Tissue engineering hydrogel materials. Functional Polymer Journal 2004;(04):689–97.

[134] Chen M. Research on processing and molding technology of polymer materials. New Industrialization 2021;11(8):174–5.

[135] Liu W. Research on molding technology of polymer materials. Science and Technology Wind 2017;(26):220.

[136] Su L. Research on processing and molding technology of polymer materials. Information Recording Materials 2017;18(5):43–4.

[137] Wang Z. Progress of polymer material processing technology. Information Recording Materials 2019;20(6):64.

[138] Dong W. Characteristic analysis of polymer extrusion foaming equipment. Chinese Science and Technology Investment 2014;(14):235.

[139] Gao F, Li H, Shen C. Development status of extrusion molding. Engineering Plastics Application 2003;31(6):52–5.

[140] Gao F, Li H, Shen C. Extrusion molding of plastic pipes and plates. Engineering Plastics Application 2003;31(7):52–7.

[141] Gao Y. Study on polymer material molding and its control. Value Engineering 2020;39(31):179–80.

[142] Structure and main components of single screw extruder, <http://www.jingdongsuji.com/news-dlgjcjjgjzylbj.html>; June 15, 2019 [accessed 28.07.23].

[143] Zhang C, Tang Y, Zhang H. Development of plastic extrusion. Application of Engineering Plastics 2004;32(2):67–70.
[144] Tong K, Wang C. Analysis and research on technical problems in the extrusion hollow blow molding process. Modern Industrial Economy and Informatization 2020;10(8):130–1.
[145] Wu H, Mei Y, Feng Y. Research on automatic production technology of plastic extrusion molding. Equipment Management and Maintenance 2021;(18):158–9.
[146] An Z, He J, Wang G, et al. Research progress in injection compression mold and injection compression molding technology. Modern Plastic Processing and Application 2019;31(6):60–3.
[147] Shen C., Chen J., Liu C., et al. Plastic injection molding processing equipment. In: Proceedings of 2000 China Engineering Plastics Processing and Application Technology Symposium; 2000, p. 84–289.
[148] Hengsheng S series/VT series HS650S horizontal injection molding machine. <https://www.tuotuo.com.cn/p/product/10001346.html> [accessed 28.07.23].
[149] Dayu Machinery TY-700DS Vertical Injection Molding Machine. <http://www.tayu.cn/zhusuji/676.html> [accessed 28.07.23].
[150] Xing W. Progress of injection molding technology. Modern Plastic Processing and Application 2002;14(4):33–7.
[151] Zhang Q, Zhang Y, Jin Z, et al. Experimental study on disc injection molding machine melting process. Plastics Industry 2021;49(9):60–3.
[152] Zhang X, Zhang S, Li Q, et al. Research progress of micro injection molding. Polymer Materials Science and Engineering 2012;28(5):148–52 156.
[153] Zheng J, Qu J, Zhou N. Research progress of water assisted injection molding technology. Engineering Plastics Application 2003;31(7):65–8.
[154] Zhang X, Zheng Y, Han J. Research progress of low-pressure injection molding process. Materials Science and Technology 2007;15(1):79–82 86.
[155] Duan Z, Xu D. Discussion on composite calendering process and equipment. Journal of Kunming Institute of Technology 1990;(03):73–9.
[156] Liu Y, Zhang Q, Yuan W, et al. Preparation and properties of lotion electrospun Meiteng fruit oil/polyvinyl alcohol nanofiber membrane. Food Science 2021;42 (17):233–40.
[157] Suzhou Peichuan Intelligent Equipment PCM Phase Change Material Calender, <http://www.szpczb.com/cp/32.html>; April 6, 2021 [accessed 28.07.23].
[158] Han X, Li M, Xing Y. Comparison of production methods of calendered and extruded rubber waterproof rolls. China Building Waterproofing 2001;(06):29–30.
[159] Ke Q. Introduction to the development trend of the plastic calendar at home and abroad. Science and Fortune 2016;8(4) 617–617.
[160] Cang Q, Du Y, Chen C. Overview of polymer material molding. Modern Vocational Education 2019;36:228–9.
[161] Zeng J, Zeng X. Study on parison wall thickness control in hollow blow molding. Plastics Industry 2010;38(9):49–52.
[162] Hao C, Wang T. Patent analysis of hollow blow molding technology. Henan Science and Technology 2016;(22):56–8.
[163] Li H, Yuan H, Wang J. Study on the blow molding PVF film process. Plastics Industry 2005;33(10):28–30.
[164] He J, Lin Y, Xiao Z. Research on design and key technology of three-layer parison head of hollow blow molding machine. Plastic Packaging 2020;30(1):28–36 40.
[165] Extrusion blow molding machine, <https://baike.sogou.com/v62079084.htm>; June 8, 2022 [accessed 28.07.23].
[166] Hu C, Feng Z, Liu L, et al. Research progress of PET bottle blow molding. Modern Plastic Processing and Application 2020;32(1):61–3.

[167] Tong K, Wang C. Analysis of hollow blow molding technology. Science and Technology 2020;(7):19.

[168] Chen B. Processing method of forming rubber mold. Electromechanical Engineering 2004;20(1):42–4.

[169] Lei X. Forming equipment and process of liquid silicone rubber products. Unique Rubber Products 2020;41(02):54–60.

[170] Dayu Machinery TY600.2R.J Vertical Injection Molding Machine. <http://www.tayu.cn/zhusuji/218.html> [accessed 28.07.23].

[171] Wang H. Processing and molding technology of rubber sealing products. Proceedings of 2008 New Rubber (Sealing) Technology Exchange and Information Conference, 2008, 234–235.

[172] Huang N, Jiang B. Rubber injection molding technology and equipment. Rubber and Plastic Technology and Equipment 2007;33(7):32–7.

[173] Ding Z, Qi L, Ye J. Compatibility of PVC/PS melts spinning system. Polymer Materials Science and Engineering 2009;25(11):65–8.

[174] Hu X, Xiao C. Research progress in the preparation of hollow fiber membranes by melt spinning. Polymer Bulletin 2008;(6):1–7.

[175] Li Z, Ying Z, Liu M, et al. Preparation of carbon nanotubes/polypropylene composite fibers by melt spinning and their tensile properties. New Carbon Materials 2005;20(2):108–14.

[176] Gao Y, Xiao C, Ji D, Huang Y. Preparation of PVDF hollow fiber membrane by melt spinning drawing method and its oil-water separation performance. Journal of Higher Education Chemistry 2021;42(06):2065–71.

[177] Hu Y, Liu T. Application of nanofibers prepared by solution jet spinning technology in tissue engineering. World's Latest Medical Information Abstracts (Continuous Electronic Journal) 2020;20(18):27–8.

[178] Zhao J, Xing C, Jiang M, et al. Construction of cyclodextrin/polyvinyl alcohol fiber membrane based on wet spinning technology and its removal performance for air pollutants. Polymer Materials Science and Engineering 2021;37(8):11–124.

[179] New Development of Foreign Chemical Fiber Testing Equipment. <http://img.album.texnet.com.cn/view/2018/11/21/56/5bf4be7ff3656.jpg> [accessed 28.07.23].

[180] Pang Y, Meng J, Li X, et al. Preparation and properties of graphene fiber by wet spinning. Journal of Textiles 2020;41(9):1–7.

[181] Wang Y, Xue J, Zhao Z, et al. Preparation of low oxygen content SiC fiber by dry spinning. Rare Metal Materials and Engineering 2008;37:24–7.

[182] Dry-spinning, <https://baike.sogou.com/v8563061.htm>; December 28, 2022 [accessed 28.07.23].

[183] Nanoscale: Biologically inspired wood like MXene@GO Coaxial fiber, <http://www.nanomxenes.com/bonews.php?id = 1108>; October 30, 2020. [accessed 28.07.23].

[184] Gao J. Study on the spinnability of PAN precursor solution in dry wet spinning and the relationship between spinning process and the structure and properties of nascent fibers. Master's thesis, Shanghai: Donghua University; 2003.

[185] Bian Y, Qi J. Control of coagulation bath concentration of carbon fiber precursor in dry wet spinning. Electronic Science and Technology 2014;27(12):31–3 36.

[186] Wang Y, Wu C. Study on solvent diffusion process of polyacrylonitrile based carbon fiber precursor in dry wet spinning. Chemical Industry and Engineering 2008;25(6):543–7.

[187] Zhang D. Dry wet spinning process of polyacrylonitrile-based carbon fiber precursor. Doctoral thesis, Harbin: Harbin University of Technology; 2016.

[188] Take you to understand electrospinning in minutes. <http://www.jinghuizdh.com/detail.php?id = 365>. March 13, 2021 [accessed 28.07.23].
[189] Xue C, Hu Y, Huang Z. Research progress of electrospinning principle. Polymer Bulletin 2009;(6):38–47.
[190] Cui Q, Dong X, Yu W, et al. Latest research progress in preparing inorganic nanofibers by electrospinning technology. Rare Metal Materials and Engineering 2006;35(7):1167–71.
[191] Qin X, Wang S. Process principle, current situation, and application prospect of electrospinning nanofibers. High Tech Fibers and Applications 2004;29(2):28–32.

CHAPTER 2

Rubber and spherical tires

Contents

2.1 Special rubber and equipment for its preparation

2.1.1 Development of natural rubber

Rubber is a copolymer organic compound with high stretch and elasticity and is considered one of the three synthetic materials, along with plastics and synthetic fibers [1]. As one of the four industrial raw materials, rubber holds a high position in the national economy. Rubber is commonly defined as an elastic material that undergoes large deformation when an external force is applied and can be restored after removal [2]. According to the American Society for Testing and Materials, rubber has Young's modulus of 1−10 MPa and can be stretched to twice its length within 1 minute at 20°C−27°C. After removing the force, the length can be

New Polymeric Products
DOI: https://doi.org/10.1016/B978-0-443-19407-8.00003-8

returned to within 1.5 times its original length. Rubber has good fatigue strength, electrical insulation, chemical corrosion resistance, and wear resistance, making it an indispensable and irreplaceable material in the national economy.

Based on its origin, rubber can be classified into two types: natural rubber (NR) and synthetic rubber (SR), accounting for two-thirds of the total production [3]. NR is derived from rubber trees, rubber grass, and other plants through gum extraction and processing, while SR is produced through polymerization reactions of various monomers. NR boasts strong elasticity, excellent insulation, flexural plasticity, water and gas barrier airtightness, tensile resistance, and tough wear resistance, making it a comprehensive elastic material with outstanding performance. It finds extensive applications in industry, national defense, transportation, people's livelihood, medicine, and health and is a crucial industrial raw material and strategic resource.

Over 2000 species of plants contain NR, with the Brazilian rubber tree (*Hevea brasiliensis*) being the most commercially valuable among them [4]. Due to its high yield, it remains the primary source of NR (Fig. 2.1). Rubber trees are cultivated in over 40 countries and regions worldwide, with the global rubber planting area reaching 12.72 million hectares in 2012. Indonesia ranks first in the world, followed by Thailand in second place, and Malaysia and China tied for third. Vietnam and India rank fifth and sixth, respectively. The top six countries account for approximately 72% of the world's total planting area. The total dry rubber production is 11.93 million tons, with a unit area yield of 1387.5 kg/ha. Southeast Asia remains the primary production area of NR, with Thailand, Indonesia, and Malaysia (ITRC member countries) producing 3.7 million tons,

Figure 2.1 Trifoliate rubber tree (*left*) and natural rubber collection (*right*).

3.26 million tons, and 0.95 million tons, respectively, accounting for 66% of global production [5]. NR remains one of the four essential industrial raw materials and plays a crucial role in the national economy and defense construction. The ANRPC association released the latest NR production and demand forecast data on December 4, 2020, predicting that global NR production will reach 13.678 million tons by 2021, up 8.6% year on year. Additionally, global NR demand is expected to reach 13.436 million tons by 2021, up 4.9% year on year.

2.1.2 Chemical structure of natural rubber and its properties

The structure of rubber and all rubber-like substances necessarily meets two characteristics: long chains and sufficient chain flexibility. Long rubber chains are generally connected in the form of cross-links, but between the cross-linking points, there are hundreds of chain segments consisting of single bonds, which are needed to achieve a high degree of stretching. When not stretched, the rubber molecular chains are in the shape of random wire clusters. In contrast, molecular chains can produce huge deformation under external forces through rearrangement. Long chains are one of the reasons for the high elasticity of rubber and are a necessary condition. Another necessary condition is chain flexibility. Deformation of rubber and recovery of deformation involve rearrangement of chain conformation, and chain flexibility directly determines the ease of rearrangement. The polymer must not be crystalline or in a glassy state to produce a large deformation. Keeping the temperature high enough or adding the right amount of plasticizer can prevent the polymer from crystallizing or being in a glassy state [6].

The fundamental characteristic distinguishing rubber from ordinary metals, nonmetals, and other polymeric materials is high elasticity, also known as rubber elasticity. Rubber elasticity combines two major characteristics of matter. First, rubber can produce large deformations without breaking, and typical rubber can elongate up to 5−10 times its original length. Second, after stress relief, the deformed part of the rubber can spontaneously return to very close to its original size without permanent deformation. These two properties are similar to a liquid and a solid. Rubber elasticity is not unique to rubber. Any long-chain polymer can be rubber elastic if certain temperature conditions are met and the polymer is properly plasticized [6].

NR has rubber hydrocarbons (also known as polyisoprene) as the main component and contains 5 wt.% of nonrubber components.

The *cis*-polyisoprene structure of the rubber hydrocarbon, the main component of NR, occupies about 97% [7]. The structure diagram is shown in Fig. 2.2.

Polymers used as rubber materials must meet the following structural requirements to exploit the high elastic properties of rubber materials fully:

1. Smaller rotational potential barriers;
2. weaker intermolecular forces;
3. A smaller density of cohesion energy within large molecular chains is required. The binding energy density of rubber polymers is generally 290 KJ/cm^3, much lower than that of plastics and fiber polymers.

As mentioned earlier, polymers only exhibit high elastic properties above *Tg*, so the temperature range for rubber materials is between *Tg* and melting temperature. Table 2.1 lists the glass transition temperatures of several rubber polymers and their application ranges.

Polymers used as rubber materials must have no or low crystallinity under the conditions of use. For example, polyethylene and polyformaldehyde tend to crystallize at room temperature and should not be used as rubber materials. However, NR ideally crystallizes in tension. It melts after removing a load because the crystalline part can play the role of

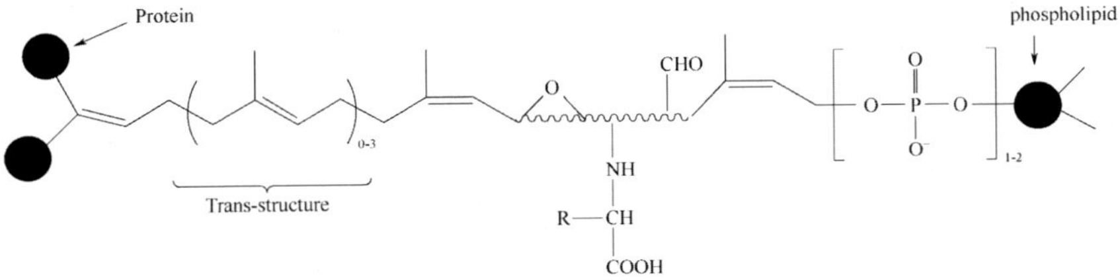

Figure 2.2 Chemical structure of natural rubber.

Table 2.1 Glass transition temperature of several types of rubber and their main temperature applicable range.

Name	T_g/°C	Temperature range/°C	Name	T_g/°C	Temperature range /°C
NR	− 73	− 50–120	NBR (70/30)	− 41	− 35–175
BR	− 105	− 70–140	EPR	− 60	− 40–150
SBR (75/25)	− 60	− 50–140	PDMS	− 120	− 70–275
PIB	− 70	− 50–150	Vinylidene fluoride-perfluoropropylene polymer	− 55	− 50–300

intermolecular cross-linking, increasing modulus and strength, and crystallization and melting after unloading do not affect its elastic recovery properties.

Additionally, there should be cross-linking sites on the macromolecular chains to cross-link and form a network structure, preventing relative slip between molecules and cold flow. Physical cross-linking methods, such as styrene and butadiene block copolymers, can also be used as rubber materials because the ends of the rubber chains can be joined together to form a network structure by styrene segments that aggregate into glassy regions at room temperature. Such rubber materials are also known as thermoplastic elastomers.

2.1.3 Production and processing of natural rubber

Rubber goes through three stages from collection to finished product: latex—raw rubber—vulcanized rubber. The latex collected from rubber trees is usually a milky white liquid consisting mainly of rubber hydrocarbons (*cis*-isoprene) and water. This liquid mixture needs to be separated and dried by centrifugation to produce a film of raw rubber. Although raw rubber is already a commodity, it does not meet the requirements of a material due to its difficult processing and poor mechanical properties. Therefore, raw rubber is often added with various compounding agents and mechanically mixed and vulcanized to make a qualified product. The general production process of rubber products is shown in Fig. 2.3.

A wide range of rubber additives is subdivided into more than three thousand types. They are mainly divided into vulcanizing agents,

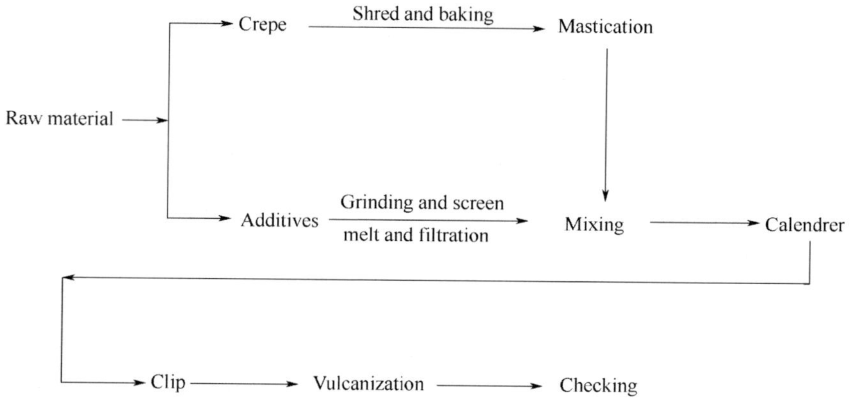

Figure 2.3 Procedure for rubber processing.

accelerators, reinforcing agents, and antioxidants. Vulcanizing agents are often sulfur, polysulfide, or peroxide compounds, while reinforcing agents are carbon black and silica. Accelerators, such as diphenylguanidine and thiazole compounds, are used to accelerate the vulcanization efficiency. Plasticizers, which are usually mineral or vegetable oils, are used to enhance the plasticity of raw rubber.

In the process of rubber production, raw rubber is first plasticized, then mixed with additives according to the formula ratio, calendered or cut according to product requirements, and then vulcanized to complete the production of finished rubber. Whether it is NR or special rubber, the basic process of production and processing is roughly similar, which is from raw materials → plasticizing → mixing → calendering → vulcanizing and shaping before becoming finished products.

The specific process is as follows:

2.1.3.1 Reparation of raw materials

The main raw materials of rubber products are raw rubber, compounding agents, fiber materials, and metal materials. Among them, raw rubber is the most basic material, while compounding agents are auxiliary materials added to improve certain properties of rubber. Fiber and metal materials are used as the skeleton materials of rubber products to improve mechanical strength and limit product deformation [8,9].

Each component must be weighed accurately according to the formula in raw material preparation. To ensure that the raw rubber and compounding agents are evenly mixed, the raw materials need to be processed as follows:

1. Raw rubber must be dried in a room at 60°C–70°C and then cut into small pieces.
2. Compounding agents, such as paraffin, stearic acid, and rosin, must be crushed if they are in block form.
3. The compounding agent must be sieved if it is in powder form and contains mechanical impurities or coarse particles.
4. If the compounding agent is in liquid forms, such as pine tar or coumarin, it needs to be heated, melted, and any water needs to be evaporated and impurities filtered out.
5. Compounding agents should be dry. Otherwise, they can easily cake, and if they are not evenly dispersed during mixing, it can lead to bubbles during vulcanization, affecting the product quality [10].

Accurate weighing and proper processing of raw materials are critical to ensuring the quality and performance of rubber products. It is important to follow the formula precisely and to use high-quality raw materials and compounding agents to achieve the desired properties of the finished product.

2.1.3.2 Plastic refining

Raw rubber is elastic and lacks the necessary plasticity properties during processing, making it difficult to process. To improve its plasticity, the raw rubber needs to be plasticized. This helps to ensure that the compounding agent is easily and uniformly dispersed in the raw rubber during mixing and also improves the permeability and molding fluidity of the rubber during the roll-forming process. The process of degrading the long-chain molecules of raw rubber to form plasticity is called plasticizing [11].

There are two methods of plasticizing raw rubber: mechanical plasticization and heated plasticization. Mechanical plasticization degrades and shortens the long-chain rubber molecules by mechanical extrusion and friction of a plastic rolling mill at a not too high temperature, changing them from a highly elastic state to a plastic state. Heated plasticization feeds hot compressed air into the raw rubber to degrade and shorten the long-chain molecules under heat and oxygen, thus obtaining plasticity [12].

Proper plastic refining is crucial in achieving the desired properties and performance of the finished rubber product. The method of plasticizing depends on the specific type of raw rubber being used and the intended use of the finished product.

2.1.3.3 Mixing (dense refining)

After plasticizing, the raw rubber is mixed with the compounding agents and other additives in the appropriate ratios to form a rubber compound [12]. This process is known as mixing or dense refining. The goal of mixing is to uniformly disperse the compounding agents and additives throughout the raw rubber to achieve the desired properties and performance of the finished product.

Mixing can be done using various equipment, including a Banbury mixer, a two-roll mill, or an internal mixer. The Banbury mixer is a commonly used equipment for mixing rubber compounds. It consists of two rotors that rotate in opposite directions and a mixing chamber. The raw rubber and compounding agents are added to the mixing chamber, and the rotors rotate, causing the rubber to be sheared and compressed between them. This mechanical action allows uniform dispersion of the compounding agents throughout the raw rubber.

The two-roll mill is another commonly used equipment for mixing rubber compounds. It consists of two horizontally placed metal rolls that rotate in opposite directions. Different compounding agents must be added to the raw rubber to achieve the desired properties and performance of the finished rubber product. Mixing disperses the compounding agents uniformly throughout the raw rubber by mechanical mixing.

During the mixing process, the filler is dispersed by the raw rubber and softeners, and the adsorption between the raw rubber and the active point of the filler is combined. The raw rubber also swells due to the softeners, and the surfactants in the mixing process promote the dispersion of the powder. Additionally, a "mechanical-chemical" reaction can occur during mixing, which plays an important role in achieving the desired properties of the finished product.

Therefore, the mixing process cannot be regarded as a simple physical mixing process. It is a complex process that involves various interactions between the raw rubber and the compounding agents. The quality of the mixing process directly affects the quality and performance of the finished product.

Mixing is an important part of the production process of rubber products. If the mixture is not uniform, it will not give full play to the role of rubber and compounding agent and affect the performance of the product. The basic requirements for the mixing process are as follows:

1. The compound must be evenly dispersed to ensure consistent adhesive performance.
2. The compounding agents, such as carbon black, must achieve the best dispersion for optimal performance.
3. The mixing time should be shortened as much as possible to reduce energy consumption and prevent overplasticization.

The rubber compound obtained after mixing is called the rubber compound. It is a semifinished material for the manufacture of various rubber products, commonly known as rubber compounds, and is usually sold as a commodity. The buyer can use the rubber to process and vulcanize it to produce desired rubber products directly. Different formulations show a series of rubber compounds of different grades, varieties, and properties.

2.1.3.4 Molding process

In the production process of rubber products, the preprocessing of various shapes and sizes by a calender or extruder is called molding. The molding process includes three main methods:

Calendering: This method is suitable for making simple sheet and plate products. It involves pressing a rubber compound into a particular shape and size film using a calender. Calendering is also used for coating textile fiber materials used in rubber products, such as tires, rubber sheeting, and hoses. The fiber material needs to be dried and impregnated before calendering, and the gluing process is done on the machine. The purpose of dipping is to improve the bonding performance of the fiber material and the rubber material [13].

Extrusion Molding: More complex rubber products, such as tire treads, hoses, and wire surface coatings, require extrusion molding. This method involves placing rubber compounds with a certain degree of plasticity into the hopper of an extruder and continuously molding them through various mouth shapes (also called templates) under the extrusion of a screw. The rubber material must be preheated before extrusion to make it soft and easy to extrude, which helps to obtain rubber products with a smooth surface and accurate dimensions.

Molding: This method makes rubber products with specific complex shapes, such as cups and seals. It involves using molding inner and outer molds to heat and mold the rubber material in the mold. The molding process is particularly useful for creating intricate shapes and details in the finished product [14].

The molding process is an important step in the production of rubber products, as it determines the final shape and size of the finished product. The quality and precision of the molding process can affect the performance and durability of the finished product.

During the molding process, different factors can affect the quality of the finished product, including the temperature and pressure used in the molding process, the quality of the mold, and the quality of the raw materials used.

To ensure the quality of the finished product, it is essential to use high-quality raw materials and compounding agents, maintain appropriate conditions during the molding process, and conduct proper quality control and testing throughout the production process. This helps ensure that the finished product meets the desired specifications and performance requirements.

2.1.3.5 Vulcanization

Converting plastic rubber into elastic rubber is called vulcanization [15]. This is done by adding a certain amount of vulcanizing agents, such as sulfur or a vulcanization accelerator, to the semifinished product made of raw rubber in the vulcanization tank. The linear molecules of the raw rubber are cross-linked into a three-dimensional mesh structure through the formation of "sulfur bridges" by heating and holding at temperature, thus turning the plastic rubber into a highly elastic vulcanized rubber [16].

The cross-linked bonds in vulcanized rubber are mainly composed of sulfur, which is why the process is called "vulcanization." With the development of SR, there are now many varieties of vulcanizing agents, in addition to sulfur, such as organic polysulfides, peroxides, and metal oxides.

The vulcanizing agent plays a crucial role in the vulcanization process, as it helps to form the cross-linked bonds that give the rubber its elasticity and strength. The amount and type of vulcanizing agent used can affect the properties of the finished product, such as its hardness, resilience, and resistance to heat, chemicals, and abrasion.

Vulcanization is a key process in producing rubber products, as it transforms plastic rubber into a highly elastic material with improved mechanical properties. The resulting vulcanized rubber is also called soft rubber or simply "rubber," and it is used in a wide range of applications, such as tires, hoses, gaskets, and seals.

Proper vulcanization is critical in achieving the desired properties and performance of the finished rubber product. The vulcanization process must be carefully controlled to ensure that the rubber is not overvulcanized or undervulcanized, which can lead to poor performance and durability.

The vulcanization process can be carried out using various methods, such as steam vulcanization, hot air vulcanization, and microwave vulcanization, depending on the specific requirements of the finished product. The temperature, pressure, and time used in the vulcanization process can also vary depending on the type of rubber and the intended use of the finished product.

In addition to the vulcanizing agent, other additives may also be added during the vulcanization process, such as antioxidants, antiozonants, and processing aids. These additives can help to improve the properties and performance of the finished product, such as its resistance to aging and weathering.

Overall, the vulcanization process is a critical step in producing rubber products. It requires careful attention to detail and quality control to ensure that the finished product meets the desired specifications and performance requirements.

Vulcanization is one of the most important processes in rubber processing, and it is necessary to vulcanize rubber products to obtain the desired properties. Unvulcanized rubber has no use value, while undervulcanization or overvulcanization can result in a decline in rubber performance.

Undervulcanization occurs when the vulcanization process is not performed long enough or at a high enough temperature to reach the optimal state. This can result in decreased strength, elasticity, and durability of the finished product. Overvulcanization occurs when the vulcanization process is carried out for too long or at too high a temperature, resulting in a decline in performance.

Therefore it is crucial to strictly control the vulcanization time and temperature during the production process to ensure that the vulcanized rubber products have the best performance and the longest service life. The vulcanization time and temperature can vary depending on the type of rubber and the intended use of the finished product.

In addition to controlling the vulcanization time and temperature, other factors can also affect the quality and performance of the finished product, such as the quality of the raw materials, the compounding agents used, and the processing conditions. Proper quality control and testing throughout the production process can help to ensure that the finished product meets the desired specifications and performance requirements.

Proper vulcanization is critical in achieving rubber products' desired properties and performance. It is important to use high-quality raw materials and compounding agents, control the vulcanization time and temperature, and conduct proper quality control and testing.

2.1.3.6 Supporting measures

To achieve the desired performance requirements for rubber products, auxiliary measures can be added during the production process [17]. Some examples of these measures include:

1. Increasing strength—Use hard carbon black and mix with phenolic resin.
2. Increasing wear resistance—Use hard carbon black.
3. High gas tightness requirements—Use fewer high-volatile components.

4. Improving heat resistance—Use a new vulcanization process.
5. Improving cold resistance—Use raw rubber without branching, reduce the tendency of crystallization, and use low-temperature-resistant plasticizers.
6. Improving flame resistance—Avoid flammable additives, use fewer softeners, and use flame retardants (such as antimony trioxide).
7. Improving oxygen and ozone resistance—Use antidiamine protectants.
8. Improving electrical insulation—Use high-structure fillers or metal powder and antistatic agents.
9. Improving magnetic properties—Use strontium ferrite powder, alnico-iron powder, iron barium powder, etc., as fillers.
10. Improving water resistance—Use lead oxide or resin vulcanization system, and add fillers with low water absorption (such as barium sulfate and clay).
11. Improving oil resistance—Fully cross-link and use fewer plasticizers.
12. Improving acid and alkali resistance—Use more fillers.
13. Improving high-vacuum performance—Use low-volatile additives.
14. Reducing hardness—Use a large amount of filling softener.

Overall, these auxiliary measures can help to improve the performance of rubber products and meet the specific requirements of different applications.

2.1.4 Production and processing equipment

The kneader and open refiner are essential equipment in producing and processing rubber products, as shown in Fig. 2.4.

The open refiner, also known as the open mill, is one of the earliest basic equipment used to process rubber products. It has been used for

Figure 2.4 Old-style knapper (*left*) and motorized single-roller knapper (*right*) [22].

over 200 years since the first single-roller knapper in 1820 [18]. The open machine's rollers are completely exposed, allowing the rubber-making process to be fully visible and convenient for staff to control the temperature of the rollers. The structure of the open machine is simple and easy to disassemble and modify, making it still used by many rubber factories. However, its disadvantage is the low degree of automation, resulting in the mixing process producing a large amount of dust pollution [19,20].

The specification of the open refiner is expressed as "working diameter of roller working length of roller," which means that the diameter of the working part of the front and rear rollers is 550 mm, and the working part of the roller is 1500 mm long. The diameter of the roller is crowned with a letter symbol to indicate the model and usage of the machine, such as XK-400, where X means rubber, K means open machine, and 400 means the diameter of the working part of the roller is 400 mm.

There are three main types of kneading machines manufactured in China [21]: Standard Kneader, Integral Kneader, and Double-Motor Drive Kneader.

The working part of the standard kneader is two parallel and relatively rotating hollow rollers with bearings on the journal of each roller on both sides. The frame is bolted to the machine base, and the upper part is connected to the cross beam. The front roller bearing can be adjusted by a distance adjustment device to control the thickness of the film by adjusting the distance between the front and rear rollers. The rear roller is equipped with a large drive gear at one end, and the motor can drive the rear roller through the reducer and small drive gear. The other end of the rear roller is equipped with a speed ratio gear that can engage with the front roller, allowing the front and rear rollers to rotate relative to each other. The rubber stopper above the bearing is designed to prevent the rubber from falling into the bearing. The emergency lever above the cross beam ensures safety and cuts off the power when necessary. The pitch-adjusting device contains safety devices, such as gaskets, to protect the rollers, frame, and other important parts.

The integral kneader is a type of kneading machine with a closed structure and can avoid dust pollution during mixing. It is suitable for processing high-viscosity rubber compounds and can achieve high mixing efficiency and uniformity.

The double-motor drive kneader is a kneading machine that uses two motors to drive the front and rear rollers separately, allowing for more precise control of the mixing process and achieving higher mixing efficiency and uniformity.

Overall, kneading machines are essential equipment in producing and processing rubber products. The choice of kneading machine depends on the specific requirements of the finished product and the production process.

The integral kneader refers to the kneader where rollers, motor, and transmission system are all installed in a single base. The advantages of the integral kneader are its compactness, easy installation, small footprint, and lightweightness. The drawback is its inconvenience for maintenance, and it is commonly used in laboratory settings.

The biggest feature of the dual-motor drive kneader is that it lacks speed ratio and drive gears. The power is slowed down by two motors using the circular arc gear reducer and then driven by the universal coupling to rotate the front and rear rollers. The speed ratio of the drum can be changed by adjusting the speed of the two drive motors. Incorporating a circular arc gear reducer improves the life and work efficiency of the kneader and also improves cost-effectiveness. Some new laboratory settings have started using dual-motor structure kneading machines with stepless frequency conversion speed control to meet the process requirements for different rubber materials [23].

The drive system of the knapper is the power source for the normal operation of the knapper, including the motor, reducer, and drive gear. The good or bad transmission system greatly influences the whole machine layout, floor space, and maintenance of the knapper. The transmission mode is generally divided into single and multiple transmissions according to the number of motors. Single drive refers to the knapper driven by one or two units. This drive is characterized by easier control of the drive, but also the most commonly used at home and abroad to drive the knapper.

A drive method in which one motor drives multiple kettles is called a multiple drives. This drive type is designed to reduce the number of motors and reducers, thereby reducing power consumption and floor space. The disadvantages of this drive method are as follows:

1. Energy consumption is wasted when the kettles do not work simultaneously.
2. Because there is only one motor drive, all the openers cannot work in case of motor failure.
3. The layout of this drive has greater requirements for plant area and process layout and is not easy to maintain.

Given the above, this kind of transmission is not generally used.

Although there are many kettling machines, the structure is similar, mainly consisting of rollers, barrels, roller bearings, frames, beams, distance-adjusting devices, temperature-adjusting devices, and brakes. The roller is the most important part of the knapper, which is directly involved in the rubber refining operation, so it has the greatest impact on the performance of the knapper. The basic requirements of the knapper roller are high enough mechanical strength and stiffness to ensure that the roller is not damaged; the roller surface should have high wear resistance and chemical resistance to prevent corrosion of the roller surface by various additives in the raw material; reasonable shape to eliminate stress concentration; the good thermal conductivity of the roller to facilitate heating or cooling of the rubber. Because of the above requirements, the general rollers are made of chilled cast iron. There are two roller structures: hollow structure and circumferential drilling structure, as shown in Fig. 2.5.

Compared with hollow rollers, rollers with circumferential drilling structures have the advantages of large heat transfer area, high heat transfer efficiency, and uniform temperature on the surface of the rollers. However the disadvantage is that the processing is complicated, and the production cost is high.

Roller bearings generally have to bear large loads, low sliding speeds, and high temperatures. Therefore, sliding bearings generally require wear resistance, high load-carrying capacity, and high life. Roller bearings use two types of bearings: sliding and rolling. Sliding bearings are simple in structure, widely used, and inexpensive; rolling bearings will reduce friction loss and have a long service life.

To meet the requirements of different refining processes, the kneader often requires a change in the roll distance. Therefore a pair of spacing devices are needed on the machine frame. The spacing range of the spacing device is between 0.1 and 15 mm. The roll distance should not be too large to not damage the speed ratio gear. Commonly used distance-adjusting devices are manual, electric, and hydraulic drive types.

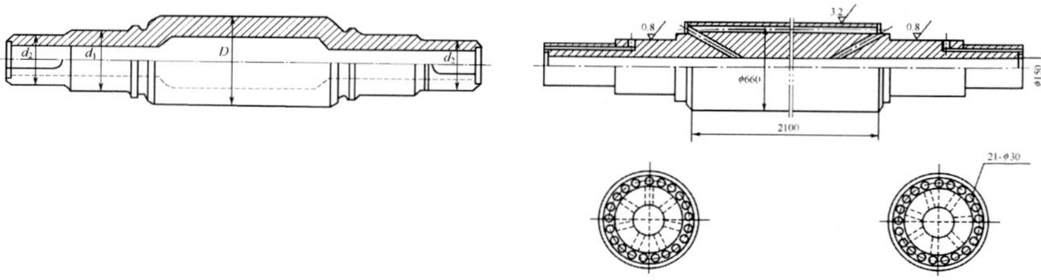

Figure 2.5 Hollow roller (*left*) and circumferential drilling roller (*right*) [21].

In using the knapper, many manual operations inevitably lead to safety or equipment accidents, so a safety brake is required. Safety braking devices generally have two parts: safety and braking. The safety device is mainly a gasket, the purpose of which is to protect the main parts of the knapper from damage due to the overload operation of the machine. A brake is an emergency stop device. Pull down the lever or press the button to cut off the power to stop the machine.

Different rubber materials have different temperature requirements when plasticizing, and the heat generated by friction during the plasticizing process can also cause the temperature of the rubber to rise. Therefore the knapper must have a device to adjust the roll temperature. Generally, rubber kettling machines use water-cooled roller cavities to adjust the roller temperature, which is simple in structure and can have a good cooling effect.

The working principle of the refiner is very simple. That is, the material is squeezed and sheared in the gap between the two relatively rotating rollers to process the material.

N is the positive pressure of the material on the roller, F is its reaction force, F_t and F_x are the tangential and horizontal components of F, respectively, and T is the friction force of the roller on the material. Obviously, when $T > F_t$, the material can enter the roller, so this is the operating condition of rubber refining. Where T is calculated by the Eq. (2.1)

$$T = F\mu \tag{2.1}$$

Where μ is the coefficient of friction between the material and the roller. And,

$$\mu = \tan\varphi \tag{2.2}$$

Where φ is the friction angle between the material and the roller.

F_t in the right triangle $F_t\,FO$ gives

$$F_t = \tan\alpha \tag{2.3}$$

Where α is contact angle.

From the operating conditions can be derived:

$$\varphi > \alpha$$

That is, when the friction angle is greater than the contact angle, the material can be pulled into the roll gap. The friction angle is affected by the plasticity of the material. Warming or changing to a more plastic rubber will increase the friction angle. The general friction angle is 38–50 degrees.

Currently, with the progress of science and technology, some new technologies have been introduced to the kneader. The control system of the kneader is programmable. It receives and sends instructions through programmable logic controller (PLC), enabling better coordination with the front and rear equipment and realizing automatic control of some functions. The application of hydraulic, pneumatic, and other technologies makes the connection between the various parts of the kneader more stable, orderly, and accurate while greatly reducing the labor intensity of workers and improving the safety of the equipment.

The development of pneumatic and electric knife devices makes the cutting and conveying a film more convenient and fast. The emergency brake device and the photoelectric switch to prevent brake reversal make the kneader work more safely to avoid threats to workers' safety. The automatic rubber turning device replaces manual rubber turning, and the quality of rubber mixing is higher and more labor-saving.

Especially the application of a hydraulic distance adjusting device not only realizes the automatic roll distance setting but also can accurately monitor the roller load in real-time during the working process. According to the programmed requirements, the hydraulic spacing device can realize the protection of the core working components such as rollers, thus avoiding damage and personal injury accidents caused by equipment overload. Furthermore, the hydraulic spacing device replaces the mechanical safety sheet, eliminating the tedious, repetitive labor of replacing the safety sheet, and ensuring the continuity of production.

Reference [24] highlights that the application of hydraulic spacing devices ensures the continuity of production while protecting the working components of the kneader. Moreover, it enables the automatic adjustment of the roll pitch after setting and accurate monitoring of the roller load in real time during the working process. Overall, the integration of advanced technologies has enabled the kneader to work more efficiently, safely, and with higher quality.

2.1.4.1 Compacting machine

The closed-type refiner, also known as a dense refiner, is mainly used for rubber or plastic refining and mixing. The closed refiner is a high-intensity, interstitial mixing equipment developed based on the open refiner. Rubber mixing with carbon black and other compounding agents was first realized using an open refiner. However, the open machine is not easy to operate, has serious dust emissions, a long mixing time, and low productivity. Since the emergence of the true sense of the Banbury machine in 1916, the advantages of the machine have gradually been recognized. In the rubber mixing process, the Banbury machine has more advantages than the opening machine, including a large mixing capacity, short time, high efficiency, less dust flying, less loss of compounding agent, as well as a good working environment, safe operation, low labor intensity, high degree of mechanization and automation, and good quality of rubber [25]. Therefore, the Banbury machine has become the most common rubber refining equipment in modern rubber factories.

However, in contrast, the Banbury machine does not allow for observing the state of the rubber material during refining, so it is impossible to adjust the rubber ratio flexibly. Additionally, the temperature is generally very high and not easy to control during refining, so it is unsuitable for mixing rubber with low-temperature requirements [26].

The Banbury machine generally consists of a compacting chamber, two rotating rotors, a pressure weight and discharge door, a temperature measuring system, a heating and cooling system, an exhaust system, a safety device, a discharge device, and pneumatic, hydraulic, and electric control systems, as shown in Fig. 2.6.

The transmission system is one of the main components of the Banbury machine, which is used to transmit power to make the rotor overcome the working resistance and rotate to complete the rubber refining operation. The transmission mode of the Banbury machine is divided

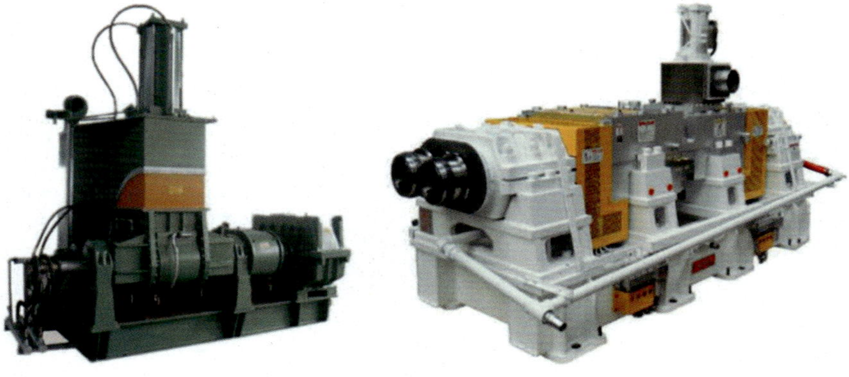

Figure 2.6 Dense refiner (*left*) and FCM continuous refiner (*right*) [27].

into a separate transmission and two linked transmissions. The single drive can be divided into large drive gear drive, no large drive gear drive, and double shaft reducer drive. The two connected drives can be divided into the left drive, right drive, and intermediate drive. The double-out reducer drive is the most commonly used transmission method for industrial Banbury machines.

The rotor is the main working part of the Banbury machine. The strength and shape of the rotor have a great influence on the efficiency, service life, and quality of the machine. In the rubber refining process, the rotor is often subject to the friction and extrusion of the rubber material, and the additives in the rubber material will also corrode the rotor. These effects are stronger than in the open refiner, and the rotor structure is relatively complex. The general material of the rotor is mostly 45 steel casting, and a layer of cemented carbide with a hardness of 55–62 HRC is welded at the protrusions. The working surface of the high-speed Banbury machine should be surfaced with 23 mm wear-resistant cemented carbide.

The elliptical rotor is divided into double-pronged and four-pronged rotors according to the number of prongs (Fig. 2.7). The practice has shown that the four-pronged rotor refiner has a higher production capacity than the double-pronged rotor refiner and consumes less power per unit of material, improving the rubber refining quality. The rotors are cavities that are cooled or heated by condensate or steam during operation. The rotor generates a lot of heat during the rubber refining, making it difficult to dissipate. Therefore, cooling the rotor is also the key to improving the quality of the rubber. Generally, there are two types of rotor cooling methods: spray type and jacket type, among which the spray type is the most commonly used.

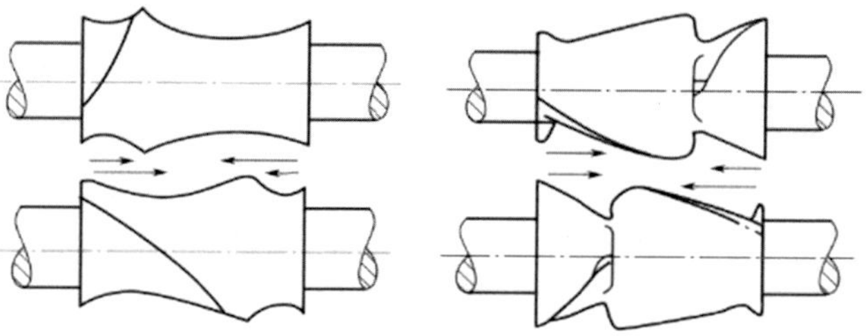

Figure 2.7 Double-protrusion rotor (*left*) and four-protrusion rotor (*right*) [21].

The spray-type rotor cooling method uses a cooling water spray nozzle to directly cool the rotor surface during operation, which has a simple structure and good cooling effect. The jacket-type rotor cooling method uses a cooling water jacket to surround the rotor. The cooling water circulates in the jacket to cool the rotor, which has a better cooling effect but a more complex structure.

The mixing chamber is also a critical component of the kneader. Similar to the rotor, it is subjected to strong friction, extrusion, and chemical corrosion of the rubber during the kneading process. These effects can cause serious wear on the wall of the mixing chamber and even affect the use of the machine. Therefore the inner wall of the high-speed kneader needs to be surfaced with a layer of wear-resistant carbide with a thickness of 2–4 mm.

The structure of the mixing chamber is diverse, using a combination of open-type and front and rear combination (Fig. 2.8). The open combination-type mixing chamber is more convenient for manufacturing and maintenance. The front and rear combination-type mixing chamber is more conducive to the heat dissipation of the mixing chamber. The combination of open-type and front and rear combination mixing chamber has the advantages of both mixing chambers, effectively improving heat dissipation and the convenience of manufacturing and maintenance.

The working principle of the kneader (Fig. 2.9) is more complicated than that of the open kneader. The rubber is kneaded in the mixing

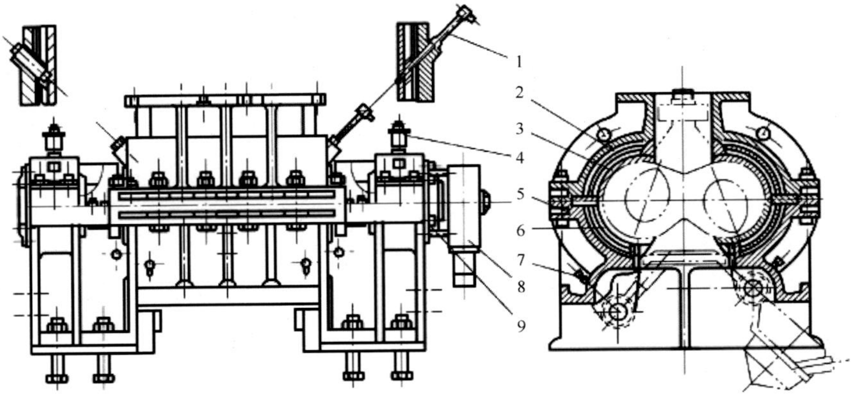

Figure 2.8 Combined kneading chamber with Folio [21]. 1—Thermocouple; 2—Upper housing; 3—Upper mixing chamber; 4—Dry oil cup; 5—Lower housing; 6—Lower mixing chamber; 7—Oil cup; 8—Cooling water tank; 9—Gland.

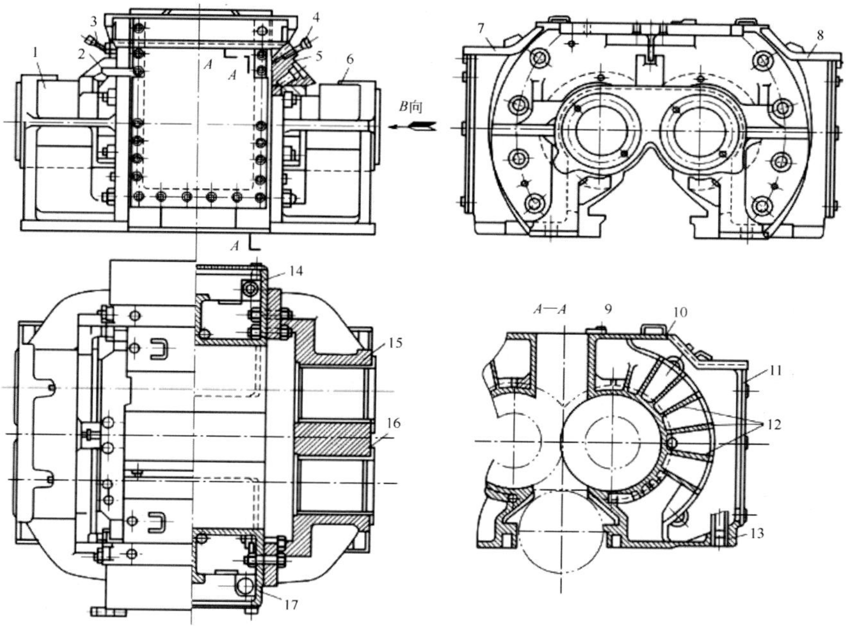

Figure 2.9 Combined front and rear mixing chamber [21]. 1,6—side wall; 2—left bracket; 3—thermocouple; 4—refueling tube; 5—right bracket; 7,8—shell cover; 9—sealing pressure plate; 10—upper cover; 11—front cover; 12—reinforcement; 13—tube; 14, 17—positive wall; 15,16—bushing.

chamber not only in the gap between the two rotating rotors but also in the gap between the rotors and the wall of the mixing chamber and in the gap between the rotors and the top and bottom bolts, which causes the rubber to be sheared and deformed.

After the raw rubber and compound are added to the mixing chamber, the material receives shearing and squeezing in the gap between the two rotating rotors, the gap between the rotors and the wall of the mixing chamber, and the gap between the rotors and the top and bottom bolts, resulting in shear deformation. The rotor has spiral ribs, which cause the rubber to repeatedly move axially when kneading, playing a stirring role and making the kneading more intense, as shown in Fig. 2.10.

When mixing a large amount of compound with raw rubber, glue is a non-Newtonian fluid in the process. The mixing process is roughly divided into two steps: firstly, the granular solids and compounds are mixed into the raw rubber to form binder blocks, called simple mixing.

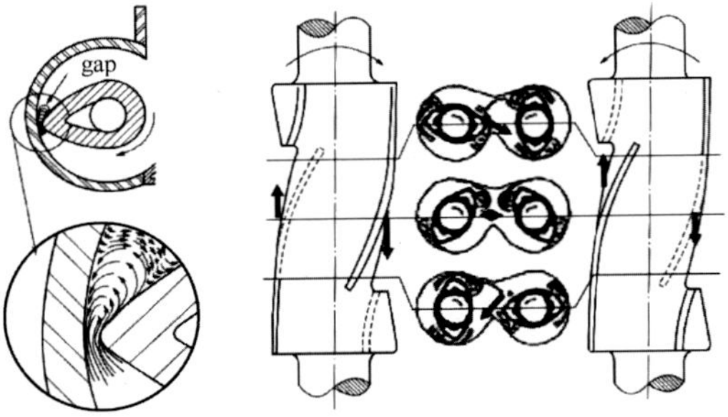

Figure 2.10 Working principle diagram of oval rotor compacting machine [21].

Then these binder blocks are further dispersed evenly, also called strong mixing. It has been proved that high shear stress is required to achieve good dispersion.

Currently, three main types of compactors are used worldwide: the American Banbury compactor, the German GK compactor, and the British K compactor. Each type has advantages and disadvantages, depending on the specific application and requirements. The Banbury compactor is the most commonly used and has high automation and efficiency. The GK compactor has a higher mixing efficiency, but its structure is more complex and requires more maintenance. The K compactor has a simple structure suitable for small-scale production but relatively low mixing efficiency.

The Banbury machine was invented by an American Farrel company and named after its inventor in 1915 [28]. Since then, it has been developed and improved from the D to the NF model, which includes two and four prongs. The D-type refiner replaces the shear-extruded circular rollers of the knapper with prismatic curved rollers (rotors), while the F-type refiner uses linearly tangential oval rollers (rotors). Due to the different shapes of the rollers, the mixing effect of the compactor and the knapper is different. The Bembridge kneader is suitable for mixing large batches of rubber, and the amount used accounts for about 1/3 of the total amount of the kneader.

In 1913, the German company W&P first patented and produced the GK-type kneader technology. The principle of the GK mixer is similar to that of the Banbury mixer. The GK-type mixer includes the GKN-type nonengaging tangential elliptical rotor mixer and the GKE-type engaging cylindrical rotor (two-sided and four-sided) mixer, which is suitable for mixing a large amount of rubber and has a large consumption, accounting for about 1/3 of the total consumption of the mixer [29].

In 1986, a Chinese company introduced the GK300 mixing machine from Germany. They also introduced GK technology from Germany and developed the GK270 machine in the same year. After more than 30 years of development, the key technology of the GK type of compacting machine is close to the advanced international level [30].

In 1934, British R. Cook invented a type of bite-in rotor internal mixer with the same shape as open mill rolls and put it into production at Francis Show Company. This mixer is a K-type internal mixer. In 1937, the Italian Pomini company improved the bite-in mixing machine. It developed a variable roll pitch K-type mixing machine with an adjustable rotor gap like an open-type mixing machine. They also developed two synchronous rotor K-type mixers with the same rotor speed. K-mixers are suitable for mixing low-temperature high-quality rubber, and their consumption accounts for about one-tenth of the total consumption of the mixing machine.

The host control system is the control core of the compactor. It includes a PLC [31] and a range of software with high integration and modularity. The main control system in Germany, the United States, and England is usually set up in a closed space isolated from the site. In contrast, the main control system of the domestic refiner is mostly located directly on the operating site. In recent years, many of these machines have been equipped with infrared scanners to facilitate the identification of raw material codes and to prevent the use of wrong raw materials. Generally, the main machine control system is not easily contaminated by dust and has a long service life.

As an upgrade of the opener, the compactor has made numerous research advances in recent years. To develop high-performance tires and improve the application of silica in compounding rubber, tandem low-temperature primary equipment has been developed to alleviate the problems of the traditional compounding process with many stages, high energy consumption, and poor dispersion of silica [32]. Recent studies have focused on improving the performance and efficiency of compactors, specifically in the areas of constant temperature mixing time on silica dispersion, the effect of adding sequence on mixing quality and physical properties, and the impact of high-speed mixing on processing performance and physical properties. These studies were conducted using imported mixing machines and provided a foundation for optimizing processing performance, improving physical properties, and enhancing industrial production efficiency [33]. The mixing chamber of a compactor is a closed and opaque structure, making it difficult to observe the flow of rubber during the mixing process visually. However, numerical simulation techniques and visualization studies have proven effective in reflecting the rubber flow in the chamber during the mixing process [34]. This technique has demonstrated its effectiveness in reflecting the flow conditions of the compounding process. New technologies, such as electromagnetic dynamics, have been introduced to improve the compounding process further [35]. The introduction of new technologies, such as dynamic electromagnetic technology, has led to the implementation of vibration fields caused by electromagnetic fields into the polymer processing and molding process. This has resulted in a series of new physicochemical effects, such as phase changes in polymer material, changes in aggregation state, and self-reinforcement mechanisms of polymers. These effects enhance the rubber mixing process, reduce energy consumption, and provide design ideas for energy-saving and consumption reduction of compacting machines.

With the rapid development of computing technology, network technology, artificial intelligence, automatic control, applied mathematics, and related disciplines, the automation and intelligence of compacting machines have become inevitable. For example, fuzzy control technology can be combined with traditional Proportional Integral Derivative—PID [36] for temperature control, achieving the goal of a low-temperature refining process. The temperature control technology can be further optimized for better control and energy efficiency.

Many domestic compactor manufacturers have recently developed automatic weighing devices for carbon black, calcium carbonate, other powders, and rubber oil based on foreign experience [37]. These devices allow for the automatic injection and weighing of powder and liquid oil through screw conveyors and pneumatic butterfly valves, all done in a closed state throughout the process. This not only reduces dust pollution but also minimizes oil waste.

2.1.4.2 Rubber injection press

The rubber injection molding process is similar to plastic extrusion and injection molding, but it is much more difficult for two reasons. Firstly, rubber is viscous and has poor flowability, requiring high-pressure injection to complete the mold, while this problem does not exist in injection molding. Secondly, after injection, plastic can be released from the mold after cooling, whereas rubber must be vulcanized. This makes rubber injection molding much more complicated than injection molding, requiring complex equipment construction, mold design, and formula composition [38].

Despite its complexity, rubber injection molding is necessary for producing higher-quality products, especially in the automotive and aerospace industries. Rubber injection machines, such as the Japanese KRV-600 type, the French Rep Company's V type, and the Italian RUTIL Company's RS-1000/150 type, are required to meet the development needs of these industries. These machines are designed to handle the unique challenges of rubber injection molding, as shown in Fig. 2.11, such as the need for high-pressure injection and vulcanization.

Rubber injection presses can typically be divided into three types based on the injection mechanism: screw type, plunger type, and screw/plunger type, as shown in Fig. 2.12. The screw type is unsuitable for the high viscosity of rubber due to low injection pressure, and the plunger type cannot achieve sufficient mixing and blending. Therefore, most rubber injection presses adopt the screw/plunger structure to combine injection and mixing functions. In addition, heating and pressurizing means are added to complete the vulcanization process.

The typical components of a rubber injection press include:

1. Injection device: This consists of a barrel, screw, screw drive, nozzle, and injection/reciprocating cylinder.
2. Heating/cooling device: To ensure that the rubber material reaches the required vulcanization temperature and subsequent vulcanization temperature, the barrel part can be heated and cooled simultaneously.

Figure 2.11 Injection molding machine for wheel fabrication [39].

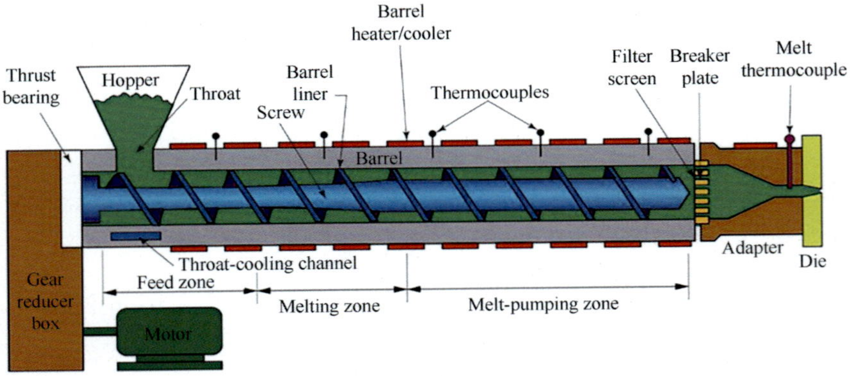

Figure 2.12 Schematic of thermoplastic injection molding machine [39].

3. Transmission and electrical system: This system includes the motor, transmission system, and electrical control system, which control the movement of the injection device and the heating/cooling device.

The electrical control part adopts a computer and programmable controller, and the control panel is equipped with a Control Receiver-Transmitter (CRT) monitor to monitor all operation actions or data. The equipment adds intelligent software, such as (1) a closed-loop control system, to improve the adjustment accuracy and repeatability. (2) The system automatically calculates the PID adjustment coefficients for the mold so that when the mold is used for the first time, these coefficients are calculated during the preheating process. Assuming a significant temperature change, the microprocessor calculates the PID adjustment coefficients to provide more accurate temperature control. (3) Modifying the curing time system causes certain process parameters to fluctuate due to external influences. Automatic adjustment of these changes allows the curing time to be corrected in each cycle. (4) Network management system. Currently, each rubber injection machine in our country is independent. The modern rubber industry is connected to multiple injections by a group of central computer equipment, forming a central management network. They will communicate with each other. The data transfer enables production management, quality control, and statistical processing control of the production process, forming a production line [40].

2.1.4.3 Calender

Calendering is a fundamental process used in rubber processing to create a film or fabric coating of uniform thickness from rubber materials on a calender, as shown in Fig. 2.13. The process is versatile and can be used for sheeting, molding, gluing, rubbing, laminates, thin channels, filtering rubber, and more. Special operations such as filtering and laminating are also commonly used in other rubber processing operations, such as extrusion.

Calendering is also vital in production lines for conveying rubber materials and connecting before and after processes. This helps to create a seamless and integrated production process, improving efficiency and productivity [35].

A calender is a large, complex, highly automated plastic and rubber processing equipment crucial to calendering production lines. With different auxiliary machines, it can produce various products such as plastic film, PVC flooring, and modified artificial leather. The number of rollers in a calender can range from two to five, and the arrangement of rollers can include horizontal, I, inverted L, S, and E types. The roller structure can either be hollow or drilled, with the roller blank made of alloy chilled cast iron by centrifugal casting. The four-roller calender can have equal or reduced diameters rollers and is equipped with different auxiliary machines before and after the calender to produce the plastic film, modified artificial leather, PVC flooring, or rubber products.

The working principle of a calender is based on the gradual thinning of the material due to extrusion and shearing as it is pulled into the two adjacent roll slits rotating in opposite directions under the action of friction. With different rotational speeds of adjacent rollers, there are different roller gaps between the rolls for products of different thicknesses. The materials rotate and flow regularly between the roller gaps, thus further plasticizing and making the internal plasticization of materials more

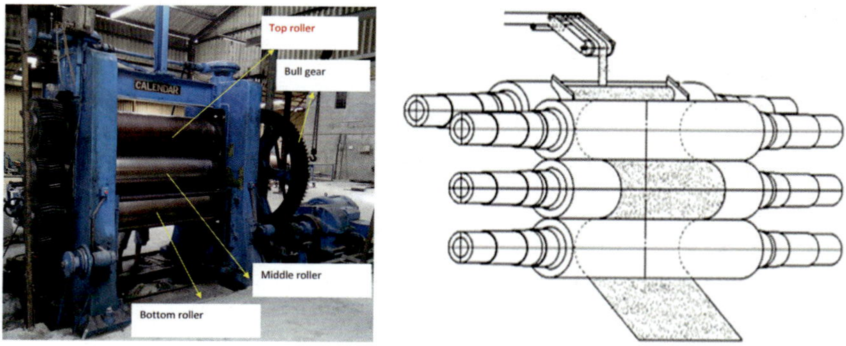

Figure 2.13 Calender [43] (*left*) and working principle diagram of calendar [42] (*right*).

adequate. This also excludes some volatile gases and mixed air. After the material is crushed and sheared by a certain number of roller gaps, the plasticization of the material is strengthened, and the required specifications and precision are achieved.

The rollers, bearings, frame, baffle, lead-off, laminating and embossing devices, lubrication, and temperature control are the main parts of calendering equipment. The rollers are the most critical component and are made of alloy chilled cast iron by centrifugal casting. The bearings are designed to handle the high loads and speeds of calendering. The frame provides the rigidity to withstand the high pressures and forces generated during the process. The baffle plate guides the material into the roll gap while the lead-off device removes the material from the rolls. The laminating and embossing devices finish the product, while lubrication and temperature control ensure that the process is consistent and accurate.

In summary, a calender is a complex and highly automated equipment that produces various products such as plastic film, PVC flooring, and modified artificial leather. Its working principle is based on the gradual thinning of the material due to extrusion and shearing as it is pulled into the two adjacent roll slits rotating in opposite directions. The number of rollers and roller gaps determines the precision and thickness of the product. The main parts of a calender include rollers, bearings, frame, baffle plate, lead-off device, laminating and embossing device, lubrication, and temperature control [41]. Calendering equipment is critical in determining the quality and accuracy of the final product. With the ongoing development of technology, the automation and intelligence of calendering equipment will continue to improve, leading to more efficient and high-quality production processes in the plastic and rubber industry.

The plastic calender production line, consisting of a calender and its associated auxiliary machines, is a fundamental piece of equipment in the production process of plastic products. It is a large, high-precision equipment. The calender has been used in plastic processing for over 100 years, during which time it has continuously improved and undergone some significant changes. As the technology for processing plastic products has advanced, and the variety and quality of plastic products have increased, the development of the plastic calendering machine production line has entered a period of rapid growth [42].

Furthermore, with the increasing requirements for environmental protection, energy saving, and safety, plastic calenders have made significant progress in increasing varieties, improving quality, saving energy, reducing consumption, and automatic control. They constantly evolve toward large-scale, high-precision, high-efficiency, and high-automation machines.

2.1.4.4 Extruder

Extruder is a common polymer processing equipment, not only limited to rubber processing. Its working principle mainly relies on the shearing and extruding action produced by the screw and the barrel during the movement. Since the screw can be heated, some chemical reactions (such as grafting, cross-linking, and blocking materials) can coincide during the extrusion process.

The body of the extruder is a barrel/screw/head/port. The barrel has a jacket to pass steam and cooling water to regulate hot and cold. The screw is the operating part of the extruder. As the material passes through the gap between it and the cylinder wall, it is stirred by strong extrusion pressure and friction. The main parameters of the screw are (1) length-to-diameter ratio L/D. The L/D of a plastic extruder is 10–15 because the latter also plays a role in mixing homogeneously, while in a rubber extruder, these tasks are done by a thermal mixer. However, the L/D ratio of rubber extruders also tends to increase to 8 to meet the needs of cold feeding. The rubber passes along the screw from back to front, successively through three zones: feeding section, compression section and extrusion section. (2)The pitch of two adjacent threads decreases from back to front to increase the pressure. (3) Thread coefficient H/B. The ratio of thread depth to grain width is usually between 0.3 and 0.5. (4) Compression ratio. The barrel-to-barrel volume ratio gradually increases from back to front, so the rubber material gradually changes from loose to tight. The best value is 2:1. That is, the rubber material from entering the barrel shape to the mouth shape of the compression force doubled. (5) Screw peak. The width should be moderate. Too narrow will decrease extrusion capacity, and too wide will cause local overheating, resulting in burning rubber. (6) Speed. There should be several speeds to meet the linkage needs with the production line [38].

The extruder head is the forming mechanism located at the front end of the machine, mainly used to install the mouth shape. The purpose is to make the rubber mix evenly and without stratification. The shape of the head varies. Most of the head geometry is cylindrical but flared and flat (for tread extrusion) to suit different product shapes. The head is also equipped with a temperature regulation device to ensure the smooth flow of the rubber. The mouth shape is divided into two types: round mouth and flat mouth. The former makes inner tubes, hoses, and rubber strips, while the latter is used for tread extrusion. The mouth shape of extruded hollow products consists of an outer shape, core shape, and bracket. The size of the mouth hole should be 1/3–3/4 of the screw diameter. Too much or too little pressure will cause burning. To speed up the extrusion speed, the cone angle of the outlet mouth should be large, and the diameter near the outlet port should be small. The inner wall of the discharge port should be smooth, streamlined, and without dead ends. If there is a big difference between the amount of glue discharge and the size of the discharge port, the flowing glue hole can be added to the mouth plate to prevent burning. The calculation of the expansion rate is essential in the design of the discharge chamber structure.

First of all, the shrinkage rate of various adhesives is different, and even if the same adhesive is used, the number of layers will be different due to the different glue content. Generally, high glue content and expansion (or shrinkage) deformation are also large. In addition, the expansion rate is also related to the amount of filler. A large amount of filler is a small amount of deformation. Finally, it is related to the type of filler, such as filling 75 FEF carbon black (fast-pressing carbon black, known as small extrusion deformation) with rubber filler. The extrusion deformation is only 5%. In contrast, the deformation rate of the same amount of carbon black by the can-changing method is 67%. Of course, there is a relationship with the plasticity of the plastic material [38].

Extrusion equipment is a plastic machine with a wide range of applications. The composition of the auxiliary machine of extrusion equipment depends on the type of products, as shown in Fig. 2.14 [44]. The auxiliary machines mainly include a shaping device, cooling device, traction device, cutting device, wire release device, straightening device, preheating device, cooling device, traction device, meter counter, spark tester, take-up device, cutter, blow dryer, printing device, etc. The screw extruder is combined with special follow-up equipment (auxiliary machine) to form a special extrusion line [45]. The following is a brief introduction to the current status and progress of extruders.

Extruders first appeared 200 years ago, invented by J. Brand. Early extruders are plunger type, belong to the discontinuous equipment, are generally used to produce lead pipe, and have low production efficiency. It was not until 1881 that the British Show Company officially launched the world's first single-screw extruder. Still, the efficiency of the early single-screw extruders was not high due to power and machining technology limitations.

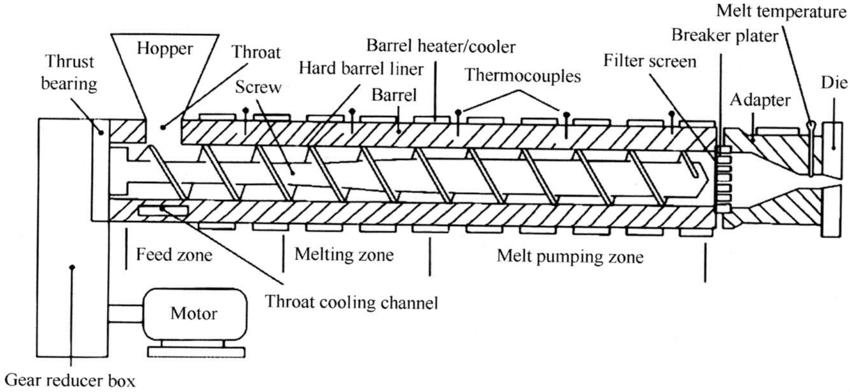

Figure 2.14 Cross section of an extruder showing principal components [44].

The modern single-screw extruder was developed by P. Troester in 1939, with barrel heating and air cooling, an automatic separate control cabinet for temperature control, a stepless variable speed drive, and an increased screw length-to-diameter ratio of 10, which significantly improved the efficiency of single-screw processing [46]. The efficiency of single-screw processing has been greatly improved since then.

The emergence of single-screw extruders has improved extrusion production efficiency, but there are still limitations. Difficulties in material conveying and high friction between the material and the equipment in the barrel lead to the limitation of material addition, which has a greater impact on the process of inorganic fillers to be added; strong reverse pressure is easily generated at the extrusion outlet die, the extrusion capacity of the screw is poor, and the extrusion rate is low; poor exhaust performance in the exhaust area, the melt cannot discharge gas in time, which is not conducive to the flow of the melt, thus reducing the quality of the extrudate. To solve these problems, screw extruders developed into twin-screw or even multiscrew (Fig. 2.15). Compared with single-screw extruders, twin-screw extruders have the following advantages: (1) twin-screw conveying performance is better, which facilitates material addition; (2) with the ability to devolatilize and self-clean the air or low molecular mass components brought into the production; (3) excellent mixing effect and easier plasticization of materials; (4) twin-screw extruders work with much lower energy consumption, usually In general, the energy consumption of twin-screw extruders is nearly 30% lower than that of single-screw

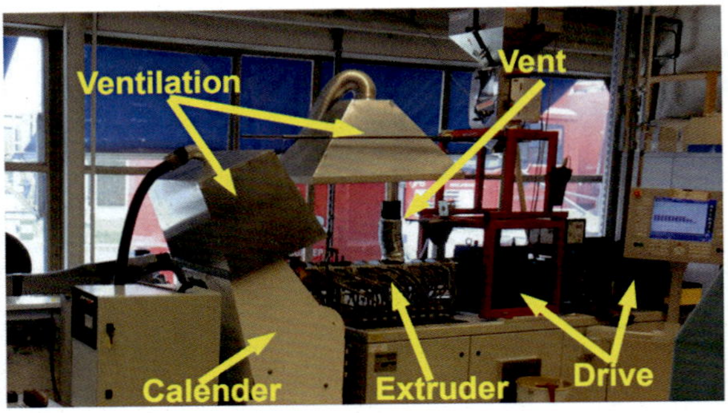

Figure 2.15 Extruder with calendar and venting system [47].

extruders during the process. It is obvious that twin-screw extruders have more advantages than single-screw extruders, so they will soon be widely used.

The current research and development of extrusion equipment focus on producing high speed, high productivity, precision, intelligence, and energy efficiency to achieve their goals—the temperature control problem of the extruder barrel based on a programmable computer controller—PCC controller developed by Bergerac. The use of Bergerac's PCC controller for temperature control experiments shows that the nonlinear PID control method and the self-turbulence controller have a significantly better control effect than the traditional linear PID control method for temperature control processes and have better robustness and self-adaptive ability to optimize the extruder barrel temperature control algorithm [48]. An intelligent computer-aided design (CAD) system for rubber extruder design has been developed based on information sharing and intelligent processing. It is significant to improve the efficiency and quality of rubber extruder product design, lower design costs, accumulate intellectual wealth in rubber extruder design, and share design resources [49]. The intermittent compactor operation makes the twin-screw extrusion press speed control have nonlinearity and time-varying problems. Fuzzy control is a systematic theory based on fuzzy mathematical logic reasoning. It is widely used in various control systems, especially where controlled objects are time-lagged, time-varying, nonlinear, or without precise mathematical models. To improve the twin-screw extrusion system's production efficiency, product quality, and service life, some scientists developed a fuzzy control algorithm to improve the screw speed control system [50]. Precision extrusion can improve the quality of products, reduce material costs, and eliminate subsequent processing means, such as multilayer coextruded composite film and precision medical catheters. Rigorous barrier function requirements, especially for multilayer coextruded films, make the thickness accuracy of each layer greatly impact the film's barrier performance. The key point of precision extrusion research and development is to improve the precision of extruded products, which involves extrusion equipment, auxiliary machines, heads, dies, inspection systems, and control systems. Using modern electronic and computer control technology, the process parameters of the whole extrusion process, such as the melt pressure and temperature, the temperature of each section of the body, the speed of the main screw and feeding screw, the feeding volume, the ratio of various raw materials, the current and voltage of the motor and other parameters, are detected online and controlled by a microcomputer in a closed loop. It is extremely beneficial to the stability of the processing conditions and the improvement of the accuracy of the products. Network control of the extrusion line is carried out by online remote monitoring, diagnosis, and control.

Realizing the innovation of extrusion processes and principles is the root of energy saving in extrusion. The screw produces heat evaporation of water in the raw material while overcoming the traditional equipment's uneven feeding and other disadvantages. It can adapt to different melt viscosity and melt temperatures of the raw material blending requirements. This process saves energy for predrying raw materials, has higher output, and achieves high productivity and energy-saving extrusion [51].

Increasing the specific flow rate and reducing specific power are the key to energy saving in extrusion. The "IKV" screw extruder has forced feeding and forced cooling and features longitudinal grooves of different shapes, depths, and widths in the bushing of the barrel feeding area. A forced cooling jacket structure is designed in the feeding area, which can improve the conveying efficiency of raw materials and increase the extrusion capacity of High-Molecular-Weight Polyethylene—HMWPE from the conventional 0.3 to 0.85. With the reasonable design of the barrier section and mixing section, the plasticizing capacity can be increased to more than 50%. The extrusion volume is stable, which can reduce the energy consumption per unit [51]. The extrusion capacity can be increased to more than 50%, and the extrusion volume is stable, which can reduce the energy consumption per unit.

2.1.4.5 Vulcanization equipment

The final process in tire manufacturing is curing. As a hot press product, the quality of the tire's appearance, the final uniformity of the tire, and the production efficiency of the enterprise are almost entirely determined by the tire curing equipment. Tire curing equipment is diverse, numbering in large quantities and occupying a large area, with its equipment investment accounting for over 25% of the tire company's equipment investment. Moreover, tire curing is the main steam energy-consuming process in tire production, with steam consumption accounting for about 80% of the entire plant [52]. Therefore tire curing equipment is often considered the most important part of the tire production process and a symbol and sign of the modern level of tire production. The optimization of the curing process, as well as the improvement and development of new curing equipment, has always been a hot topic.

There are four tire-curing equipment categories: press-type curing machines, container-type curing tanks, closed curing chambers, and open curing tanks. Curing tanks have been replaced by curing presses and are not discussed here. After nearly 60 years of development, tire-shaping curing presses have developed three types of curing presses: mechanical, hydraulic, and electric spiral. While the latter injection molding vulcanization is suitable for the continuous automatic production of a few varieties, plate vulcanization remains the dominant vulcanization operation process for rubber products. The plate vulcanizer is a vulcanizing equipment by temperature and pressure, and its heat source can use steam, hot water, or electricity. Based on the way of press, plate vulcanizing machines can be divided into hydraulic and hydraulic types while using manual or electric machinery such as screws or cranks. Hydraulic plate vulcanizers are mostly used today because of their excellent performance in pressure stability, the parallelism of the hot plate, and the degree of mold damage. Mechanical vulcanizing presses can be used for mass production occasions because the mold is easy to release. Flat-plate vulcanizing presses are generally divided into column type, side plate type, window frame type, and jaw type based on the shape of bearing clamping force parts [53]. The column type comprises an upper beam, plunger frame, pillar, and nut, which can be operated from all sides of the hot plate with few accessories, such as piping and good operation performance. The side plate type consists of upper and lower beams and steel plates on both sides, which have excellent heat preservation performance and high thermal efficiency. It can reduce the width of the flat plate machine and be used as a small flat-plate vulcanizer. The window-frame type is the whole welded structure of the window-frame type, and the hydraulic cylinder is configured in the middle of two frames. It is mostly used for large-sized plate vulcanizing presses. The jaw type is an integrally welded structure with an open front, left, and right, suitable for flowing operation from both left and right sides.

As shown in Fig. 2.16, the flat-plate vulcanizing machine is mainly used for vulcanizing flat tape, such as conveyor belts and transmission belts. It has advantages such as a large pressure per unit area of the hot plate, reliable operation, and low maintenance. Its specifications are generally expressed by the "width × length × number of layers" of the heating plate in mm, such as XLB-D350 × 350 × 2 type. The X indicates rubber machinery, L means a vulcanizing machine, B means plate structure, D means electric heating (Q means steam heating), 350 mm wide and 350 mm long hot plate, and 2 means the number of heating layers. The structure of the flat-plate vulcanizing press is shown in Fig. 2.17 and is mainly composed of a plunger, working cylinder, heating plate, and clamping and stretching device. With sealing devices and flanges, the plunger and working cylinder form parts for transmitting pressure, while the heating plate pressurizes and heats the products during vulcanization. Since the temperature has the greatest influence on the vulcanized products, the uniformity of the temperature distribution of the heating plate directly affects the quality of the vulcanized products [15]. Therefore the uniform temperature distribution of the hot plate is crucial for the quality of the vulcanized products.

Figure 2.16 Photo of press vulcanizer [54].

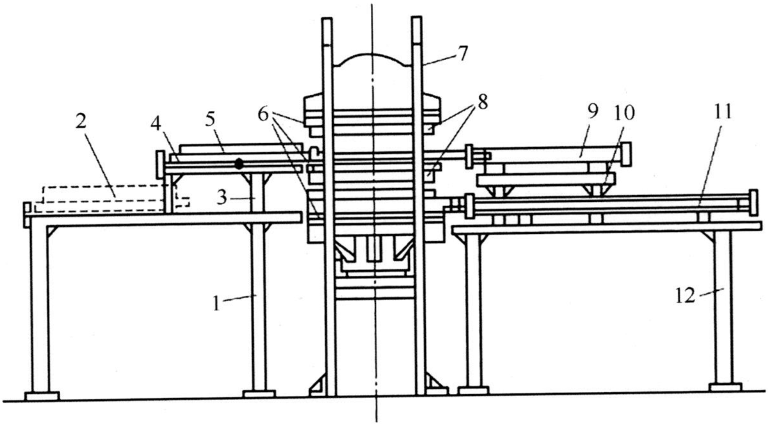

Figure 2.17 Structure diagram of 250/600 × 600 pumping type double-layer hydraulic electric plate vulcanizing machine [15]. 1—Lower mold changing table; 2—lower mold; 3—upper mold changing table; 4—extraction plate; 5—lower mold of the upper model; 6—heating plate; 7—frame; 8—upper mold fixed to heating plate; 9,11—upper and lower hydraulic cylinders of mold changing device; 10,12—bottom frame of upper and lower hydraulic cylinders.

Although the basic structure of the present plate vulcanizing press, as shown in Fig. 2.17, is the same as in the past, its pressure capacity has been increased, and the parallelism and temperature distribution have been greatly improved with the computer-controlled, vacuum-tensioned structure. The computer control can automatically set the time and interval of exhaust, holding pressure, and mold release and shorten the time for the vacuum press to reach the required vacuum level. Induction heating is also used for curing to reduce the heating time of the mold. Injection molding machine molding and curing can also be realized as an automatic molding and curing method. The principle of the flat-plate vulcanizing machine is simple. Still, for newly introduced machines, it is important to confirm the capacity, pressure, clamping force, and matching of the product, as well as the correspondence between the size of the hot plate of the platen vulcanizing machine and the size of the product, the parallelism of the hot plate, the temperature distribution of the hot plate, the vacuum stretching structure, the installation of the mold changing device, and the control system way and precision [55].

Cutting-edge research on vulcanizing presses has focused on how to make the temperature distribution of the vulcanizing process uniform and how to control the temperature flexibly. Shang Wenlu of the Beijing

University of Chemical Technology has studied electromagnetic induction heating vulcanization technology [56]. This method uses the electromagnetic induction heating method to generate heat in the vulcanization mold of the product, and the heat is transferred from the mold to the rubber and from the outer surface of the rubber to the center. This method has advantages such as fast temperature rise, low energy consumption, easy arrangement of electromagnetic coils, and easy adjustment of electrical and thermal parameters. By using pulse heating, heat is supplied to the vulcanizing device intermittently so that the heat from the rubber surface has enough time to be transferred to the center. Yu has applied PLC to the vulcanizer for the control of the "three elements" of the vulcanization process, namely vulcanization time, vulcanization temperature, and vulcanization pressure, which has effectively improved the quality of vulcanization products and saved energy [57].

Vulcanization is a crucial process in tire manufacturing that transforms rubber properties from malleable to elastic, imparting fine surface patterns and excellent performance to uncured tires. However, incomplete swelling, the asymmetric structure, and vapor condensation deposition in the existing vulcanization technology poorly affect the uniformity and dynamic balance performance of tires. Qiu et al. designed a novel tire curing equipment called the Vulcanizing Inner Mold Drum (VIMD) to improve the conventional automotive tire curing process, as shown in Fig. 2.18 [58]. The design of the drum tiles, telescoping mechanism, and transmission mechanism, as well as the finite element simulation analysis and reliability check of the key components, were carried out. The shrinkage rate can be significantly improved by increasing the drum tiles and adopting the structural design method of cross-symmetrical arrangement of the linkage mechanism. The clever design of the double-cam linkage mechanism allows for orderly movement and effectively avoids complicated control designs. The self-locking structure design greatly reduces the force of the connecting rod and improves the overall dimensional accuracy. The results show that VIMD has excellent dimensional accuracy and structural stability.

The development of curing equipment depends not only on the technological innovation of the curing equipment itself. Still, it is also closely related to the innovation of the tire curing process and the overall development of science and technology. Moreover, it must align with the development trend of the whole society and relevant national policies. With the further development of electronic and control technology, the

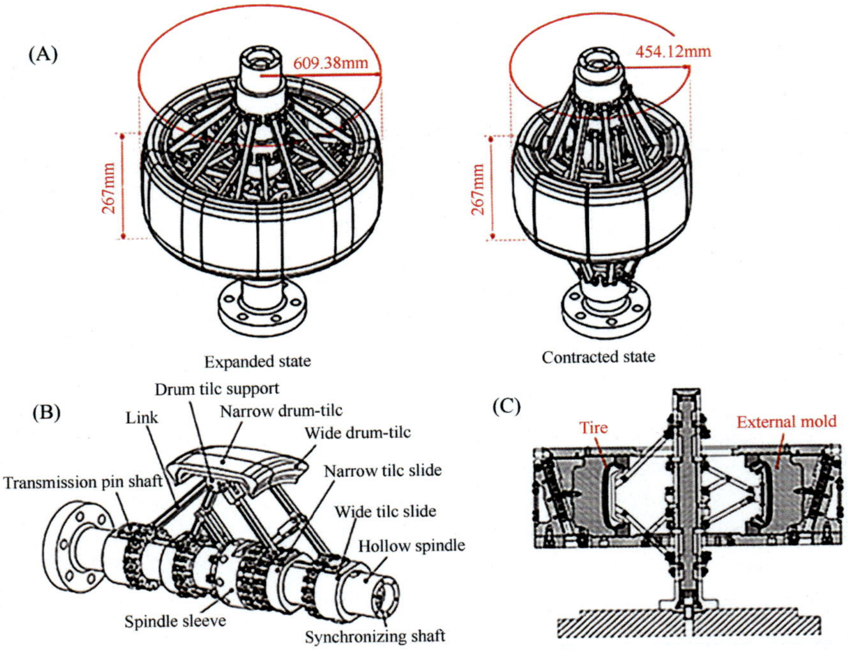

Figure 2.18 (A) Schematic diagram of expanded and contracted states of the VIMD. (B) Structural diagram of the VIMD. (C) Schematic diagram of vulcanization through the cooperation of the VIMD and external mold [58].

tightening of land resources and energy, the increasing awareness of environmental protection, and the changes in the tire market, new requirements for tire curing equipment will be constantly put forward. Therefore it is essential to keep up with technological advancements and adapt to changing demands and trends to improve the efficiency, quality, and sustainability of tire production.

2.1.5 Development of synthetic rubber

NR has been used since the 11th century, and SR was developed in the 1930s, with dozens of different kinds of SR being developed. However, it is challenging to find SRs with good overall performance compared with NR. Therefore the consumption of NR still accounts for about 40% of the total rubber consumption. More than 2000 species of plants contain rubber components in nature, but the Brazilian rubber tree is the most widely used due to its high gum content, the best quality of gum, highest yield, and ease of collection. More than 98% of the world's NR production comes from the Brazilian rubber trees. Brazilian rubber trees are suitable for growing in tropical and subtropical high-temperature wetland areas. More than 90% of global NR production comes from Southeast Asia, mainly from Malaysia, Indonesia, Sri Lanka, Thailand, India, southern China, Singapore, the Philippines, and Vietnam.

The raw materials used in the production of tires include NR and SR. Synthetic methods make SR, and different raw materials (monomers) can be synthesized into different types of rubber. General rubber is the most widely used rubber in SR. General rubber is the part or various kinds of rubber used to replace NR, such as butadiene rubber, cis-butadiene rubber, isoprene rubber, and neoprene rubber.

2.1.5.1 Styrene–butadiene rubber

Styrene–butadiene rubber (SBR) is the most efficient general rubber to produce and is prepared by copolymerization of butadiene and styrene. It includes emulsion styrene–butadiene rubber (ESBR), soluble styrene–butadiene rubber (SSBR), and thermoplastic rubber (SBS) [59–61]. The ESBR process is more advantageous regarding cost savings, and about 75% of the global SBR plant capacity is based on the ESBR process. High-end SSBR and rare earth paraben rubber are necessary to develop green tires to meet the global demand for low rolling resistance. SSBR is easily modified due to its unique structure of molecular chain ends, which is a breakthrough to achieve product differentiation and becomes a new research direction. The functionalization modification of SSBR has blunted the free macromolecule ends and improved the affinity with carbon black. The product has low rolling resistance, antislip, and excellent physical and mechanical properties.

Currently, SSBR research is focused on further improving the wet slip resistance of its tread rubber and reducing its rolling resistance. The hysteresis loss of tread rubber has an important influence on the rolling resistance of tires. The free macromolecular chain ends in rubber are an important factor affecting its hysteresis loss. Therefore the number of free macromolecular chain ends can be reduced by end group modification, thus reducing the rolling resistance of tires and improving their antislip property [62]. The Beijing University of Chemical Technology patent discloses a method to functionalize SSBR end groups, resulting in SSBR with no linear macromolecule, double-ended modified, and all-star structure. This modification has been shown to improve the properties of SSBR and can be used to produce green tires that meet the requirements of low fuel consumption and environmental performance.

According to the analysis of China Rubber Network, there was a further surplus of production and sales of common ESBR in 2013, while there was an obvious shortage of high-end SSBR products and environmental protection ESBR products [63,64]. Environmentally friendly SSBR products are produced using environmentally friendly additives and aromatic oils that pose no harm to the environment and human health. China National Petroleum has produced environmentally friendly SSBR products SBR1500E, SBR1502E, and SBR1723 using environmentally friendly additives that do not contain nitrosamines and can generate nitrosamines preparations and foreign environmentally friendly aromatic oils. Japan Synthetic Rubber Corporation, the United States Goodyear, and others have achieved industrial production, and the output is increasing yearly.

2.1.5.2 Cis-butyl rubber

The polymerization of butadiene solution prepares para butadiene rubber and has excellent cold resistance, abrasion resistance, elasticity, and aging resistance. It is mainly used in the production of tires, with a small portion used in the production of cold-resistant products, cushioning materials, tapes, and rubber shoes. Maleic butadiene rubber has poor tear and wet-slip resistance, and rare earth butadiene rubber is used in the tread and sidewall of tires to reduce rolling resistance, improve fuel efficiency, and improve the safety, durability, and economy of tires.

The development of rare earth catalysts in China started early, but technical bottlenecks still exist in achieving industrial mass production. Lanxess Chemical's patent mentions a method for preparing high relative molecular mass, bimodal, neodymium-catalyzed polybutadiene [65]. The catalyst components include an alcohol salt, phosphonate, hypophosphate, or phosphate, carboxylate, complex compound of rare earth metal with a diketone, or an addition compound of a halide of a rare earth metal with an oxygen or nitrogen donor compound, preferably neodymium terephthalate; a dialkyl aluminum hydride, preferably diisobutyl aluminum hydride; a diene, preferably butadiene or isoprene; and at least one organometallic halide, preferably sesquioxoethylaluminum chloride. The polybutadiene produced with the formulated catalyst has a high cis-1, 4-structure and a low 1,2-vinyl content. It includes straight and long-chain branching polymeric fractions, satisfying low rolling resistance and resilience, low processing viscosity, and high solubility.

According to a disclosure by the Dalian University of Technology, a method for preparing rare earth catalysts that can produce star-branched polybutadiene has been discovered [66]. The catalyst is aged by sequentially introducing nitrogen into the dry catalyst reactor and adding rare earth organic compounds, diolefins, and alkyl alumina. The reaction is carried out at 0°C–60°C for 1–20 minutes before the addition of chloride, and the reaction is carried out at 0°C–60°C for 5 minutes to 24 hours. The resulting polybutadiene product has a heavy average molecular weight of (1–10) $\times$ 10^5, a branched-chain heavy average molecular weight of (0.5–20.0) $\times$ 10^4, a mass fraction of cis-1,4-polybutadiene of 80%–98%, and a total mass fraction of 1,2-polybutadiene and trans-1,4-polybutadiene of 2%–20%. The rare earth catalytic system produces a star-shaped branched structure with a high cis structure.

To produce the polybutadiene, components A, B, and C are mixed at −20°C to 80°C (preferably 0°C–40°C) for 5–600 minutes (preferably 10–120 minutes) and cooled to below −10°C, preferably below −30°C. Finally, component D is added to the mixture. The polybutadiene produced with the formulated catalyst has a high cis-1,4-structure with a mass fraction >95% and a low 1,2-vinyl content with a mass fraction <1%. The product includes straight and long-chain branching polymeric fractions, which satisfy low rolling resistance and resilience, low processing viscosity, and high solubility. The rare earth polybutadiene rubber obtained from this method has a star-shaped branched structure with a high cis structure and a heavy average molecular weight of (1–10) $\times 10^5$, with the branched-chain heavy average molecular weight of (0.5–20.0) $\times$ 10^4, and a mass fraction of cis-1,4-polybutadiene of 80%–98%.

The catalyst used in this method is composed of various components, including an alcohol salt, phosphonate, hypophosphate or phosphate, carboxylate, complex compound of a rare earth metal with a diketone, or an addition compound of a halide of a rare earth metal with an oxygen or nitrogen donor compound, preferably neodymium terephthalate; a dialkyl aluminum hydride, preferably diisobutyl aluminum hydride; a diene, preferably butadiene or isoprene; and at least one organometallic halide, preferably sesquioxoethylaluminum chloride.

The method can produce polybutadiene rubber that satisfies low rolling resistance, high resilience, low processing viscosity, and high solubility. The rare earth polybutadiene rubber obtained from this method can be used in various applications, such as tire production, to improve the safety, durability, and economy of the final product.

2.1.5.3 Isoprene rubber

Polyisoprene rubber, or synthetic isoprene rubber, is produced through the solution polymerization method. Similar to NR, isoprene rubber exhibits excellent elasticity, abrasion resistance, heat resistance, and chemical stability. While the strength of raw isoprene rubber (before processing) is lower than that of NR, its quality uniformity and processing performance are superior. Therefore isoprene rubber can replace NR in producing truck tires, off-road tires, and various other rubber products.

In China, the production capacity of NR is inadequate, and the country heavily relies on imports to meet its demand. In light of this capacity gap, the development of synthetic isoprene rubber with similar performance characteristics has become an important area of focus. Despite the high cost of isoprene rubber and the low prices of NR, developing new technologies that reduce costs and cope with price fluctuations in the international NR market remains a priority in the long run.

Various catalytic systems are used to produce isoprene rubber, including lithium, titanium, rare earth, and metallocene catalytic systems. Shell has pioneered lithium catalytic systems, while most other companies use titanium catalytic systems. However, only one company in the Soviet used rare earth catalytic systems [63].

Rare earth–catalyzed polyisoprene products have several advantages, such as high cis-structure content, regular structure, and good processing performance. The Changchun Institute of Applied Chemistry, Chinese Academy of Sciences, has a patent advantage in rare earth–catalyzed isoprene rubber [67]. The patent involves preparing a sulfonic acid rare-earth catalyst for polyisoprene preparation, with a monomer conversion of over 80% and a heavy average molecular weight of $(1.8-12.0) \times 10^5$. The resulting cis-1,4-structured polyisoprene rubber has a mass fraction of over 92%. This catalytic system is simple to prepare, with no third component added, thus simplifying the process and reducing costs.

The Changchun Institute of Applied Chemistry has also published a method for preparing polyisoprene using naphthalene sulfonic acid rare earth catalyst [68]. The main catalyst is a naphthalene sulfonic acid rare earth compound, and the cocatalyst is alkyl aluminum. The catalytic system catalyzes the polymerization of isoprene resulting in polyisoprene products with $>80\%$ conversion, $>92\%$ mass fraction of cis-1,4 structure, and heavy average molecular weight of $(2-15) \times 10^4$. This rare earth catalyst is low in cost, widely available, and easy to prepare, and it has opened up new fields of application for naphthalene sulfonic acid compounds.

2.1.5.4 Ethylene propylene rubber

Ethylene propylene rubber (EPR) is a SR produced using ethylene and propylene as the primary raw materials. This type of rubber has excellent aging resistance, electrical insulation, and ozone resistance. EPR can be filled with oil and carbon black in large quantities, making it a low-priced product.

The chemical stability of EPR is good, with abrasion resistance, elasticity, and oil resistance closely related to that of SBR. EPR is widely used in tire sidewalls, rubber strips, inner tubes, and auto parts. It can also be used for wire and cable winding and high-voltage and ultrahigh-voltage insulation materials. Additionally, EPR can produce rubber shoes, sanitary products, and other light-colored products.

EPR has several advantages, including ozone resistance, aging resistance, chemical corrosion resistance, and high-temperature resistance. It has a small density and excellent insulation properties, allowing it to be highly oil-filled. EPR is currently one of the most promising varieties of rubber [69,70].

Currently, research on EPR primarily focuses on catalyst systems and adding third components. The catalytic system of EPR is mainly based on vanadium-based compounds such as VX3, VX4, VOX3, and VO(OR) X3 − n (where X is a halogen, mainly chlorine atoms, R is an alkyl group of C6−C12, and n is 0, 1, 2). However, preparing these substances involves using chlorine gas through VO2−5, which is toxic, unstable, and inconvenient for storage and transportation.

To address this issue, the Changchun Institute of Applied Chemistry has published a new vanadium complex catalytic system for preparing EPR [71]. The method uses stable performance, long shelf life, and high-activity vanadium complexes. Rubber products have a wide relative molecular mass distribution and good mechanical properties.

The choice of the third monomer in EPR is also important, as its type and structure can impact its vulcanization rate, affecting processing, application, and technoeconomic aspects. Currently, the production mainly uses diene-dicyclopentadiene, ethylene norbornene, and 1,4-hexadiene [72]. However, these third monomers have various disadvantages, including toxicity, slow vulcanization, difficulty in synthesis and separation, and high cost.

The Changchun Institute of Applied Chemistry has also published a method of preparing EPDM rubber using di-olefin liquid oligomer as the third monomer [73]. The third monomer is liquid polybutadiene containing a 1,2-structural mass fraction of 20%–80% or liquid polyisoprene with a 1, 4-structural mass fraction of more than 90%. The EPDM produced has good vulcanization characteristics and excellent processing performance, with a relative molecular mass distribution of 20,000–100,000.

2.1.5.5 Neoprene

Neoprene is a SR made mainly from chloroprene as the raw material. It can be produced through homopolymerization or copolymerization with several other monomers. Compared with NR, butadiene rubber, and para butadiene rubber, neoprene exhibits high tensile strength, excellent heat resistance, light resistance, aging resistance, and oil resistance. It also has high flame retardancy, good elasticity, chemical stability, and water resistance.

However, neoprene has some disadvantages, including poor electrical insulation, cold resistance, and unstable raw rubber during storage. Despite these drawbacks, neoprene is widely used in manufacturing transportation and conveyor belts, packaging materials for wires and cables, oil-resistant rubber hoses, gaskets, and chemical-resistant equipment linings. Its unique properties make it a precious material in many industrial and commercial applications.

The main imports of chlorobutyl rubber products, such as the United States Exxon 1066 grade products, are expensive. Domestic rubber enterprises should increase their research and development efforts in halobutyl rubber to seize the domestic market.

The Beijing Institute of Petrochemical Technology has published a patent for a halobutyl rubber production process [74]. This process involves four steps in sequence, including the synthesis of butyl rubber, removal of chloromethane, precipitation separation, and halogenation of butyl rubber. The method includes the preparation of solution butyl rubber and halogenation of it.

Specifically, under low temperatures, in the mixture of alkane and chloromethane, isobutylene and isoprene prepare solution butyl rubber through a positive ion copolymerization reaction. After removing chloromethane and sink separation, the rubber liquid enters the halogenation stage and finally obtains halogenated butyl rubber. The advantages of this process are that the halogenated butyl rubber is cross-linked, which improves the vulcanization speed and covulcanization performance with unsaturated rubber and improves the adhesion with metal.

Furthermore, using the solution method avoids the switching process of solvent and the dissolution process of rubber in the slurry method, which reduces the production cost [75,76]. Through this process, halobutyl rubber can be produced domestically, reducing the reliance on expensive imports and supporting the development of the domestic rubber industry.

While SR has many advantages, NR still outperforms it in many aspects, making it a preferred material for high-grade tires. However, certain chemical additives need to be added to NR to meet the performance requirements for tire manufacturing. These additives help enhance the rubber's properties and ensure sufficient performance to meet various requirements.

Carbon black or silica can be added to the rubber to improve elasticity, increasing its cross-linking density and strength. Silica can also be used as a reinforcing agent to reduce rolling resistance and improve fuel efficiency. Aromatic oils are added to the rubber to reduce heat generation, while fillers such as clay or talc improve wear resistance.

To improve puncture resistance, NR is often blended with SRs or thermoplastic elastomers to improve toughness and resistance. High load capacity can be achieved by increasing the strength and stiffness of the rubber through reinforcing fillers or fibers. Lastly, oils can be added to the rubber to ensure a comfortable ride, improve damping and reduce noise and vibration.

By carefully selecting and blending different chemical additives, tire manufacturers can produce tires that meet various performance requirements.

2.1.6 Chemical structure of special rubber and its properties

Special rubber is known for its superior performance, wide variety, high technical content, and versatile applications. However, despite its potential, special rubber products account for a very low proportion of total rubber products, resulting in a small market share.

Fortunately, with the increasing application of high-tech in the rubber industry, particularly the widespread use of material modification technology and product formulation design technology, the special rubber industry has been presented with unprecedented opportunities. The development of special rubber is full of prospects and hope.

As these technologies continue to advance, the performance and properties of special rubber products can be improved even further. This will allow for creation of new and innovative products that can meet the specific needs of various industries. By leveraging these opportunities, the special rubber industry can grow and expand its market share, fulfilling its potential as a key player in the rubber industry.

Special rubber, or special SR, is a type of SR similar to general rubber. However, the main difference is that special rubber often exhibits one or more extreme resistances. For example, some types of special rubber are resistant to high temperatures of up to 300°C, strong erosion, ozone, light, climate, radiation, and oil. Others can withstand temperatures as low as −100°C and as high as 260°C while exhibiting low-temperature dependence, low viscous flow activation energy, and physiological inertness. Special rubber can also resist high temperatures, solvents, and oil and have good electrical insulation properties.

There are many types of special rubber, such as fluoroelastomer, polyurethane (PU) rubber, polyether rubber, chlorinated polyethylene (CPE), chlorosulfonated polyethylene (CSM), propylene oxide rubber, and polysulfide rubber. Each type of special rubber has excellent and unique properties that can meet the special requirements of various industries, such as national defense, industry, cutting-edge science and technology, and medical and health care.

The manufacturing process and equipment used for special rubber can vary depending on the specific type of rubber being produced. However, the production process typically involves mixing raw materials, polymerization, compounding, and vulcanization. The equipment used for special rubber production can include kneaders, mills, extruders, and presses.

In summary, special SR exhibits unique properties and extreme resistance. It has a wide range of applications in various industries, and the manufacturing process and equipment used can vary depending on the specific type of rubber being produced [77].

2.1.6.1 Silicone rubber

Silicone rubber is a linear polymer with Si—O units as the main chain and monovalent organic groups as side groups, a polymeric elastic material with both inorganic and organic properties in the molecular chain [77].

Especially R in silicone rubber refers to an organic group such as methyl, vinyl, or phenyl. The integer degrees of polymerization, m and n, can be varied in a wide range. These compounds are initially called "silicones" because their main chain structure is similar to the ketone group. However, they are later called polyorganosiloxanes because they can form structures like (RR'SiO)*n*.

1. The properties of silicone rubber vary depending on the substituent, but they are similar to those of polydimethylsiloxanes, which are ordinary silicone rubbers. The Si—O backbone of the silicone rubber molecule determines that it has unique properties, including:

a. High- and low-temperature resistance: Silicone rubber has the widest working temperature range among all types of rubber, ranging from −100°C to 350°C. It can resist instantaneous high temperatures of thousands of degrees and be used as a heat-proof coating for the inner wall of rocket nozzles.
b. Aging resistance: Silicone rubber is resistant to ozone, light, and climate aging. Even after exposure to outdoor sunlight for several years, its vulcanized rubber maintains its performance.
c. Good electrical insulation performance: The vulcanized rubber of silicone rubber does not change insulation performance drastically when exposed to moisture or temperature rise. The SiO_2 generated by burning silicone rubber is also an insulator that protects electrical equipment.
d. Nontoxic and harmless: Silicone rubber is nontoxic and has no negative impact on human health.
e. Hydrophobic surface: The surface of silicone rubber is hydrophobic and nonsticky to many materials, making it suitable for isolation applications.
f. Good permeability: The permeability of silicone rubber is tens of times greater than ordinary rubber, and the permeability of different gases varies widely.
g. Good resistance to radiation, oil, and combustion.

However, silicone rubber has some major defects, including low tensile and tear strength (much lower than NR), poor acid and alkali resistance, and high cost.

According to its different vulcanization methods, silicone rubber can be divided into two categories: high-temperature vulcanized silicone rubber and room-temperature vulcanized silicone rubber. High-temperature vulcanized silicone rubber refers to turning polysiloxane into elastomer, which is vulcanized and shaped by high temperature (110°C−170°C). It mainly takes high-molecular-weight polymethyl vinyl siloxane as raw rubber, mixed with reinforcing filler, vulcanizer, etc., and vulcanizes the elastomer under heating and pressure. The difference between room-temperature vulcanized silicone rubber and high-temperature vulcanized silicone rubber mainly lies in that it is based on smaller-molecular-weight polysiloxane (such as polydimethylsiloxane), which can be vulcanized into an elastomer at room temperature or slightly heated under the action of cross-linker and catalyst. Room-temperature vulcanized rubber generally has a low shrinkage rate and antiadhesive characteristics, usually used to produce soft mold and pattern. The following section focuses on high-temperature vulcanized rubber.

As mentioned earlier, the different types of silicone rubber are mainly due to the differences in substituents R and R_1. When both substituents are methyl, it is the oldest rubber of silicone rubber species, namely polydimethylsiloxane rubber (polydimethylsiloxane rubber) or methyl silicone rubber (methyl silicone rubber). The raw material of dimethylsiloxane rubber is chloromethane and silicon powder. The two substances catalyze the synthesis of dimethyl dichlorosilane and then hydrolyze to produce dimethylsiloxane monomer and then condense to get the finished rubber. However, the mechanical properties of this rubber are poor, with low vulcanization activity, poor process performance, and products in the vulcanization of bubbles. Only a small number of fabric coatings are used. Other types of silicone rubber have replaced the rest of the product. Although dimethylsiloxane rubber has its advantages, the defects are also worrying. On this basis, methyl vinyl silicone rubber (polymethyl-vinyl siloxane rubber) was born. Structurally, only a small amount of ethyl [generally 0.07%–0.15% (molar ratio)] is introduced into the side group of the dimethyl silicone rubber. Still, it significantly improves the vulcanization activity of methyl silicone rubber, allowing the use of less reactive organic peroxides, increases the hardness of the product, and reduces compression deformation. Methyl vinyl silicone rubber is obtained by hydrolysis of dimethylsiloxane into octamethylcyclotetrasiloxane and then catalytic ring-opening copolymerization with tetramethyltetravinylcyclotetrasiloxane. Methyl vinyl silicone rubber is widely used in oil seals, pipe sealants, etc.

To improve the cold resistance of silicone rubber, polymethyl-phenyl-vinyl silicone rubber was obtained by introducing diphenylsiloxane links into the molecular chain of methyl-vinyl silicone rubber. The introduction of the phenyl group has destroyed the structural regularity of the molecular chain, significantly reduced the T_g and crystallinity, and given the rubber excellent cold resistance properties. The best low-temperature resistance is achieved when the molar ratio of phenyl to silicon in the molecular chain is 6%, which can maintain flexibility at $-100°C$. When this ratio reaches 15%–20%, the rubber is flame-retardant; at 35% or more, it has good radiation resistance. However, the increase in phenyl introduction also means the molecular chain rigidity increases, making processing difficult. And because the molecular chain regularity is destroyed, the mechanical properties of the products are not high. The rubber and methyl vinyl silicone rubber uses are similar, not repeat.

Fluorosilicone rubber is a polymer made by introducing fluorosilane or fluoroaryl group on the side chain of methyl vinyl silicone rubber molecule with the following structure.

There are wide varieties of silicone rubber, but only methyl-vinyl-γ-trifluoropyl silicone rubber (methyl-vinyl-γ-trifluoropyl silicone rubber) is widely used. The most outstanding feature of fluorosilicone rubber (polymethyl-vinyl-γ-trmuoropropyl siloxane rubber) is its oil and solvent resistance. It has good stability in aliphatic, aromatic, and chlorinated hydrocarbon solvents and various fuel oils and lubricants at room and high temperatures. However, the high- and low-temperature resistance of this rubber is reduced compared with the previous three, and fluorosilicone rubber is more expensive. It is generally used only for gaskets, seals, or adhesives with high resistance to oil or solvents.

2.1.6.2 Fluoroelastomer

Fluoroelastomer is a synthetic polymer elastomer that contains fluorine atoms on the carbon atoms of the central or side chains [77]. The strong electronegativity of the fluorine atoms on the main chain or side chain gives fluororubber high heat and oxygen resistance due to the large C—F bond energy (485 kJ/mol). The C—F bond energy increases with the increase of fluorine atoms in the molecular chain, making the C—F bond difficult to be destroyed. In addition, the covalent radius of the fluorine atoms is 0.064 nm, nearly half of the C—C bond length. As a result, the fluorine atoms can shield the C—C main chain and ensure the high stability and chemical inertness of the C—C bond.

Fluoroelastomer is known as the "King of Rubber" due to its comprehensive and excellent physical and mechanical properties, high-temperature resistance, oil resistance, electrical insulation, and weather resistance. The excellent performance of fluoroelastomer makes it suitable for a wide range of applications. It can be used as the outer skin of electric wires, anticorrosion lining, rocket seals, oil pipes, and electrical line sheathing in various military, chemical, and aviation products.

Fluoroelastomers were first introduced in the 1940s and have developed rapidly, with wide varieties of branches, including fluorinated diene rubber; fluorinated polyacrylate rubber; fluorinated polyester rubber, and so on, according to chemical composition. Despite the huge production volume, two types of fluoroelastomers are mainly produced: type 23 fluoroelastomer and type 26 fluoroelastomer. Type 26 fluoroelastomer is made by copolymerizing vinyl chloride and hexafluoropropylene emulsion, while type 23 fluoroelastomer is made by suspension polymerization of vinyl chloride and trifluorochloroethylene. The overall performance of fluoroelastomer is outstanding, which is reflected in the following points:

1. High-temperature resistance. Fluoroelastomer can be used at 250°C for a long time, and the limit temperature is 300°C. In contrast, IIR, EPM, and ACM have an ultimate use temperature of 150°C and NBR only 130°C.
2. Resistant to chemical reagents, oil, acid and alkali. Through the immersion of fluorine rubber, silicone rubber, and acrylate rubber in toluene and gasoline, it is found that the volume change rate of fluorine rubber is the smallest, only 3% in gasoline and 6% in toluene; the volume change rate of silicone rubber in the two liquids is 130%, and acrylate is 55% and 300%, which shows that the oil resistance of fluorine rubber is far better than both.
3. Excellent weather resistance and ozone resistance.
4. Very low compression permanent deformation.

Although fluorine rubber has the above advantages, there is still a lack of low-temperature resistance to this defect. Fluorine rubber at −20°C will lose the rubber elasticity, and at −34°C will become hard and brittle, so it cannot be used at very low temperatures. Fluorine rubber has balanced performance and is generally used as seals and oil seals for ships, automobiles, and chemical equipment.

2.1.6.3 Polyurethane rubber

PU rubber is an elastomeric material containing many urethane groups in the main chain of the polymer, referred to as PU elastomer. In addition to carbamate groups, polymer chains generally have ester groups, ether groups, aryl groups, and aliphatic chains. They are usually made by reacting oligomeric polyols, polyisocyanates, and chain extenders. The traditional classification of PU elastomers is based on the processing method. It is divided into cast PU elastomers, blended PU elastomers, and thermoplastic elastomers, with the cast type producing the most, the thermoplastic type the second, and the blended type the least. Depending on the raw materials, synthesis equipment, and production purpose, they can be divided into injectable PU elastomers and liquid dispersion PU elastomers [77].

Although there are some differences in the performance of PU elastomers due to different raw materials and manufacturing methods, they also have a series of common properties due to the same structure. The modulus of elasticity of PU elastomer is between rubber and plastic, and the most important characteristic is that it considers high hardness and high elasticity. In addition, good abrasion resistance is an excellent feature. Therefore PU elastomers can be used not only as rubber but also as plastic. PU elastomers are divided into two categories: polyether and polyester. The polyester type is generally low in hardness but good in heat and oil resistance, while the polyether type is high in hardness, cold resistance, and antibacterial properties. The mechanical strength of PU elastomers vulcanized with diamines is affected by intermolecular hydrogen bonding, and the hydrogen bonding force decreases at higher temperatures.

General PU elastomer long-term use temperature is 80°C–90°C. Polyester-type PU embrittlement temperature is −40°C. In comparison, the polyether types up to −70°C and polyether-type PU resistance to low-temperature performance are more substantial. PU elastomers have better oil and solvent resistance and good swelling resistance to lubricants and fuel oil. PU elastomers have one major drawback: poor water resistance. Water has two effects on PU elastomers. One is that the inhaled water forms hydrogen bonds with the polar groups in the PU polymer, weakening the hydrogen bonding between the molecules of the polymer itself and thus reducing the physical properties. This process is reversible, and the physical properties are restored after drying. Usually, polyester polyurethanes are about 10%–20% lower than polyether ones; secondly, water makes PU rubber hydrolyze, which is irreversible. Polyether PU is completely decomposed by immersion in water at 100°C for 8 weeks. The polyether type loses 2% of its weight, which shows that different groups have a greater influence on water resistance PU [77].

The tensile strength of PU is generally 28–42 MPa, of which the highest is 42 MPa for cast PU, and the lowest is 28 MPa for blended PU. Elongation is generally 400%–600%, and the maximum is 1000%. Although the PU elasticity is high, the hysteresis loss is also large, so the heat generation is large and often damaged under multiple bending or high-speed loads. PU elastomer products have a wider range of hardness than any other rubber, with a minimum Shore hardness of 10, and most PU products have a hardness of 45–95. PU vulcanized rubber has high tear strength, up to 62 kN/m. However, as the experimental temperature increases, the tear strength decreases rapidly. The abrasion resistance of PU elastomer is much better than that of general rubber, 9 times that of NR and 1–3 times that of SBR. However, due to the high heat generation caused by hysteresis, PU elastomers can generally only be prepared in thin products. The practical use of PU is mainly to use its abrasion resistance, and the main products are solid tires, rubber rollers, tapes and shoe soles, etc. It can also manufacture seals, oil, and valve gaskets [78].

2.1.6.4 Acrylate rubber

Polyacrylate rubber, or acrylic rubber, or ACM, is a copolymer of an alkyl acrylate monomer and a small amount of monomer with a cross-linking group. The main chain of the polymer is saturated and contains polar ester groups, so acrylate rubber is endowed with excellent resistance to oxidation and ozone and has outstanding resistance to hydrocarbon oil swelling. Many acrylic esters can be divided into three main varieties: ethyl acrylate/oxyethyl vinyl ether, ethyl acrylate/acrylic acid, and butyl acrylate/acrylic acid by monomer. Among them, butyl acrylate/acrylic clear is the best application performance, the largest amount. Its processing process is easy, so as the representative of acrylic ester. The saturated structure of the main chain of acrylate rubber gives it heat and oxidation resistance, and the polar ester group on the side group gives it good oil resistance. In SR, only fluorine rubber, acrylate rubber, and nitrile rubber have heat and oil resistance. Although fluorine rubber can withstand 300°C high temperatures, the mechanical properties are poor and expensive, nitrile rubber cannot exceed 150°C, and ozone resistance is poor. ACM heat resistance performance is second only to fluorine rubber, the highest use temperature up to 180°C, but only one-tenth of the price, in some applications, can replace it. With the development of the automobile industry, the demand for acrylate rubber is increasing. Currently, ACM is mainly used as gaskets, radiator hoses, ignition wires, spark plug covers, and rollers.

2.1.6.5 Polysulfide rubber

Polysulfide rubber is a polymer elastomer containing sulfur atoms on the main chain (—S—C—, or—S—S—). Because the molecular bonds are saturated, these rubber products have good oil resistance, solvent resistance, aging resistance, and material bonding properties. Polysulfide rubber has three types: solid rubber, liquid rubber, and water dispersion. Three types of liquid rubber production are the largest, about four-fifths of the total production. The molecular structure of polysulfide rubber varies depending on the organic dichloromonomer used. Dichloro compounds commonly used in industry are dichloroethane, dichloropropane, dichlorodiethyl ether, dichloroethyl acetal, etc. Sometimes, a few trifunctional substances, such as trichloropropane, are added to form a cross-linked structure. Solid-state polysulfide rubber is condensed from organic dichloromonomer and inorganic sodium polysulfide. The commonly used organic dichloromonomer is 2,2-dichloroethyl acetal [77].

Solid-state polysulfide rubber has poor compression deformation resistance. The cross-linked structure makes polysulfide rubber solvent-resistant, and the saturated backbone makes it excellent in oxidation resistance. Liquid polysulfide rubber is a new type of oil-resistant rubber, mainly made of dichloroethane and 1,2-dichloropropane copolymerized with sodium polysulfide as a polymeric aqueous dispersion, which can be used directly and can maintain low viscosity even if it contains more solids. Therefore it has important applications in coatings. Polysulfide latex is very important for coatings, and good results have been achieved in oil-resistant anticorrosion coatings. Liquid polysulfide rubber of vulcanized rubber has good low-temperature performance and can be used at −60°C. In addition, vulcanized rubber also has good solvent resistance and chemical stability, and the performance is related to the sulfur content [78].

2.1.6.6 Polyether rubber

Polyether rubber is a hydrocarbon polyether elastomer obtained by ring-opening copolymerization of cyclic ether compounds (epoxy alkanes) containing epoxy groups, whose main chain has an ether-type structure with no double bonds, and its side chains are generally with polar groups or unsaturated bonds [77]. At present, there are mainly the following types of polyether rubber: chloroether rubber, copolymerized chloroether rubber (ECO), propylene oxide rubber, and unsaturated propylene oxide rubber.

Chloroether rubber (epiehlorohydrin rubber) is an amorphous polymer elastomer of epichlorohydrin polymerized by ring-opening under the action of a coordination initiator.

The main chain does not contain double bonds, so it has the characteristics of saturated rubber; the main chain contains ether bonds, so it has low temperature and oil resistance; the side groups are strongly polar chloromethyl, which makes it not easy to burn; the side chains are cross-linked, so it has good chemical stability. Thus it has the characteristics of oil resistance, high temperature resistance, ozone resistance, etc.

Propylene oxide rubber is a polymeric amorphous elastomer obtained by ligated negative ionic polymerization of propylene oxide under the action of a complexing initiator.

Epoxy propane rubber is a saturated aliphatic hydrocarbon polyether with a molecular side chain of methyl. This main chain formed by the ether chain will give the rubber excellent low-temperature resistance (brittle temperature of −65°C) and flexural properties. It does not contain double bonds, and the oxygen atom has a polarity. Hence, epoxy propane rubber has good elasticity, excellent ozone, and heat and oil resistance.

2.1.6.7 Chlorinated polyethylene rubber

CPE is a modified polymer made by the substitution reaction of polyethylene with chlorine. The chlorine content defines the use of the polymer: <15% chlorine content is plastic; 16%–24% chlorine content is thermoplastic elastomer; 25%–48% chlorine content is rubber; and >73% chlorine content becomes a brittle resin. CPE rubber is a noncrystalline saturated elastomer (CM) whose main chain structure is the same as polyethylene, with the difference that chlorine atoms partially replace the central chain carbon atoms, and there are no unsaturated bonds. These two structures give CM excellent heat resistance, ozone resistance, and oil resistance; it is compatible with various polar and nonpolar polymers and maintains the chemical stability and electrical properties of polyethylene. The most important feature of CPE rubber is that the performance can be controlled by adjusting the chlorine content, and as the chlorine content increases, the T_g rises. The common CM rubber chlorine mass fraction is about 36%. CPE rubber alone is mainly used to manufacture wire and cable, and other rubber mixed with the manufacture of acid and oil-resistant hose and sealing products, but also through modification so that it has a broader range of applications.

2.1.6.8 Hydrogenated NBR

Hydrogenated nitrile rubber (HNBR) is a rubber obtained by hydrogenating the double bonds contained in the main chain of NBR polymer to improve NBR's heat and weather resistance. HNBR has improved chemical stability and mechanical strength compared with NBR and heat and weather resistance.

HNBR is currently prepared mainly by solution hydrogenation. HNBR is produced by selective hydrogenation of the butadiene units in the NBR chain segments, and the unsaturated double bonds are hydrogenated to form C—C single bonds. HNBR can work under an oil medium at 150°C for a long time, and the working life of HNBR vulcanized by peroxide is 1000 hours at 150°C. The embrittlement temperature of HNBR is 7°C–10°C lower than that of NBR, and the abrasion resistance is more than two times higher. HNBR is mainly used to make various high-temperature oil-resistant rubber parts for automobiles, aerospace, and the oil field, such as fuel hoses, automobile oil seals, engine seals, and flame-retardant cable sheaths.

2.1.6.9 Chlorosulfonated polyethylene rubber

CSM, known abroad as Hypalon, is a special rubber made by chlorination and chlorosulfonation of polyethylene. It is usually prepared by dissolving polyethylene in carbon tetrachloride, tetrachloroethylene, or hexachloroethane and treating it with azo diisobutyronitrile as a catalyst or with a mixture of chlorine and sulfur dioxide under ultraviolet (UV) light or with SO_2Cl.

The performance of CSM is related to the molecular weight of polyethylene and the relative content of chlorine and sulfur, but the molecular weight of polyethylene is the decisive factor. If the molecular weight of polyethylene is too low, the rubber is dense, and the tensile strength is low. High molecular weight will improve the mechanical strength of CSM, but there is a maximum molecular weight limit. If the molecular weight is too high, the viscosity of the system increases, which will cause the rubber to be difficult to be processed. Generally, the molecular weight of polyethylene in CSM is 20,000–100,000. Introducing chlorine atoms in polyethylene molecules can maintain the excellent performance of polyethylene while eliminating the crystallinity of molecules and getting a soft and easy-to-process elastomer. In addition, the chlorine content also affects the rubber properties. A chlorine content of 27% has been proven to be optimal. High chlorine content reduces the resistance to permanent deformation; low chlorine content reduces the mechanical strength.

The role of sulfur dioxide is to form—SO_3CI groups in CSM, thus forming a cross-linked structure. If the sulfur content is too high, rubber burning will occur during the rubber processing, and the general sulfur content is 1.5%. CSM rubber is a low-strength, viscous polymer. Density is about 1.1 g/cm^3; soluble in aromatic and halogenated hydrocarbons; insoluble in acids, aliphatic hydrocarbons, mono- and di-alcohols. CSM has excellent ozone and chemical resistance due to its fully saturated chemical structure. CSM contains more CI in its structure, so it has a good heat resistance and continuous use temperature of 120°C–140°C. CSM has excellent dielectric properties of vulcanized rubber but poor low-temperature resistance. Compared with NR, CSM has good thermoplasticity and can be processed directly without plasticizing. The industrial application of CSM is mainly the use of its ozone resistance, oil resistance, and weather resistance characteristics of aging, such as the manufacture of building waterproof layers, cable insulation materials, and automotive devices such as hoses.

2.1.6.10 Application of special rubber in the automotive industry

The ideal rubber parts should have excellent working performance and the same service life as the car. It can be seen that the requirements for rubber parts for automobiles are quite demanding. For the engine, transmission system, and fuel injection device of the car, its rubber parts are functional, requiring the material to have stable dynamic and static mechanical properties under severe conditions such as high temperature, low temperature, and oil medium [79,80].

According to the current development trend of the automobile industry, with the reduction of cabin space and the adoption of auxiliary devices such as catalytic converters and turbochargers, the temperature in the cabin keeps rising. According to the design principle of automobile aerodynamics, it can meet the needs of high-speed and safe vehicle operation. At the same time, to improve the combustion efficiency of fuel and the appearance of the vehicle, the cooling window of the engine compartment will be reduced. Some models will also use a fully enclosed cabin, increasing the cabin temperature. This requires engine rubber products with high heat resistance material, generally requiring the rubber parts in the cabin in—40°C−150°C effective work. The United States, Japan, and some European car manufacturers put forward more stringent requirements for engine rubber parts to make the car adapt to cold and high-temperature areas. The working temperature range of rubber parts for engines in the United States and Japan proposed is—40°C−180°C. Some European automakers even require the working temperature range of engine rubber parts to be—45°C−200°C. The rubber parts of the transmission system should maintain good physical and mechanical properties in gear oil from—40°C to 170°C. In addition, automotive lubricants and fuels are constantly being improved to meet high performance and environmental requirements. For example, gear oils containing wear-resistant polar additives, fuels with high aromatic hydrocarbon content and alcohol fuels, peroxide-resistant acid gasoline, etc. This also requires automotive rubber materials to resist increasingly demanding fluid media. Some of the rubber parts and materials for automotive engines and drive trains are shown in Table 2.2.

Table 2.2 Rubber parts and corresponding materials for automobile engines and transmission systems [79].

Main production	Performance requirements	Rubber
Engine crankshaft, oil seal	Heat, oil and abrasion-resistant	Fluoroelastomer
Stem seal	Heat and compression deformation resistant	Fluoroelastomer
Oil pan gasket	Heat and compression-resistant	Acrylate rubber
Gearbox rubber parts	Heat and wear-resistant, additive-resistant	Acrylic rubber
Exhaust valve rubber parts	Heat-resistant	Fluoroelastomer
Vacuum regulator diaphragms and seals	Oil and temperature-resistant	Fluorosilicone rubber
Engine timing gear belts	Abrasion resistance and dynamic fatigue strength	Hydrogenated nitrile butadiene rubber

2.2 New bio-based rubber and its preparation process

2.2.1 Bio-based rubber

The rapid development of modern society is attributed to the prosperous petrochemical industry. However, the development of the petrochemical industry inevitably brings a series of problems. Fossil resources are nonrenewable, and with hundreds of years of exploitation, they have gradually failed to meet the growing energy demand of people. The overexploitation of fossil resources has caused damage to the ecological environment and produced many greenhouse gases, contrary to the current ecological and environmental protection and sustainable development concepts. People have begun to explore new sustainable development paths to cope with the energy and environmental crisis [81].

Biomass energy is converted from solar energy and is used by plants to synthesize carbon dioxide and water into biomass through photosynthesis, generating carbon dioxide and water, forming a material cycle with theoretically zero net carbon dioxide emissions [82]. Biomass is also considered the only solar energy that can be stored and has unparalleled advantages in replacing fossil fuels. By 2020, China's biomass energy consumption is expected to account for 20% of the total oil consumption, limiting the country's foreign dependence on oil to 50%. Using biomass resources to refine biomass fuels and chemicals, the U.S. Department of Energy (DOE) plans that biofuels will provide 30% of energy demand and bio-based chemicals will provide 25% of total organic chemical demand in 2025, and it is expected that by 2050, bulk chemicals derived from renewable materials could reach about 113 million tons, accounting for 38% of all organic chemicals [83].

The use of bio-based chemicals has already been successful. Bio-based chemicals have been successfully used to manufacture such products as polylactic acid (PLA) [84], starch-based polymers [85], polyhydroxyalkanoates (PHAs) [86], 1,3-propanediol-based polymers [87,88], and polybutylene [89] materials. Dow Chemical has built a 140,000-ton-per-year PLA plant and plans to increase capacity to 450,000 tons/year. Toyota uses sugarcane and potatoes to make plastic car parts, and Fujitsu uses corn starch to make plastic computer cases. Although bio-based plastics such as these still have disadvantages such as low heat resistance temperature and poor toughness, biological resources have replaced fossil resources in the field of plastics to obtain new polymer materials. At the same time, efforts should continue to develop environmentally stable bio-based polymers to meet high-performance needs. The molecular chains and the aggregated structure of rubber give it high elasticity under large deformation, so it is widely used in defense, military, medical transportation, and daily life.

The vast majority of rubber still comes from natural and petroleum-based SR. NR is mainly from Brazil rubber trees, which can only be grown in specific tropical areas due to their harsh growing conditions. China's demand for NR is great, but the lack of self-production and heavy reliance on imports and imports are increasing yearly. The development of petroleum-based SR has filled the gap between the supply and demand of NR. Still, at the same time, it has brought problems such as the rapid consumption of fossil resources and excessive carbon emissions. Compared with bio-based plastics and fibers, there has been less research and development of bio-based elastomers for engineering applications. Developing new bio-based elastomers has become particularly important to meet the growing demand [90].

The main source of current chemicals is petrochemical feedstock, and the bulk chemicals are methanol, ethylene, propylene, butadiene, benzene, toluene, xylene, etc. [91] Although there are still technical and cost issues in replacing such petroleum-based chemicals with bio-based chemicals, there is still a huge potential. Bio-based chemicals are chemicals produced from grains, legumes, cotton straw, and other biomass as raw materials [92]. With the maturity of biofermentation technology, conversion costs have been greatly reduced, and mass production is possible. Due to the wide variety of bio-based chemicals, industrial production and scientific research focus on a few high-value-added bio-based chemicals. In 2004, the U.S. DOE launched the first batch of high value-added bio-based chemicals in a report called the "TOP 12" chemical platform, including succinic acid (fumaric/malic acid), 2,5-furan-dicarboxylic acid, 3-hydroxypropionic acid, aspartic acid, glucaric acid, glutamic acid, itaconic acid, levulinic acid, 3-hydroxypropionic acid, and gluconic acid. Itaconic acid, levulinic acid, 3-hydroxybutyl lactone, glycerol, sorbitol, and xylitol (arabinitol) [93]. Subsequently, in 2010, a new "TOP 10" chemical platform was introduced, including ethanol, furans, glycerol, biohydrocarbons, lactic acid, succinic acid, hydroxypropionic acid/aldehyde, levulinic acid, sorbitol, and xylitol [94] Most of them are already in commercial production.

2.2.2 Conventional bio-based rubber

SBR, EPR, butadiene rubber, and other traditional elastomers are widely used in industry, agriculture, national defense, and daily life. However, they rely heavily on fossil resources, which is not conducive to the sustainable development of the rubber industry. Using renewable resources to replace fossil resources is significant for sustainable development in the SR industry. How to use renewable biomass monomers to synthesize valuable rubber materials is very important to us [90].

2.2.2.1 Isoprene rubber

Williams first produced isoprene in 1860 through the pyrolysis of NR. Today, most isoprene products come from fossil sources. In the Soviet, isoprene is usually obtained by dehydrogenating isoprene and synthesis of isobutene with formaldehyde. In the United States, extractive distillation separates isoprene directly from the C5 stream [95]. In March 2010, Genencor and Goodyear announced an alliance to develop an integrated fermentation, recovery, and purification system to produce isoprene from sugar to isoprene rubber. Amyris and Michelin also signed an agreement in September 2011 to collaborate on developing renewable isoprene. Amyris plans to commercialize isoprene for use in tires and other specialty chemicals such as adhesives, coatings, and sealants. Amyris' technology is already being used for the commercial-scale production of farnesene and for converting plant-based sugars to isoprene [96,97], as shown in Fig. 2.19.

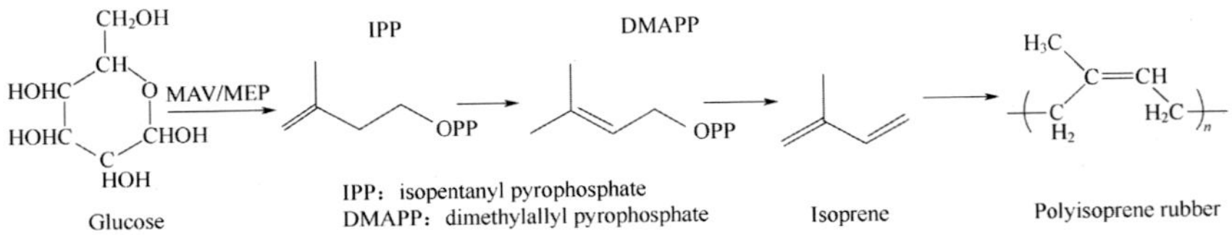

Figure 2.19 Synthesis process of plant-based sugar conversion to isoprene rubber [90].

2.2.2.2 Ethylene propylene rubber

Traditionally, EPDM rubber is synthesized from ethylene and propylene extracted from petroleum. With the continuous development of biomass monomers, bio-based ethylene and propylene have also been successfully used to produce EPDM rubber. The German company LANXESS extracts ethanol from Brazilian sugar cane and then dehydrates bio-based ethylene as a raw material for producing ethylene-propylene rubber [98,99]. Braskemsa supplies bio-based ethylene via pipeline to the Lanxesse EPDM production facility in Triunfo, Brazil. 2011 saw the announcement of Lanxess' world's first bio-based EPDM product under the trademark name KeltanEco. 2017 saw the announcement of Arlanxeo's "Telstar18" World Cup ball for Adidas. "Telstar18" World Cup ball features the best rebounding bio-based EPDM soccer pad [90]. Arlanxeo announced the first bio-based EPDM soccer pad for the Adidas Telstar18 World Cup.

2.2.2.3 Sisal rubber

The key to the preparation of bio-based butadiene rubber is how to obtain bio-based butadiene monomers. In recent years, the production of 2,3-butanediol by biofermentation of glucose has become one of the new focuses in bioenergy. Bio-based 2,3-butanediol can be further converted to 1,3-butadiene (Tsukamoto et al.) on SiO_{2}- loaded cesium dihydrogen phosphate catalyst for highly selective dehydrated 2,3-butanediol to produce bio-based 1,3-butadiene, and this type of single-bed catalyst system has the highest 1,3-butadiene selectivity ($>90\%$) [100], Fig. 2.20 shows the route for the preparation of bio-based butadiene.

2.2.3 New bio-based synthetic rubber

New bio-based SR is to prepare new SR from existing bulk bio-based chemicals and develop its engineering applications in rubber. It should follow the following criteria: (1) the raw materials used do not depend on

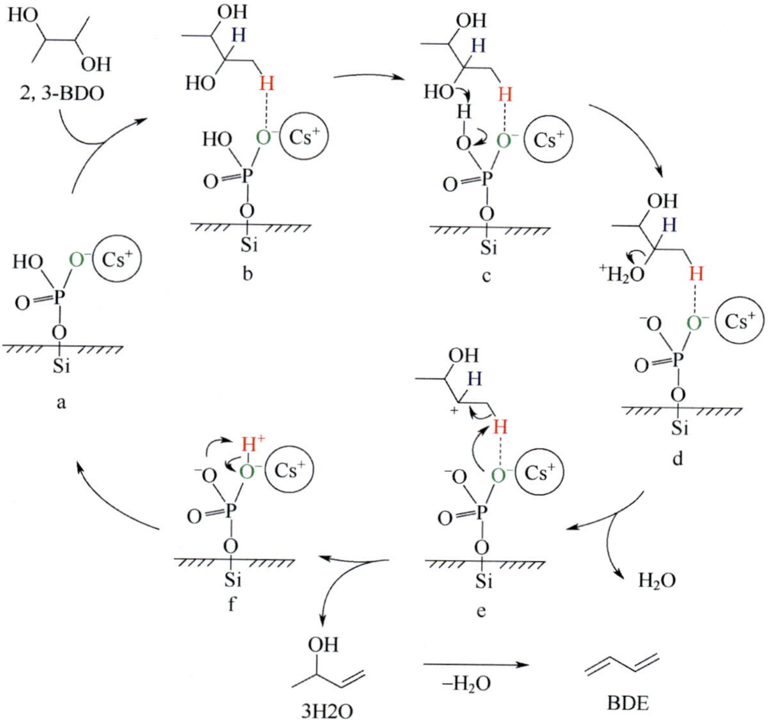

Figure 2.20 Preparation route of bio-based butadiene rubber [90].

fossil resources, mainly prepared by renewable biological resources, easy-to-obtain monomers, and low prices; (2) have good environmental stability, such as low water absorption rate and low degradation rate; (3)should have good compatibility with traditional rubber processing and molding processes and can be processed and molded using traditional rubber processing processes, such as mixing and vulcanization processes; (4) have physical and mechanical properties similar to those of traditional SR. Several new bio-based SR preparation methods have been introduced [90].

2.2.3.1 Polyester-based bio-based synthetic rubber

Polyester materials are made by condensation polymerization of polyol and polyacid and are widely used in various packaging materials, agricultural films, biopharmaceuticals, information electronics, and other fields. Such as PET, PBT, is a high-performance polyester material with excellent overall performance. Due to their high crystallinity and use as plastics, elastomers need long soft chains and cross-linked networks. To meet the above requirements, various biomass monomers are selected to prepare amorphous structures and types of polyester bio-based SR by molecular hybridization to inhibit crystallization. The monomers include butanedioic acid (SA), 1,3-propanediol, 1,4-butanediol, sebacic acid, which provides long-chain flexibility, and itaconic acid, which provides cross-linking point. The synthesis formula is shown in Fig. 2.21.

Bio-based polyester costs about $27,000/ton, and petroleum-based polyester is $17,000/ton. Currently, the price of bio-based monomers is still higher than petroleum-based monomers. With the rapid development of fermentation technology, this phenomenon is improving.

At present, for polyester elastomer conventional vertical reactor, when the molecular weight of polyester elastomer increases to a certain level, due to the influence of viscoelasticity, polyester elastomer experiences axial forces, climbing leverage effect, resulting in polyester elastomer that cannot continue to stir and further increase the limit molecular weight. The molecular weight of polyester elastomers synthesized in vertical reactors is generally below 40,000. However, lower molecular weights lead to poor processing and mechanical properties of polyester elastomers, making it necessary to increase their molecular weights. The current methods to increase the molecular weight are mainly chain expansion and the introduction of horizontal reactors and viscosity reactors to continue stirring

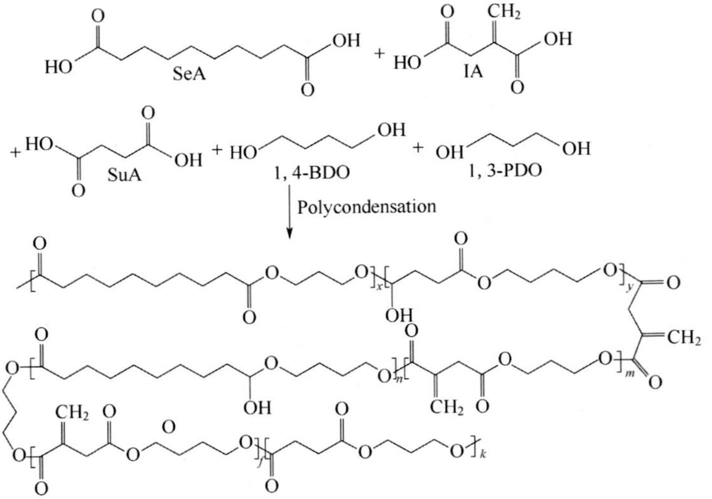

Figure 2.21 Polyester-type rubber synthesis reaction [90].

polyester elastomers. In the study of chain expansion of polyester elastomers, end-hydroxy polyester elastomers were first synthesized, and then diisocyanate (MDI) was used to react with the hydroxyl group to achieve an increase in molecular weight.

By adjusting the molar ratio of monomer, the amount of catalyst, the reaction temperature, and the vacuum, it is possible to make the invisible molecular weight (Mn) greater than 35,000, and the molecular weight (Mw) greater than 128,000, the specific dispersion index (PDI) is about 3.2, and the glass transition temperature of polyester bio-based SR (PBEE) is −56°C. Compared with conventional petroleum-based rubbers, pBEE has a relatively low molecular weight, but its glass transition temperature is lower. This is mainly due to the introduction of a long carbon chain sebacate structure in the main chain, which makes the molecular chain of the polymer very flexible. In addition, using multiple monomers for copolymerization disrupts the regularity of the molecular chains, inhibits crystallization, and makes them amorphous at room temperature. This linear, double-bonded cross-linked elastomer can be processed secondarily, and different kinds of rubber composites can be prepared using traditional rubber processing methods, thus improving the mechanical properties of the elastomer and broadening its application range. Polyester-based SR can also be used as a bio-based plastic toughening agent to toughen PLA with excellent toughening effects. In addition, researchers have explored potential applications of polyester-based synthetic elastomers in oil-resistant seals, shape memory, electrodeformation, and 3D printing materials. A 100-ton polyester bio-based engineering elastomer pilot production line has been built, as shown in Fig. 2.22.

2.2.3.2 Bio-based itaconic ester rubber

Itaconic acid has two carboxyl groups and one double bond, and because the carboxyl group affects chain growth during emulsion polymerization, only low-molecular-weight polymers can be obtained. Polymerization of itaconic acid esters usually results in high-molecular-weight polymers. The isopentyl acetate monomer with high-molecular-weight bio-based itaconate rubber, referred to as PDII, is prepared by fermentation production of the bio-based itaconate esterification reaction through the bio-based monomer itaconate and isopentyl alcohol, followed by copolymerization of isopentyl acetate emulsion with isoprene. Fig. 2.23 shows a schematic diagram of the PDII preparation process.

Figure 2.22 100 tons of bio-based polyester synthetic rubber pilot production line [90].

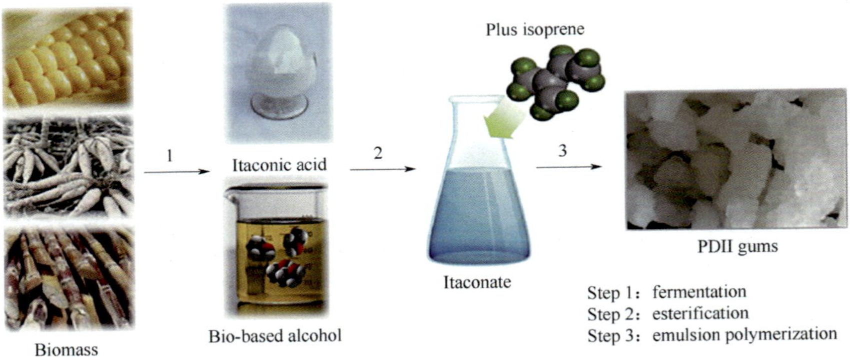

Figure 2.23 Preparation process of PDII [90].

In the study of PDII, it was found that the side groups of itaconic acid esters have a significant effect on the properties of PDII. To analyze the effect of side groups on the properties of PDII, PDII raw materials with different side groups were synthesized. Itaconic acid diesters with different side group lengths were firstly prepared from itaconic acid and monomeric alcohols such as methanol, ethanol, *n*-propanol and *n*-butanol, and then dextrinated PDII with different side group lengths were prepared from copolymerization of itaconic acid diesters with isoprene. Fig. 2.24 shows the reaction equation of PDII with different side group lengths. PDIIs with different side groups were synthesized by initiating emulsion polymerization with a redox initiation system under similar polymerization conditions. The average molecular weight of PDII ranged from

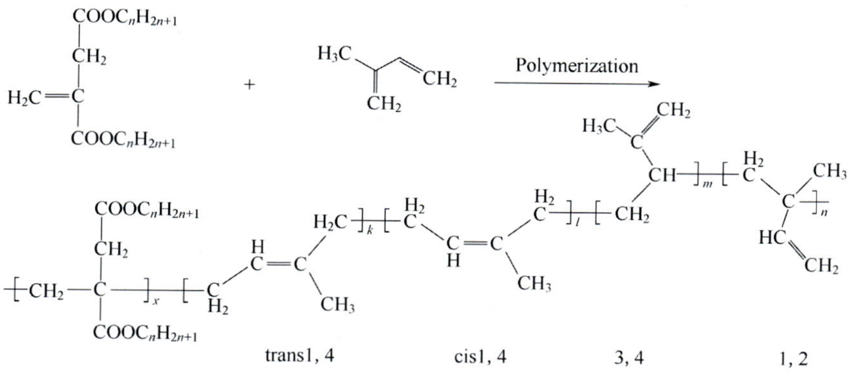

Figure 2.24 Synthesis equation of different sides of PDII [90].

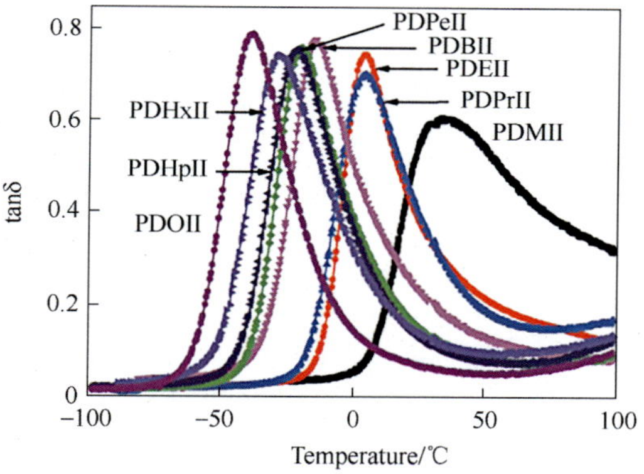

Figure 2.25 DMTA curves of different side PDII/silica composities [90].

100,000 to 300,000, with a molecular weight distribution coefficient of about 3.0. When the feed ratio of itaconate to isoprene copolymer was controlled at 2:3, it was found that the polymer's glass transition temperature decreased gradually with the increase of side group length. For example, the glass transition temperature of the dimethyl itaconate/isoprene copolymer was about 15°C. The glass transition temperature of the itaconic acid dimethyl *n*-decyl ester/isoprene copolymer was −68°C.

Fig. 2.25 shows the DMTA curves of PDII with different side groups. From the DMTA curves, it can be seen that the flexibility of PDII molecular chains gradually increases with the increase of the length of side groups. This is because the increase of the side group length increases the spacing between PDII molecular chains and improves the free volume of PDII. The results show that: for the DMTA curves of rubber composites, the 0°C tanδ value can reflect the wettability of the material. The higher the tanδ value at 0°C, the better the resistance of the material to wettability.

Similarly, a tanδ value of 60°C can reflect the rolling resistance of the material. The lower the tanδ value, the lower the rolling resistance of the material. PDII is expected to be an ideal tread material as it must balance rolling and wetting resistance in practical applications. In addition, to further improve the performance, poly(di-*n*-butyl butadiene/butadiene) elastomer (PDBIB) was made by copolymerizing di-*n*-butyl itaconate with di-*n*-butyl itaconate instead of isoprene.

The prices for preparing bio-based itaconic acid ester rubber monomer are as follows: itaconic acid is about 8000–9000 yuan/ton, *n*-butanol is about 4500–6000 yuan/ton, and isoprene and butadiene are about 9000–11,000 yuan/ton. By comprehensive calculation, the production cost of bio-based itaconic acid ester rubber is 10,960–12,550 yuan/ton, and the market price of soluble polystyrene butadiene rubber is about 16,000 yuan/ton. There is a certain price advantage. Because silica can reduce rubber's rolling resistance without sacrificing its moisture resistance, it is an ideal filler for "green tires." However, there are many hydroxyl groups on the surface of silica, with strong hydrophilicity and easy agglomeration, resulting in poor dispersion of silica in rubber and poor compatibility with the rubber matrix. Compared with petroleum-based soluble butadiene rubber, its dynamic mechanical properties still have a certain gap. To solve this problem, the rubber matrix can be functionally modified to have some functional groups that can react with the hydroxyl group to enhance the interaction between the rubber matrix and silica and improve its dynamic mechanical properties.

2.2.3.3 Soybean oil-based elastomers

As a natural renewable resource, soybean oil is the production of vegetable oil. However, the preparation of elastomeric materials using soybean oil was studied mainly because of the structure of soybean oil with triglycerides. Hence, using polymers to prepare soybean oil is a thermosetting reticulated material that cannot be further rubberized. The bio-based elastomeric poly(epoxidized soybean oil-decylenediamine) was prepared by Wang et al., as shown in Fig. 2.26. On the one hand, decanediamine reacted with the epoxy group of epoxidized soybean oil to form polymer chains; on the other hand, decanediamine reacted with the ester group of epoxidized soybean oil to destroy the triglyceride structure and

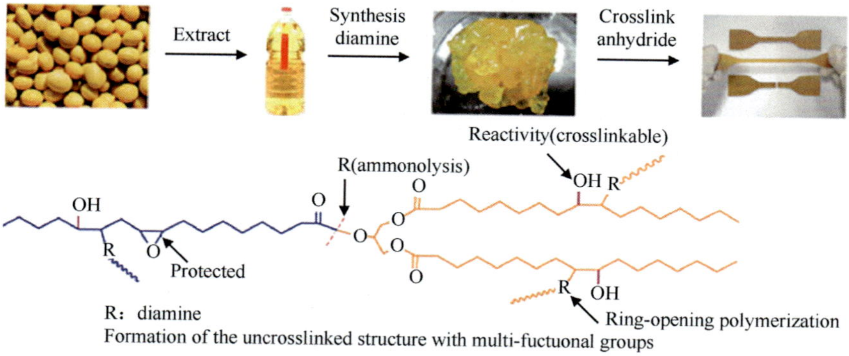

Figure 2.26 Preparation of bio-based elastomeric polyepoxy soybean oil—decylenediamine [90].

obtained a linear processable soybean oil-based bioengineered elastomer. By adjusting the molar ratio of epoxidized soybean oil to sebacic acid, the glass transition temperature of soybean oil—based bioengineered elastomers is −17°C to −30°C. Soybean oil—based bioengineered elastomers can be processed using existing rubber processing technologies. After cross-linking soybean oil-based bioengineered elastomers with succinic anhydride, unreinforced soybean oil—based bioengineered elastomers achieved a mechanical strength of 8.5 MPa and an elongation at a break of 200%. Goodyear's research and development center found that using soybean oil—based bio-based tires could extend the life of tread rubber by 10%, potentially reducing fuel consumption by 7 million tons per year. Also, soybean oil elastomers can be easily blended with silica, improving plant efficiency and reducing energy consumption and carbon emissions.

2.2.3.4 Bio-based polyurethane elastomers

PU elastomers are widely used in biomedical engineering and industrial products because of their excellent mechanical properties, abrasion resistance, resilience, chemical resistance, and structural designability. The monomers used to synthesize PU include diisocyanates and polyols, etc. The primary raw materials for bio-based polyurethanes are vegetable oils, xylose, starch, etc. Research on bio-based PU elastomers has focused on the copolymerization of vegetable oils as soft segments with various isocyanates. Aziz et al. synthesized bio-based PU elastomers using castor oil and poly(3-hydroxybutyrate)-diol with nontoxic 1,6-hexamethylene diisocyanate. An et al. prepared sustainable polyurethanes from castor oil and diisocyanates due to highly cross-linked and flexible structures resulting in very low strength and toughness. They enhanced castor oil-based polyurethanes by adding the hard component isosorbide (IS) to strengthen and toughen the network stiffness and reduce the cross-link density. As the IS content increased, the cross-linkedness decreased while the strength, modulus, ductility, and heat resistance increased significantly. This bio-based PU showed excellent thermal stability with an onset decomposition temperature above 280°C. These studies provide ideas for designing and manufacturing high-performance, sustainable polymers from other vegetable oils. In terms of commercialized products, some major companies in the world have developed bio-based products to meet the market demand. Such as, Cargill, USA, developed a soy-based polyol with the trade name BiOH and built a production unit for BiOH in Chicago. Merquinsa of Spain developed the Perlthene ECO series of bio-based PU elastomers. Bayer of Germany has also launched vegetable oil—based foamed PU materials accordingly.

2.2.4 Brief introduction of bio-based rubber

After World War II, NR as a strategic material declined considerably. Its price gradually became cheaper, thus significantly increasing its value as a fundamental industrial raw material. Although the development of SR has brought considerable pressure on NR, it still cannot shake its position as a significant primary industrial raw material. The NR industry has been dominated by trilobal rubber in the past decades, and the main growing countries of this plant are mostly concentrated in a narrow area near the equator. Still, its rubber output has to be worldwide, which is unreasonable from the viewpoint of industrial layout. At the same time, pests and diseases also greatly impact the trilobal rubber tree. The most prominent of these is the South American leaf blight [101,102], a fungal disease caused by this cotyledon fungus that causes the leaves of the trilobed rubber tree to fall off and even die in severe cases. Geographical and natural constraints have led to an oversupply of triticale rubber worldwide, a major impediment to the development of the global rubber industry.

In the past few years, China's economy has experienced a period of rapid development, that time due to the rapid development of the automotive industry and thus led to the rapid development of the tire industry. Take 2012 as an example [103]. That year, China's tire production reached 470 million, accounting for more than 1/3 of the global wheel tire production, with the proportion of exports at 50%. The fast-developing tire industry requires large miles of raw rubber materials. That year, China's total consumption of rubber raw materials was 7.3 million tons, of which 3.51 million tons of NR, accounting for 48.1%. In comparison, the self-produced NR is only 795,000 tons, and NR's foreign dependence is close to 80%. The high external dependence on NR has constrained the development of China's NR industry and its downstream industries. Therefore, seeking an alternative plant that can produce NR becomes urgent. At present, more than 2500 kinds of plants can produce polyisoprene substances, but only a few can produce valuable rubber. A large part of the reason is that most of the molecular weights to reach the length of the molecular chain of NR are in the hundreds of thousands or even millions. Most plants produce polyisoprene with low molecular weights and have no practical value as rubber. Among these known plants, the two most promising rubber-producing plants are Taraxacum Kok-saghyz (TKS) [104] and Guayule [105]. These two rubber-producing plants have comparative advantages in rubber production, rubber quality, and harvesting methods, and both have the potential as NR replacement plants. The other part of the plant yields trans-isoprene polymer, the representative one is gutta-percha [106]. This rubber cannot be considered as an elastomer due to the regularity of the trans structure, which is in the crystalline state at room temperature. Therefore, it cannot be used as an alternative to NR trees. The following briefly describes each of these three types of rubber.

2.2.4.1 Silver gum daisy rubber

Silver gum daisy rubber is a NR extracted from silver gum daisies (as shown in Fig. 2.27) and is an important source of NR. It has the same chemical structure as trilobal NR and is the same as trilobal rubber in terms of quality and performance [107]. The research, development, and application of silver gum rubber began in the early 20th century, and the main research efforts were concentrated in the United States. Still, there were many interruptions due to various factors such as yield, cost, and politics. The cultivation of silver gum chrysanthemum has made a breakthrough, as shown in Fig. 2.27. Silver gum chrysanthemum's gum content and rubber yield have made a qualitative leap, and the average yield has reached $1-1.5$ t/hm^2. Some companies have tried to commercialize it on a small scale and built several pilot plants, laying the foundation for the commercial development of silver gum chrysanthemum rubber.

Figure 2.27 Silver gum chrysanthemum [109].

The cultivation, harvesting, and latex production of silver rubber chrysanthemum can be fully mechanized compared with the production of labor-intensive trifoliate rubber. However, processing silver rubber chrysanthemum shrubs is technically complex and requires significant investment costs and operating expenses. This is because its rubber is present as micron-sized particles in the thin-walled cells of the bark and cannot shed latex-like rubber trees. The plant body must be destroyed to release the rubber particles from the individual cells to obtain the rubber. Subsequently, rubber can be separated by latex extrusion, centrifugation, extraction, purification of latex, and solvent extraction methods. A problem that needs to be solved in further developing silver rubber chrysanthemum rubber is the presence of proteins that are difficult to separate by solvent extraction steps. This problem has not been solved in a good way [108].

In recent years, companies such as Bridgestone and Goodyear have reopened the research and development of silver rubber tires. Other countries have also researched and developed silver rubber chrysanthemums at different historical periods, such as the former Soviet Union,

Mexico, and parts of Europe, but have not yet achieved outstanding results [107]. However, no outstanding results have been achieved. At the end of 2013, SGB and Yulex collaborated to improve the breeding and selection of silver gum chrysanthemums using a non-GM technology platform combined with genomic technology, including high-energy genotyping and genome-wide trait association. The genome selection and proprietary plant redomestication methods were used to improve the breeding and selection of silver gum chrysanthemum and to double the rubber yield of silver gum chrysanthemum [110].

In terms of applications, the fact that silver chrysanthemum rubber contains essentially no or only a small amount of allergenic proteins and avoids allergic reactions to trilobal proteins in people with allergies determines that it has a huge application in the healthcare sector [111]. In April 2008, Yulex received the first marketing approval from the U.S. Food and Drug Administration to produce examination gloves made from silver latex [112]. In 2013, Yulex collaborated with 4D Rubber in the United Kingdom for the first time to use silver chrysanthemum latex as a rubber barrier material for dental applications [113]. In 2013, Yulex partnered with 4D Rubber in the United Kingdom to produce the first dental rubber barrier material using silver chrysanthemum latex.

Another important area of research is using silver rubber in automotive tires. In 2009, at the United Nations Climate Change Conference in Copenhagen, the Goodyear Tire Company presented the world's first concept tire using bio-rubber technology, manufactured in collaboration with Janenko [114]. Bridgestone began researching the production of silver rubber in 2012 by establishing a research farm in Arizona, USA, which was opened in September 2013. Bridgestone launched its Bio-Rubber Process Research Center a year later to establish all the processes needed to develop Agaricus NR for tire applications, including R&D, experimental production, and manufacturing. In 2015, Bridgestone successfully produced passenger car tires using 100% NR components derived from Agaricus blazei. In the production of these tires, all of the major NR components of the tire, including the tread, sidewall, and sidewall core, were replaced with NR derived from the silver gum chrysanthemum grown and harvested by Bridgestone, for which it was awarded the 2015 Edison Gold Medal. The company's silver gum chrysanthemum rubber tires are currently in the testing phase, and it plans to achieve commercial production of silver gum chrysanthemum tires by 2020. In the meantime, the Bridgestone Group will continue its research on Agaricus and various other raw materials to move toward 100% sustainable tire materials by 2050 in a long-term environmental effort [115].

Goodyear Tire United States has developed a concept tire with several components made entirely of silver chrysanthemum rubber and has tested it on dry, wet, and off-road courses. [116] In 2017, a research team led by Goodyear Tire, including Clemson University, Cornell University, PanAridus, and the USDA's Agricultural Research Service, produced more than 450 concept car tires using silver chrysanthemum rubber instead of all SR. They conducted extensive evaluations, including rigorous wheel and road tests. The overall performance was equal to that of tires made with natural and SR from the trefoil tree. In particular, it is significantly better than conventional tires in terms of rolling resistance, wet handling performance, and wet braking. The study took 5 years and cost $6.9 million in federal funds [117]. Pirelli has produced its first ultrahigh-performance tire with a NR component derived from the silver chrysanthemum plant. Developing this tire sample took 2 years and was the first basic materials research project between Pirelli and the Italian chemical company Visaris [117]. The project was the first basic material research project between Pirelli and the Italian company Visaris Chemicals. Pirelli chose a Maserati test car that could guarantee the operation of the tire under high-load conditions and completed a runway test of the silver chrysanthemum high-performance tire. The results showed no difference in performance between the silver chrysanthemum tire and the normal tire in wet and dry conditions. The company will test the tire's performance in winter [118].

In the 1970s, China introduced silver gum chrysanthemums from Mexico and the United States [119]. We have obtained five high-yielding systems through screening, and the highest gum content can reach 20% in 4 years of cultivation. However, due to the lack of understanding of the utilization value of silver gum chrysanthemum, the research on silver gum chrysanthemum in China has been in the preliminary exploration stage. There are only a few reports of related applied research [120].

2.2.4.2 Dandelion grass rubber

TKS is the common name for the Soviet dandelion. First discovered by Soviet botanist Rodin in 1931 in Kazakhstan, the genus Taraxacum of the lettuce family of Asteraceae is very similar in appearance to the common dandelion, as shown in Fig. 2.28. Still, the mass fraction of rubber components in the roots of TKS reaches more than 0.2 (dry weight of perennial),

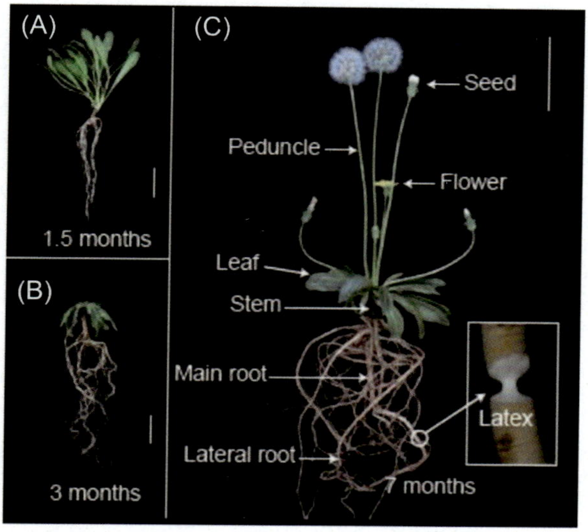

Figure 2.28 Morphologies of Taraxacum kok-saghyz at different growth stages and tissues for transcriptome [124].

and the rubber quality is excellent. Its value for industrial use is no less than Brazilian tree rubber [121]. After the discovery, the Soviet scientific team returned the rubber grass to the country and quickly started researching and promoting its industrial production. By 1941, the planting area of rubber grass in the Soviet Union had expanded to 67,000 hectares, and rubber grass rubber accounted for 30% of the entire Soviet Union's NR consumption. And in 1943, it successfully produced 3000 tons of TKS rubber [122]. The origin of rubber grass is located between approximately 42°–43°N and 79°–82°E. During World War II, rubber grass was widely planted in the former Soviet Union, the United Kingdom, the United States, Spain, Germany, and other countries. Rubber grass is highly adaptable and grows well in arid and saline areas and high mountains, slopes, and plains. Compared with rubber trees, dandelion rubber grass is a plant that can be harvested in the same year to extract rubber. The rubber production cycle is short and has a lighter texture [123]. The rubber tree must be cut manually, dandelion rubber grass can be planted intensively, and the harvesting can be automated, making it possible to obtain a large yield quickly. In addition, products such as gloves made from traditional tree rubber can cause some personal allergic reactions. In contrast, dandelion grass rubber is more biocompatible and antiallergic rubber.

Grass rubber is present only in the latex tubes and vascular bundles of grass rubber roots. Fresh roots break with white latex containing the rubber particles, while dried roots pig off with a clear pulling phenomenon. The value of dandelion rubber lies not only in its rubber tubes containing powerful antiallergy latex but also in its plant tissue containing solid rubber filaments.

The chemical structure of dandelion grass rubber is similar to that of triticale rubber, being the same cis-polyisoprene, but the form of existence is very different. The ratio of latex to solid rubber filaments varies according to season, climate, growing conditions, and time of growth [104]. Research showed that the genes of some rubber of dandelion grass are the same as those of Brazilian tree rubber, which is cis-polyisoprene. The heavy average relative molecular mass of grass rubber can reach 2.18 million, the average relative molecular mass reaches 1.21 million, and the relative molecular mass distribution index is 1.8.

In terms of performance, tires made of dandelion grass rubber can reach the same level as trilobal rubber tires [121]. In 1940, American research reported that tires made of dandelion rubber grass rubber could meet the same requirements as trilobal rubber tires. 1950, Beijing Industrial Laboratory made dandelion grass rubber into finished bottle stoppers, trademarks, shoe heels, and rubber rings. They found that raw rubber was soft in quality because it contained more resin and grease. And more sulfur and zinc oxide were needed to enhance hardness. In 1953, Xinhua News Agency reported that the rubber extracted from dandelion rubber grass was made into bicycle tires, rubber shoes, machine belts, and automobile tires by Lanzhou Industrial Laboratory. The physical properties met the requirements of the national standards at that time, and the tires had good results for Beijing Public Transport [108].

Dandelion rubber grass has excellent biological properties, but how to extract rubber from it is the key. Depending on the morphology of the extraction target products, the extraction methods can be divided into latex extraction, solid rubber extraction, and mixed latex–solid rubber extraction. In the field of latex extraction, A. U. Buranov et al. [125] invented the mixed extraction method by improving the mixing method for extracting silver rubber chrysanthemum latex. Soviet scientists proposed the flow method for extracting dandelion rubber grass latex. In solid gum extraction, an attempt was made to extract dandelion rubber grass solid gum by analogy with the solvent sequential extraction method of silver rubber chrysanthemum solid gum [125]. In solid latex gum mixed extraction, P. Stramberger and R. K. Eskew et al. [126,127] proposed a wet grinding method to extract rubber from dandelion rubber grass using a gravel mill; Chunyan Du et al. [108] first invented the solvent extraction method through their research, and the average rubber content obtained by this method was 27.48%.

Goodyear Tire & Rubber Company recently participated in the NR Alternatives Advantage program, which seeks new sources of NR within the United States. Over the next 4 years, the PENRA project team will use dandelion grass rubber products on a small scale to manufacture tires and conduct comparative tests with NR, including tire traction, durability, rolling resistance, treadwear, etc. According to the PENRA project team, dandelion grass rubber products have the potential to replace NR with 30% of the market demand [108]. The dandelion rubber grass project was proposed by Delta in 2006 [128]. The project includes: (1) improving the seeding of dandelion rubber grass to increase its rubber content; (2) exploring the experience of planting on a large scale; and (3) establishing a processing plant to extract rubber from the harvested dandelion rubber grassroots. On the other hand, dandelion rubber grass is an herbaceous plant with a short growth cycle. So we can try to establish the genetic transformation system of dandelion rubber grass and use dandelion rubber grass as a carrier to study the functional genes of the Brazilian rubber tree. In 2010, Zhang Liqun of the Beijing University of Chemical Technology began conducting in-depth research on dandelion rubber grass, including screening and breeding high-quality varieties of dandelion rubber grass. The project includes selecting and cultivating high-quality dandelion rubber grass varieties, developing and designing efficient rubber extraction methods and equipment, determining the characterization and performance of dandelion rubber, and applying dandelion rubber in radial tires. In collaboration with the Rubber Research Institute of the Chinese Academy of Tropical Agriculture and Shandong Linglong Tire Co., Ltd, the research project has devised an efficient and rapid rubber extraction method and a systematic and comprehensive characterization of dandelion rubber. This research is currently in the leading position in China and has also accelerated China's progress in second NR resources research [108].

2.2.4.3 Eucommia gum

Eucommia gum is another kind of bio-rubber, mainly found in the leaves, bark, and seeds of Eucommia trees (Fig. 2.29). It is an excellent polymer material with dual rubber and plastic properties, broadly divided into

Figure 2.29 Photo of eucommia trees.

natural and synthetic juniper rubber. Both are chemically identical to the NR produced from the trilobal rubber tree, but their molecular structures are different: juniper is trans-polyisoprene, while NR is cis-polyisoprene. Due to the orderliness of the trans structure of its molecular chain, juniper rubber is easy to crystallize and is a crystalline polymer. At room temperature and without applied pressure, the intrinsically crystalline juniper rubber generally forms two stable crystalline forms, namely the orthogonal B-crystalline form and the monoclinic a-crystalline form. In addition, the flexible, double bond and trans structure of the eucommia rubber itself meet the conditions of shape memory material, so it has shape memory properties. Still, it cannot be considered an elastomer.

Eucalyptus rubber is a leathery and tough material at room temperature. Pure rubber is colorless, but commercial rubber has various colors. The pure rubber crystallizes at 10°C, shows elasticity and elongation at 40°C–50°C, and softens at 100°C. It has plasticity and can recover its original property after cooling. Its T_g and plasticity vary with the resin content: when the resin content is 2%–6%, T_g is 62°C–63°C; when the resin content is 7%–12%, T_g is 59°C–60°C. The relative density of eucommia rubber is 0.95–0.98; Mn is 160,000–173,000 and degrades after mixing. Eucommia rubber has good corrosion resistance and impact resistance. Experiments show that 20–25 parts of synthetic eucalyptus rubber can save about 2.5% of fuel. If 20 parts of high vinyl polybutadiene rubber are mixed, fuel consumption is reduced, and antiskid performance will significantly improve.

Eucommia is a natural plant endemic to China, and more than 95% of the world's eucommia resources are in China. Eucommia has been cultivated in China for more than 2000 years, and after the founding of the People's Republic of China in 1949, it was introduced on a large scale, and the cultivation range was extended to 27 provinces and autonomous regions in China, except Heilongjiang, Tibet, Guangdong, and Hainan. Japan, the United States, and the Soviet have conducted a series of research and practice on the cultivation of Eucommia [129]. Studies and practices on juniper cultivation have been conducted in Japan, the United States, and the Soviet. Hitachi has applied for patents in China and the United States on extracting eucommia biopolymers, and Osaka University in Japan has conducted many research projects on eucommia applications. Some companies in the United States have also applied for patents in China for using Eucommia gum in dental materials.

Domestic research on eucommia gum mainly focuses on eucommia extraction and purification process development, as well as eucommia gum processing process and application technology development [130–132]. Around 1982, Yan Ruifang, a researcher at the Institute of Chemistry, Chinese Academy of Sciences, processed trans-polyisoprene, which used to be only as plastic, into a high elastomer, thus opening a new chapter in the use of eucommia gum. Later, he developed three other materials with different applications: low-temperature functional plastic, thermoelastic shape memory material, and elastic rubber material. Yan Ruifang also found that blending with plastic and rubber can develop new materials. He applied for a series of patents. The Shenyang University of Chemical Technology has introduced many dynamic and reversible ions or chemical bonds into eucalyptus rubber to give it a self-healing function [133]. This self-healing elastomer material can be applied to robots, electric vehicles, lithium-ion batteries, and artificial muscles.

Three Rubber Co., Ltd. has developed a new process of blending juniper rubber with NR, solving the critical technology of extrusion and molding of blended rubber, which can be molded at room temperature without additional molding equipment. It was the first time to realize the application of blended rubber in existing aviation tires. The project prepared the juniper rubber aviation tires through the highest speed level dynamic simulation experiments. The results show that the performance of all indicators meets the requirements. Skid distance, takeoff and landing load, and takeoff and landing speed are the highest level of the same specification aviation tires. The technology is now available for industrialization and promotion conditions [134]. Li et al. [135] found that grafting of maleic anhydride on dulcimer makes a chemical link between dulcimer and asphalt. It establishes a reaction network between vulcanized rubber powder and asphalt, significantly improving the mixture's performance, and the modification effect reaches the level of foreign products. Long et al. [136] tried to apply eucommia gum to the dental root canal filling material and achieved a better encapsulation effect. Other scholars used the high elastomer and nontoxicity of Eucommia gum to prepare chewing gum and teething stick to achieve oral health care and promote the growth and development of children's teeth and mouth.

2.3 Tire production process and spherical tires

2.3.1 Tire structure

Tires usually consist of an outer, inner, and cushion belt. Some tires do not need an inner tube, such as tubeless tires, which have an airtight rubber layer on the inside and must be equipped with special rims. The tire structure is developing in the direction of the tubeless, radial structure, flattening (small tire cross-sectional height to width ratio), and lightweight.

The outer tire comprises tread, side, and bead. The outer tire slices can be divided into several independent areas: crown area, shoulder area (tread slope), flexion area (side area), reinforcement area, and bead area.

2.3.1.1 Tread

The tread is the outermost rubber layer of a tire in contact with the road. Usually, the outermost rubber crown of the tire, the shoulder, the edge of the crown on both sides, the side of the tire, and the reinforcement area are collectively called tread rubber. The tread is used to prevent mechanical damage and early wear of the tire body, to transfer the traction and braking force of the car to the road, to increase the grip between the tire and the road (soil), and to absorb the vibration of the tire during operation. In the normal driving process, the part of the tire tread in direct contact with the road is called the driving surface. The driving surface consists of different shapes of tread blocks and tread grooves. The protruding part is the tread block. The tread block's surface increases the tire's grip and the road (soil) to ensure that the vehicle is resistant to side slippage. The tread grooves' lower layer, the tread base, cushions the impact and shock.

2.3.1.2 Placenta

Tire body usually refers to the force structure of a pneumatic tire, composed of one or more layers of rubber curtains and beads suspended at certain angles. The tire has a certain strength, flexibility, and elasticity to withstand the complex stresses and multiple deformations of the tire and to relieve the vibration and impact of the external road. The tire line layer consists of tire lines suspended side by side in the tire body. It is the stress skeleton layer of the tire. It supports all the tire components and holds the tire profile to ensure strength and dimensional stability.

2.3.1.3 Tire bead

The bead is a part of the tire mounted on the rim. Its purpose is to hold the tire tightly on the rim and withstand the various forces acting between the outer tire and the rim. It can bear the tension caused by internal pressure, overcome the lateral force acting on the tire when it turns, and keep the outer tire from coming out of the rim. Therefore it must have high strength, and the structure should be compact, firm, and difficult to deform.

The bead consists mainly of the traveler, the triangular filler, and the bead package. The ring is a rigid ring made of copper-plated steel wire and is the main component that holds the tire to the rim. Its structure is divided into four cross-sectional forms: square, hexagonal, semicircular, and circular. The filler rubber is a semirigid rubber body, which is the transition of the elastic sidewall of the rigid rover and is also a compound filler rubber composed of two formulations.

The inner tube is a flexible round rubber cylinder, as shown in Fig. 2.30. When inflated, the inner tube can be stretched and attached to the inner

Figure 2.30 Inner tube [137].

cavity of the outer tube to seal the air. The rubber used to make the inner tube should have high elasticity, tear resistance, fatigue resistance, and good air tightness. Usually, butyl rubber can meet these requirements. The outer dimensions of said inner tube are slightly smaller than the inner cavity of the said outer tube. The thickness of each part of the inner tube is usually thicker than the other parts due to the different conditions of use, such as inflation.

According to their structure, inner tube valve nozzles can be divided into three categories: rubber—metal valve nozzles (mainly for inner tubes of automobiles), rubber cushion valve nozzles (mainly for inner tubes of truck tires), and water and gas dual-use valve nozzles (special tools for inner tubes). The valve nozzle for tubeless tires is fixed directly to the rim.

The cushion belt is placed in the contact area between the inner tube and the rim to protect the inner tube from the wear and tear of the rim assembly. The cushion belt can be divided into three types according to its structure: ordinary, no, and flat belt. The cushion belt has a thin edge, smooth surface, and heat resistance.

2.3.2 Tire classification

Different classifications of tires are made from different perspectives. Commonly used classification methods include skeleton structure classification, skeleton structure classification, and usage classification.

2.3.2.1 Classification by skeleton

According to the different structural designs, that is, the arrangement of the cord in the carcass, it can be divided into bias-line and radial tires. The fundamental difference between radial and bias tires is the tire body, as shown in Fig. 2.31.

The oblique tire carcass is a bias cross-curtain ply, and the radial tire carcass is a polymer multilayer cross material. The top layer of oblique tires is a steel belt cord woven by steel wire, which can reduce the probability of the tire being punctured by out objects.

The cord of the bias tire is arranged in a biased, horizontal direction, hence the name. It is characterized by high tread and sidewall strength but high sidewall stiffness and poor comfort. Due to the movement and friction between the plies of the cord at high speed, it is unsuitable for high-speed transmission. With the continuous improvement of radial tires, bias tires will be eliminated.

The radial tire cord finding direction is consistent with the radial cross section of the tire. The radial tire cord layer is equivalent to the basic

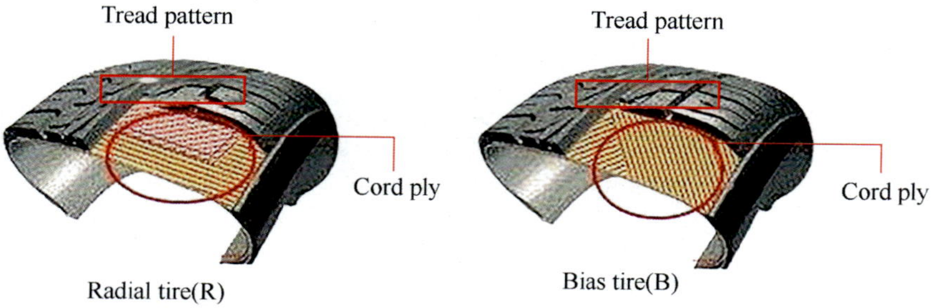

Figure 2.31 Structure of radial and bias tires [137].

skeleton of the tire. To ensure the stability of the cord, several layers of band plies (also called hoop plies) are made of high-strength, nonstretchable materials outside the tire. The direction of the cord is at a greater angle to the radial cross section. Oblique tires have many design limitations, such as the strong friction of the horizontal cord making the carcass easily heated, accelerating tread wear, and the arrangement of the cord does not provide good handling and comfort. In contrast, radial tire steel belts have better elasticity, adapt to the irregular impact of the road, and are durable. This structure also means much less friction than biases when driving a car, resulting in longer tread life and better fuel economy.

At the same time, radial tires have characteristics that make it possible to achieve tubeless. There is a recognition of the advantages of tubeless tires: when a tire is punctured, it suddenly does not look like a biased and tubeless tire (which is very dangerous). Still, it keeps the tire air pressure periodically, improving the car's safety. In addition, radial tires have better grip than bias tires.

Radial tires have greater elasticity, better wear resistance, less rolling resistance, better adhesion, better cushioning performance, higher carrying capacity, and less penetration than ordinary bias tires. The disadvantages of radial tires are that the side of the tire is prone to cracking, slightly poorer lateral stability of the vehicle due to high lateral deformation, high manufacturing process, and high cost.

2.3.2.2 Classification by carcass structure

Car tires can be divided into pneumatic tires and solid tires, as shown in Fig. 2.32, according to the structure of the carcass.

Most modern cars use pneumatic tires. Pneumatic tires can be divided into high-pressure, low-pressure, and ultralow-pressure tires according to

Figure 2.32 Photo of solid tire [137].

Figure 2.33 Wheels with and without tubes [137].

the level of air pressure inside the tire. Low-pressure tires are widely used in all kinds of automobiles. Pneumatic tires can be divided into tubular and tubeless tires according to their composition and structure. Tubeless tires are usually used for automobiles, as shown in Fig. 2.33.

2.3.2.3 Classification by use and international standards

According to international tire standards, tires can be divided into the following categories:

A car tire is a type of tire that is mounted on a car. It is mainly used for high-speed driving on good roads. The maximum speed can be more than 200 km/h. It requires a comfortable ride, low noise, good maneuverability, and stability. Tire structure mainly adopts radial structure. According to the requirements of driving speed, it can be divided into different series. In the standards and instructions, the 95 and 88 series are bias tires, and the 80, 75, 70, and 65 series are radial tires.

Light truck tires have a rim diameter of 16 inches and a cross-sectional width of 9 inches or more. These tires are mainly driven on highways and can generally reach speeds of 80–100 km/h.

Truck and bus tires are usually truck, dump truck, special tires, and trailer tires with a rim diameter of 18–24 inches and a cross-sectional width of 7 inches or more. Its road surface is more complex, with good asphalt roads, poor gravel roads, dirt roads, mud roads, ice and snow roads, or even no road conditions. Its driving speed is generally not more than 80 km/h.

Construction tires are installed on special construction vehicles such as loaders, bulldozers, excavators, straighteners, rollers, and stonework machinery. Travel speed is not high, but the road conditions and load-bearing performance are demanding. The tire structure is mainly bias tires, but the French company—Michel (Cantonese, Michelin in Mandarin) also uses radial structure. According to the tire section width classification, it can be divided into two series: standard tires and wide base tires.

The front and rear wheels drive off-road vehicle tires. Off-road vehicle tires are mainly driven on desert, muddy ground, soft soil, or other bad roads without road surface, which requires the high passing performance of tires. Off-road tires mostly use low-pressure tires; some use pressure-regulating tires to adjust tire pressure according to road conditions. To improve off-road passability, generally take measures to increase the tire section and rim width, reduce the rim diameter, etc., to increase the grounding area and reduce the grounding pressure.

Agricultural and forestry tires—Agricultural tires are mainly used for tractors, combine harvesters and farming vehicles. Forestry tires are installed on forestry tractors and machinery for cutting, skidding, shoveling, and digging operations in forestry. Low-speed requirements characterize these two types of tires, but the use of harsh conditions. They are often driven on field roads in poor conditions, hard stubble or rocky mountain roads, or even roads without pavement, and the tires are easily scratched or cut. Another characteristic is that tires have short mileage but long service life. Therefore, tires must resist bending, cracking, and antiaging. The main structure of the tire is a bias structure, but a radial structure is also used.

Industrial vehicle tires are mainly pneumatic, semisolid, and industrial vehicle solid tires. Distribution of battery tires, forklift tires, and air leakage tires. Motorcycle tires, including motorcycle tires, booster tires, and small-diameter motorcycle tires.

Air Tires—Pneumatic tires used for aircraft. Special vehicles were tires including gun tires, tank tires, armored tires, desert tires, explosion-proof vehicle tires, etc. Wheel Tires—Pneumatic tires for bicycles, tricycles, and wheelbarrows.

2.3.3 Tire production process

The tire production process is shown in the following diagram (Fig. 2.34).

2.3.3.1 Procedure 1: crafting

The compacting process combines carbon black, natural/SR, oil, additives, accelerators, and other raw materials in an internal mixer to produce "rubber." All raw materials must be tested before entering the mixer and released before use. The weight of the mixer is approximately 250 kg per tank. The compounds used for each rubber component in the tires have specific properties. The composition of the compound depends on the performance requirements of the tire. Also, the composition of the compound varies depending on the demands of the supporting manufacturers and the market, mainly from traction, driving performance, road conditions, and the tire itself. All rubber materials are tested, and rubber parts are prepared before moving on to the next process.

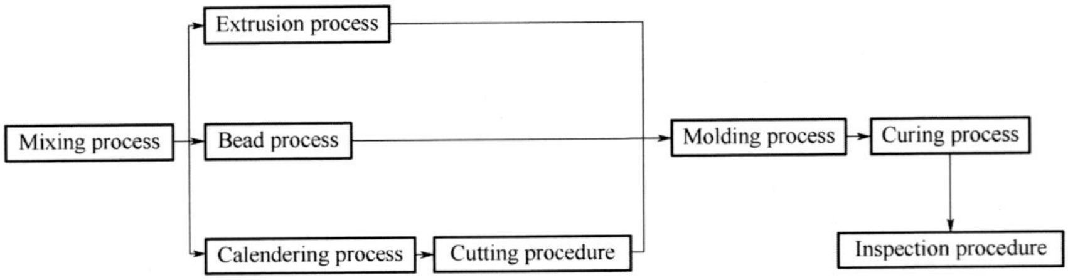

Figure 2.34 Flowchart of tire production process [137].

2.3.3.2 Procedure 2: preparation of rubber parts

The process of preparation of rubber parts consists of six main stages. In this process, all semifinished rubber parts that make up the tire are prepared, some of which have already undergone initial assembly. These six parts are:

1. *Section I: Extrusion*
   a. The rubber is fed into the extruder head, and different semifinished products are extruded: tread, sidewall/subport, and a triangular rubber strip.
2. *Section II: Rolling light*
   a. The raw material thread is passed through the calender, and a thin layer of rubber is hung on both sides of the thread. The final product is called "cord." The raw materials for the cord are mainly nylon and polyester.
3. *Section III:Bead forming*
   a. Steel rings are made by winding multiple steel wires together and hanging glue. The compound used for the beads has special properties. When vulcanized, the compound and the steel wire can be tightly bonded.
4. *Section IV: Cutting curtains*
   a. The cord will be cut to the proper width and fit during this process. The variation of the cord width and angle depends mainly on the tire specifications and the requirements of the tire construction design.
5. *Section V: Triangle tape*
   a. In this process, the extruded triangle tape from the extruder is manually applied to the bead. The triangle tape plays an important role in the running performance of the tire.
6. *Section VI:Forming of the belt layer*
   a. This process is the production of the tape layer. In the spindle chamber, many wires come out through the threading plate, while the tape is hung on both sides of the wire with glue through the template. After hanging the tape, the bundle plies are cut to the specified angles and widths. The width and angle depend on the tire specifications and structural design requirements.

All rubber parts will be transported to the "tire molding" process for tire molding.

2.3.3.3 Procedure 3: tire molding

The tire molding process is to assemble all the semifinished products on the molding machine into a tire, which is an uncured tire. After the check is born, it is transported to the curing process.

2.3.3.4 Procedure 4: vulcanization

After the proper time and conditions, the raw tire is loaded on the curing press and cured in the mold into a finished tire. The cured tire is no longer plastic but has strong mechanical properties and the appearance of a finished tire tread/font and pattern. The tire will then be sent to the final inspection area.

2.3.3.5 Procedure 5: final inspection and product performance testing

In this area, tires first pass a visual appearance test, and then a homogeneity test is done by a "homogeneity tester." The homogeneity tester measures radial force, lateral force, cone force, and fluctuation. After the homogeneity test, the tire is tested on the "dynamic balance tester" for dynamic balance. Finally, the tires are X-rayed and shipped to the finished product warehouse for shipment.

Designing new tire sizes requires extensive tire testing to ensure that tire performance meets government and supporting plant requirements.

Once the tires are in production, we continue to do tire tests to monitor the quality of the tires, the same as we do when new tires leave the factory. The machine used to test the tires is the "mileage test," which usually includes high-speed and durability tests.

2.3.4 Tire production technology

2.3.4.1 Michelin technology (command + control + communication and manufacturing − C3M)

C3M [138] is an integrated system that combines command, control, communication, and manufacturing. The system has five technical points: (1) continuous low-temperature mixing; (2) direct extrusion of rubber parts; (3) weaving/winding of the skeleton ply on the molding drum; precuring of the ring tread; and (4) electrothermal curing of the tire.

The key equipment for C3M is special weaving machines and extruders. The C3M technology can be realized by properly configuring a special weaving machine and an extruder with a forming drum as the core. The special weaving machine weaves the unconnected looped carcass cord layer and tape layer around the forming drum and winds the wire around the forming drum to obtain the wire. The extruder continuously mixes the rubber at a low temperature (below 90°C) and extrudes rubber parts such as sidewalls and triangular rubber strips.

2.3.4.2 Integrated manufacturing precision assembly cellular technology (IMPACT)

IMPACT [139] is Integrated Machining Precision Forming Cell Technology, which has four elements (also known as four cells): (1) thermoforming; (2) improved control technology to increase productivity; (3) automatic feeding; and (4) cell manufacturing. These four elements can be used individually or in combination as one element or a whole system and can be tightly integrated with existing tire processes. IMPACT will not be incompatible with existing systems like new-generation tire manufacturing systems.

2.3.4.3 Pirelli technologies (modular integrated robot system)

Modular integrated robot system (MIRS) [140] is a building block integrated automation system. The essence of MIRS is that: drum production is centered on the drum; multiple extruders are combined with remotely controlled robots for direct molding from the extruded rubber to the molding drum, and the inner layer of the embryo shell replaces the bladder.

There are only three processes in MIRS: (1) preprocessing, (2) molding, and (3) vulcanization. There are several extruders in the preprocessing process. Each extruder has a winding frame with dimensions of 1 m × 1.5 m, a hanging roll or a rope dipping roll; many wires or ropes from the frame enter the extruder. Right-angle heads are extruded with rubber to obtain reinforcing tapes for downstream processes. The forming process has three sets of eight extruders and three pairs of remote-controlled robots divided into three stations. The forming drum is foldable and hollow, and the drum body is made of eight 20 mm thick aluminum plates with small holes that connect the drum surface to the drum cavity. The forming drum is preheated into the first station and rotates around the shaft; the extruder extrudes the rubber onto the forming drum, and the robot repeatedly rolls the rubber to extrude air so that the rubber gets gas close to the drum surface. The dense layer, the drum surface is hot, and the compound is prevulcanized. Then the forming drum enters the second station, where the second pair of robots wind the various reinforcing strips produced by the performing process onto the drum, and the second group of extruders extrudes the rubber onto the drum. The machines gradually perform crossover operations to form the carcass layer, beads, etc. Then the molding drum enters the third station, and the third pair of robots is prefabricated with the strip ply. The extrusion unit directly extrudes the separation rubber, side rubber, and tread rubber onto the molding drum, and the complete embryo is obtained through compaction and molding. The green tire enters the curing process with the molding drum, and the curing machine is mounted on a six-station disk conveyor belt column. The first pair of robots puts the blank of the molding drum into the curing machine, closes the mold, and sends high-pressure nitrogen into the molding cavity, where the nitrogen escapes to the surface of the drum through the ventilation holes in the drum wall, causing the embryo to expand. Thus, it separates from the drum surface and clings to the vulcanizing mold's inner wall so that the precured carcass's inner layer acts as a capsule. As with normal vulcanization, steam passes through the cavity. After 15 minutes of curing, the pan conveyor reaches station 6, where a second pair of robots opens the mold and removes the tire along with the molding drum, which is folded to obtain the finished tire. The forming drums are assembled and sent back to the second process for recycling. This completes a production cycle.

2.3.4.4 *Digital tire technology (digital rolling simulation)*

The so-called digital tire simulation technology [141] refers to various simulation experiments by simulating a rotating tire model on a supercomputer. Mainly by the tire tread noise simulation, air pressure change simulation, wire external force absorption simulation, rubber formula simulation, wear energy distribution simulation, real car driving simulation, gas penetration simulation, tire mud road simulation, road environment simulation, and other technical components. The simulation better solves the disadvantage that the tire cannot collect the tire ground data when rotating at high speed and shortens the tire design and production cycle.

2.3.5 History of tire development

The earliest tires were made of wood or iron and can be seen on ancient Chinese chariots and foreign gentlemen's cars, as shown in Fig. 2.35 [62]. Later, when the explorer Christopher Columbus explored the New World for the second time in 1493–96 and reached Haiti in the West Indies, he was surprised by a rubber block played with by local children. Later, he brought this wonderful object back to his homeland. A few years later, rubber was widely used, and the wheels were gradually changed from wood to hard rubber. But the rubber tires at this time were still solid, uncomfortable to walk on, and noisy.

Figure 2.35 Early wooden tires [62].

2.3.5.1 The birth of hollow wheels and pneumatic tires

The first hollow wheel was invented in 1845 by the British Robert Thomson [62]. He suggested using compressed air-filled elastic airbags to reduce vibration and shock during motion. Although tires at the time were made of leather and rubber canvas, such tires showed the advantage of low rolling resistance. Although Robert Thomson invented the hollow wheel, the person who put the wheel on the tire was Dunlop. In 1888, John Dunlop made a hollow rubber tire. At that time, to increase the comfort of the carriage, he put an inflated rubber tube on the wheel to reduce vibration, but Dunlop was still a veterinarian. Thomas then made a hollow rubber tire with a valve switch, but unfortunately, it could not maintain a certain section shape and width due to the lack of canvas in the inner layer.

The advent of rubber tires determined the direction of automobile travel [62]. When it comes to rubber, people naturally think of Charles Goodyear, the father of rubber. In 1834, inspired by steelmaking, he began experimenting with soft rubber hardening. After countless failures, he found that the vulcanized rubber was not sticky and elastic after heating, so hardened rubber and the rubber tire manufacturing industry were born. The name Goodyear Tire is now given in honor of the father of rubber, Charles Goodyear (Fig. 2.36).

Pneumatic tires were widely developed in 1895 with the advent of the automobile. 1895 saw the first samples of automobile tires appear on the French Peugeot model. It was a single-tube tire made of plain canvas. Although initially pneumatic tires were used and made of rubber, there was no tread pattern.

Figure 2.36 John Boyd Dunlop (*left*) and Charles Goodyear, the father of rubber (*right*) [137].

2.3.5.2 The birth of patterned tires

It was not until 1908–12 that significant changes were made to the tire [62]. Namely, the tread rubber had a tread pattern that improved the tire's performance, thus opening up the history of tire treads and increasing the tire's section width, allowing for lower adoption. The internal pressure ensures better cushioning performance. In 1892, the British Burry Mill invented the curtain, which was used in production in 1910. This achievement not only improved the quality of tires but also expanded the variety of tires and made them moldable. As the quality of tires improved, so did the quality of fabrics, and cotton thread was replaced by rayon.

In 1904, Matt invented a carbon black reinforced rubber [62]. The mass use of tread-reinforced rubber came after using tires because the canvas was damaged faster than tire tread before. The amount of rubber used in rubber compounds increased rapidly. In the 1930s, only about 20 carbon blacks were used per 100 raw rubber parts. At this time, carbon black was mainly used for tread, and carcasses were not used. It has now reached more than 50 parts. Before the carbon black was mixed in the tread, the tires were polished after about 6000 km. The mileage of the tires improved rapidly after the carbon black was mixed. Now a set of truck tires can travel about 100,000 km, even on good roads, up to 200,000 km.

From 1913 to 1926, the invention of cord and carbon black tire technology laid the foundation for the development of the tire industry. With the standardization of the outer edge of the tire and the gradual improvement of the manufacturing process, the production speed is higher than before. With the development of the automotive industry, tire technology has been continuously improved. For example, from the early 1920s to the mid-1930s, tires transitioned from low-pressure to ultralow-pressure tires. In the 1940s, tires gradually transitioned to wide wheels. In the late 1940s, there were no tubes.

2.3.5.3 The birth of radial tires

However, the advent of many new technologies was not much better than the tire without in-line construction pioneered by Michelin in France in 1948. This tire revolutionized the tire industry by significantly improving service life and performance, especially fuel savings while driving.

A radial tire is a new type of tire in which the tire body cord is arranged along the radial direction, and the padding is arranged along the circumferential or near circumferential direction. It comprises six main parts: tread, carcass, sidewall, belt ply, bead, and sealing ply. According to the different cord materials used in the carcass and belt, radial tires can be divided into three types: all-steel radial tires, semisteel radial tires, and all-fiber radial tires.

The radial tire cord finding direction is consistent with the radial cross section of the tire, and the radial tire cord layer is equivalent to the basic skeleton of the tire. To ensure the stability of the cord, there are several layers of band plies (also called hoop plies) made of high-strength, non-stretchable materials outside the tire. The direction of the cord is at a greater angle to the radial cross section. Radial tires have high elasticity, good wear resistance, low rolling resistance, good adhesion, good cushioning performance, and high load-carrying capacity. They are not easy to penetrate compared with bias tires. The disadvantage of radial tires is that one side of the tire is easy to crack, and the lateral stability of the vehicle is slightly worse due to the large lateral deformation, high manufacturing process requirements, and high cost.

2.3.6 Spherical tires

2.3.6.1 Introduction to spherical tires

It's hard to imagine life without cars now. Perhaps it is impossible to imagine this part of our daily life in the future. One of the key elements of their design is the installation of an elastic membrane on the edge of the vehicle's wheels. In other words, it is a tire. An inflatable tire is filled with air or any gas under pressure. Depending on the sealing method, pneumatic tires are produced in indoor or tubeless versions. The quality and condition of the tire affect the car's operation, stopping distance, speed, load, and rolling resistance. The load index is the car's ability to carry the weight of cargo, passengers, and itself. The speed index determines the speed limit at which the tire can safely operate. During the vehicle's movement, the tire consumes a certain amount of energy due to deformation caused by the point contact displacement of the tire. This energy is subtracted from the kinetic energy. Rolling resistance can amount to 25%–30% of energy consumption. However, this percentage depends to a large extent on the speed of the vehicle. The index is low at high speeds. The main cause of accidents is the car's loss of control, which is directly related to the tires.

According to a recent study by Navigant Research, 85 million self-driving cars with features are expected to be sold worldwide annually by 2035. To meet the needs of self-driving cars, international tire manufacturer Goodyear has developed a concept tire called the Eagle-360. While the tire is pure concept development, it offers a glimpse of what the innovation might look like in the future [142], as shown in Fig. 2.37 and Table 2.3.

Spherical tires can move in all directions to help the body go around obstacles without changing its direction of travel. This feature of spherical tires greatly improves safety while driving, providing a smoother driving experience and increased comfort for future drivers. In addition, multidirectional tires can help cope with future space constraints when parking in cities.

The Eagle-360 program relies on magnetic levitation to connect to the body. A dielectric magnetic layer generates a magnetic field that enables levitation between the tire and the body. The physics behind it is that: If the media are exposed to an external magnetic field, they will generate a magnetic field of the same polarity so that the two fields repel each other, thus avoiding mechanical contact between the tires and the car, avoiding unnecessary vibrations, and reducing the noise inside the car, thus further improving the comfort of the passengers. Because in a car using Eagle-360, they will no longer need to drive but can spend their time working, sleeping, or using the multimedia system embedded in the car [142]. In the mobility world of the future, smooth, quiet travel will be essential. The company predicts that the market for self-driving cars will reach about

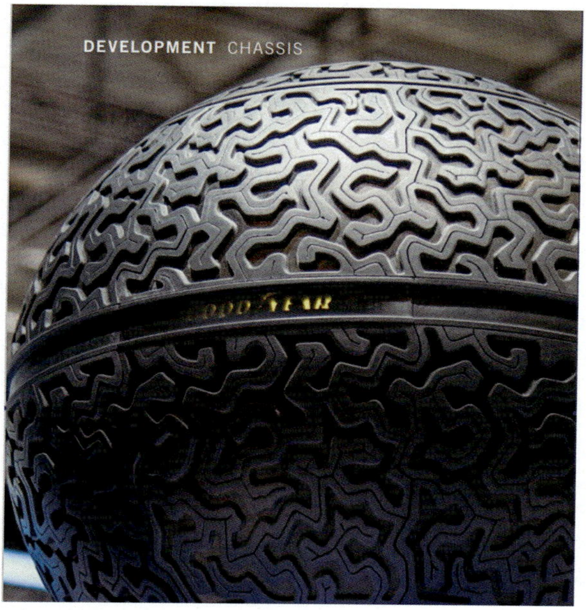

Figure 2.37 Goodyear 360 concept tire [142].

85 million per year by 2035. Based on individual designs, they will be produced using additive manufacturing technology (such as 3D printing). Their attributes will be set according to the car's location and the owner's driving style. Each tire is a separate mechanical device consisting of an electric motor and a battery, with the gaps in the tire filled with reinforced foam. They will rotate around the axle to adapt and react to different weather conditions and road changes (driving on gravel or ice). In addition to increasing smoothness control and maneuverability, this tire shape allows the car to move sideways. The main idea of this concept is to provide the tires with artificial intelligence and the ability to exchange data with the car, other road vehicles, and the environment. The Eagle-360 will integrate sensors to monitor road conditions and react to possible obstacles to improve driving efficiency and safety. In addition, the tires will transmit data on the level of aging and damage, as well as information on the level of aging and damage, to the central control system of the self-driving vehicle, and the tires can receive real-time traffic and location data from the vehicle. The tire can select the best course of action and make decisions because it can combine different data sources and process their parameters efficiently using a neural network based on deep learning algorithms [143].

Table 2.3 Key features of goodyear Eagle-360 concept tire [144].

Spherical tire design	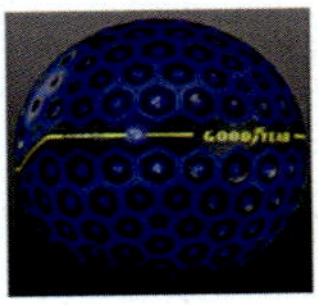	The spherical tire design supports the 360-degree steering of the vehicle, which makes driving safer and makes it easier to navigate through city streets or park in tight spaces
Magnetic levitation connection technology		The concept tires use magnetic levitation technology, a contactless connection between the tires and the vehicle, allowing for a smoother, quieter ride
Sensor technology		Embedded sensors feed road and weather conditions to the vehicle control system to improve driving safety. It also monitors tread wear and tire pressure to help improve tire efficiency and life
Bionic technology	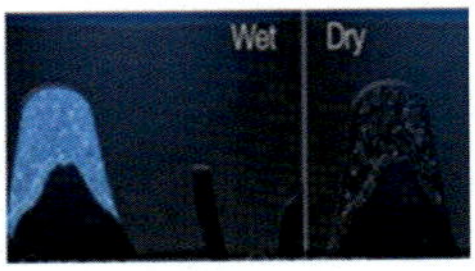	The 3D-printed tread of the tire mimics the shape of brain coral and can stretch like a natural sponge. On dry roads, the tire automatically hardens to reduce friction; on wet roads, the tread softens to reduce slippage

Eagle-360 has a self-learning function, which means that, with the help of artificial intelligence technology, it is possible to optimize the sequence of responses to the same conditions based on the stored information of actions performed before. This tire can deform, that is, to expand and compress, because it is covered with a bionic membrane made of a superelastic polymer similar to human skin, as shown in Fig. 2.38. The foam packing maintains its flexibility under this membrane and is strong enough to withstand the weight of the vehicle. The sensors detect tire holes if the tire tread's bionic membrane is damaged. When damage is

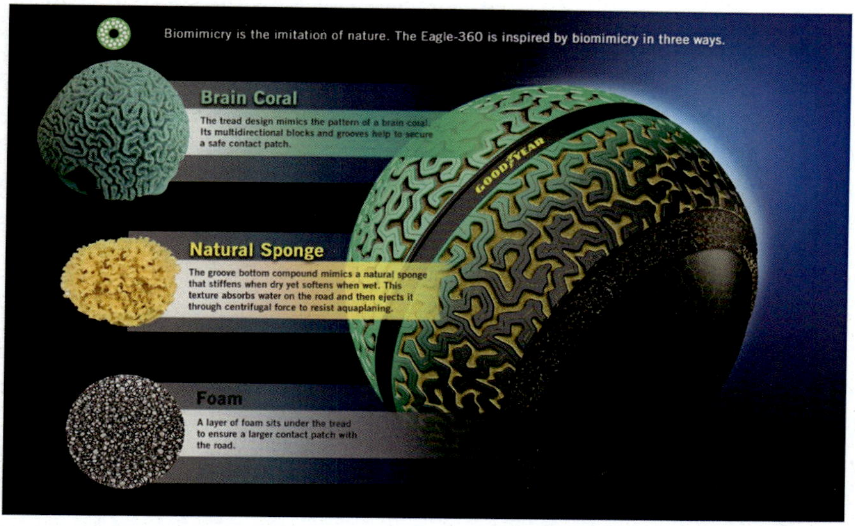

Figure 2.38 Biomimicry is the imitation of nature in technology—the Eagle-360 tire is inspired by this in three ways [142].

detected, the tire returns so that the punctured tire does not fall into the area of contact with the road. In addition, the tire can delay opening independently due to the innovative properties of the material. The tread pattern of spherical Goodyear tires resembles the natural surface structure of brain coral. The tread void is filled with a special material with natural sponge properties. It absorbs water and then throws it out by centrifugal force, preventing a water-slip effect, that is, a complete or partial loss of grip on wet surfaces. A special system drives that works like human muscles by moving individual parts of the tread on the tire's outer layer. Thanks to their action, the tire surface changes according to the weather conditions. Moisture is detected, the number of grooves increases, and the film is "smooth" with a dry coating. Since sphere-shaped wheels can rotate 360 degrees, future technology may shrink parking spaces because cars with spherical tires need less space to maneuver [143].

2.3.6.2 3D printing and spherical tires

3D printing is a manufacturing process that creates parts layer by layer. The technology eliminates the need for tools or molds that mold the desired part into shape. In addition, 3D printing allows for more flexibility in product design than traditional manufacturing methods. The design of the product is done through CAD or medical imaging and can be printed immediately, resulting in a solid part. Soon, 3D printing will also enable the creation of products that cannot be manufactured with current technologies, also known as "impossible products." The 3D printing market segments typically include enterprise printers, individual user printers, materials, and services. The main service offered is the production of custom parts, which is the most economically efficient and dominant market segment. Regarding regional distribution, the North American market currently holds more than 50% of the market share. Automotive and aerospace are the main application areas for 3D printing, while consumer goods and healthcare are seen as the fastest-growing applications. According to leading market research firms, the market for 3D printing is forecast to exceed $20 billion by 2020. At present, although 3D printing has begun to be applied to small batch direct manufacturing, to printing time, accuracy, and material limitations, 3D printing is not yet possible to replace the traditional manufacturing industry, but only to take advantage of research and development, design, and development, for high-volume manufacturing as an aid. The most pressing current need is to prototype or achieve small batch production of products at a faster rate, often referred to as additive manufacturing or rapid production. However, it is generally accepted that custom product design and high geometric complexity components will be more important in the future [145].

Researchers at the Singapore University of Technology and Design announced on February 13, 2017, that a family of highly stretchable and UV cured (SUV) elastomers, with elongation up to 1100%, has been developed for UV-cured 3D printing technology. High-resolution 3D printing with SUV elastomer compositions can directly create complex 3D lattices or hollow structures with large deformations. The manufacturing time of this SUV elastomer can also be greatly reduced. The elastomer development project is the result of a collaborative effort between researchers at the Centre for Digital Manufacturing and Design at the University of Technology and Design, Singapore, the Campus Research Enterprise for Technological Excellence (CREATE), and the Hebrew University of Jerusalem (HUJI), both of which received funding from the National Research Foundation of Singapore. The research surrounding the project has been published in the Journal of Advanced Materials. The joint team developed 3D-printed elastomers that, once cured, can elongate by up to 100% resulting in the most elongated 3D-printed material. Qi Ge, a Professor at the Centre for Digital Manufacturing and Design at Singapore University of Technology and Design, one of the project's coleaders, said that the world's most elongated 3D printed elastomer had been developed. The new elastomer can extend 1/100%, more than five times the elongation at break of any commercial elastomer, and is suitable for UV-cured 3D printing technology. The researchers explain that they were able to successfully 3D print complex 3D lattices and structures using SUV elastomer compositions and a high-resolution 3D printing process that is no longer limited by the design limitations of traditional manufacturing processes. 3D printing SUV elastomers could mean huge advances in many areas where elastic materials are needed.

The high excellence of these 3D printed elastomers, in addition to simplifying complex structures and geometries, can also significantly reduce manufacturing time: Compared witj traditional molding and injection molding methods, 3D printing of SUV elastomers with UV curing can significantly reduce manufacturing time from hours, or even days, to minutes or hours, as complex and time-consuming manufacturing steps, such as mold making, molding/unmolding, and part assembly, are replaced by complex and time-consuming manufacturing steps, such as mold making, molding/unmolding, and part assembly. Mold making, molding/demolding, and part assembly are replaced with a single 3D printing step. SUV elastomers have also shown good mechanical repeatability, suitable for flexible electronic components. This property has been demonstrated by researchers with 3D-printed elastomeric switches for Buckyball lights, which were tested to work properly after 1000 presses. Research project coleader Shlomo Magdassi, a Professor at HUJI and CREATE, concluded that, overall, it could be concluded that SUV elastomers combined with UV-cured 3D printing technology will greatly enhance the ability to make soft and deformable 3D structures and facilities, including soft drives and robots, flexible electronics, acoustic metamaterials, and many other applications [145].

At the Geneva International Motor Show on March 7, 2016, Goodyear unveiled its first 3D-printed spherical concept tire, the Eagle-360 (Fig. 2.38). Goodyear said the tire would be a great solution for self-driving cars and greatly improve driving safety. Joseph Zekoski, Goodyear's chief technology officer, said, "The future of self-driving cars will gradually reduce the driver's control, and tires will play an increasingly important role as the contact parts between the car and the road." It is said that spherical tires can move flexibly in multiple directions, effectively reducing sideslip and making lateral movement smoother. It can also effectively avoid obstacles when not changing driving direction sharply. This 3D-printed tire can make the car's entry easier for users with difficulty parking. In addition, this spherical tire is equipped with a magnetic levitation connection to improve ride comfort. It can also automatically change the tread softness according to wet and dry road conditions. Because the tread is 3D printed, the tire will have a customizable tread pattern. Owners can customize the tire tread pattern according to the specific conditions of their region or road [146]. The tread pattern can be customized according to the region or road conditions.

In today's world, environmental and resource problems are very prominent. The development of green, energy-saving, safe, and high-performance tire products represents both the future development trend and the inevitable choice for building a resource-saving and environment-friendly society. To meet the development trend of the industry and solve the problem of green development, in recent years, Linglong Tire has cooperated extensively with many domestic and foreign academic institutions, universities, and other scientific research units around the research topics of green, energy-saving, low-carbon new materials, and new technologies to develop dandelion tires, graphene tires, organosilicon rubber compound elastomer materials, etc. They launched the "Green Energy-saving High Performance Tire." On July 12, 2017, the 3D-printed PU tire was jointly developed by Shandong Linglong Tire Co. (Fig. 2.39).

Compared with traditional rubber tread, the successful development of 3D printed PU tires using thermoplastic PU material has lower heat generation and rolling resistance. Tires made of PU material are simple, safe, durable, and environmentally friendly. Waste tires can be recycled or used to manufacture other industrial products. The cost is significantly lower than that of traditional rubber tires and is expected to become the next

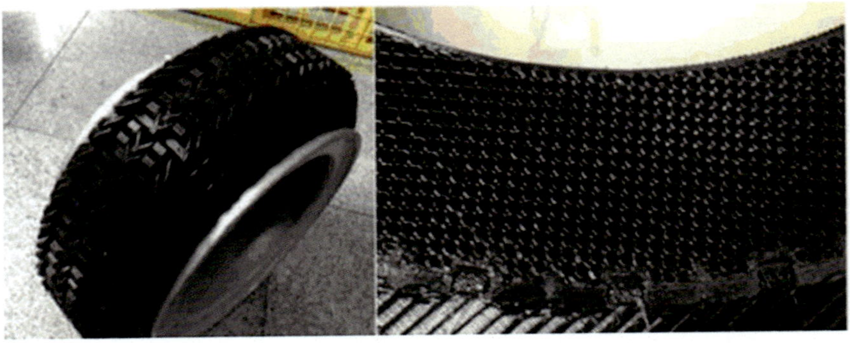

Figure 2.39 The first 3D printed polyurethane tire in China [145].

generation of green tires. It is expected to be the main material of the next generation of green tires. Meanwhile, this study combines PU material with 3D printing technology, and 3D printing is completed by the fused deposition method. The tire has an internal hollow structure of a positive hexagon, realizing a one-piece tire forming. Low rolling and high wear resistance tires with various tread structures can be prepared without molds. Its process is short, fully automatic, and can be manufactured on-site, making the manufacturing faster and more efficient. The successful development of China's first 3D-printed PU tire is a powerful practice of Linglong Tire. And Linglong will take this opportunity to continue to promote technological progress and structural adjustment, speed up the research and development speed of high quality, high performance, and green tires, and improve the level of green development of the enterprise and the industry [145].

Linglong will also take this opportunity to continue to promote technological progress and structural adjustment, speed up the research and development of high-quality, high performance, and green tires, and improve the level of green development of the company and the industry.

References

[1] Xiong JP. Modern rubber selection and design. Beijing: Chemical Industry Press; 2014. p. 41–2.

[2] Hua YQ, Jin YL. Polymer physics. Beijing: Chemical Industry Press; 2013. p. 25–6.

[3] Threadingham D, Obrecht W, Wieder W, et al. Synthetic rubbers, introduction and overview. Ullmann's Encyclopedia of Industrial Chemistry 2011;56–8.

[4] Jones KP, Allen PW. Historical development of the world rubber industry. Developments in Crop Science 1992;23(4):1–25.

[5] Department of Science. Technology and education, ministry of agriculture. China's agricultural industry technology development report. Beijing: China Agriculture Press; 2012. p. 64.

[6] Flory PJ. Principles of polymer chemistry. New York: Cornell University Press; 1953. p. 104–6.

[7] Tanaka Y. Structural characterization of natural polyisoprenes-solve the mystery of natural rubber based on structural analysis. Nippon Gomu Kyokaishi 2001;74.

[8] Zhu Jia. Import and export of major rubber raw materials and rubber products in China from January to November. Rubber Science and Technology 2018;16 (1):57–60.

[9] Zhu Jia. Import and export of major rubber raw materials and rubber products in China from January to February. Rubber Science and Technology 2018;16(4):56.

[10] Yu Q. Handbook of rubber raw materials, 2nd ed., Chemical Industry Press, 2007, 23–26.

[11] Wang GZ, Lu BY. Mechanisms and methods of rubber plasticizing. Special Rubber Products 2007;28(6):47–9.

[12] Wang Yanqiu. Rubber plasticizing and compounding. Rubber Plasticizing and Compounding 2005;31–4.
[13] Zhou Y, Dong Z, Chen F, et al. Advances in rubber calendering process and mechanical equipment. China Rubber 2003;23:25–7.
[14] Liang GZ, Gu YJ. Molding technology. Beijing: Chemical Industry Press; 1999. p. 49–51.
[15] Hou Yahe, You Changjiang. Rubber vulcanization. Beijing: Chemical Industry Press; 2013. p. 78–80.
[16] Coran AY. Science and technology of rubber. Massachusetts: Academic Press; 1978. p. 21–2.
[17] Zhang YL, Li P. Rubber varieties and selection. Beijing: Chemical Industry Press; 2012. p. 50–1.
[18] Zhang A, You C. Rubber plasticizing and compounding. Beijing: Chemical Industry Press; 2012. p. 81–2.
[19] Shi WX. Design and simulation study of fully automatic rubber turning mechanism for rubber knapping machine. Master's Thesis, Shandong: Qingdao University of Science and Technology, 2022, 72–73.
[20] Huang S. Evolution and trends of kneader development. The 14th China Tire Technology Symposium 2006;37–8.
[21] An S. Development and experimental study of a fully automatic rubber knapper. Master's Thesis, Shandong: Qingdao University of Science and Technology, 2013, 41–42.
[22] Liu WH. Development of experimental openers. Master's Thesis, Heilongjiang: Harbin Institute of Technology. 2020, 45–46.
[23] Fan HG. A new type of two-roll opener with double drive. CN Patent, CN201611045024.4; 2016.
[24] Huang SL. The development history and trend of domestic refiner. Rubber Industry 2007;54(7):440–3.
[25] Zhou Y, Zhou YZ. Technical progress of rubber compactor. Rubber Science and Technology 2017;15(3):38–43.
[26] Yu QX. A peek into the development trend of rubber processing equipment. World Rubber Industry 2002;29(2):10.
[27] Liu WH. Development of experimental kneader. Master's Thesis, Heilongjiang: Harbin Institute of Technology, 2020, 61–62.
[28] Cheng YJ, Gai ZY. China's plastic machinery development status and outlook. Plastics industry 2007;70–1.
[29] Li B. Introduction to the manufacturing technology of GK-type rubber compacting machine (I). Rubber Technology and Equipment 1987;03:9–13.
[30] Li S, Yi Y. GK compacting machine won the title of provincial famous brand. China Rubber 2005;21(1):31.
[31] Chen D. Application of PLC in the modification of control circuit of dense refiner. Electric World 2020;717(11):49–53.
[32] Li Q. Research and application of tandem compacting machine process. Master's Thesis, Shandong: Qingdao University of Science and Technology. 2019, 91–93.
[33] Sun MZ. Research on the mixing process of BB430 compacting machine. Master's Thesis, Shandong: Qingdao University of Science and Technology. 2017, 101–103.
[34] Zhai TJ. Simulation and visualization of the mixing process of the compacting machine. Master's Thesis, Shandong: Qingdao University of Science and Technology. 2020, 27–28.
[35] Ai C. Simulation study of electromagnetic dynamic compacting machine. Master's Thesis, Shandong: Qingdao University of Science and Technology, 2014, 38–39.

[36] Zhao JF. Design of rubber temperature control system for tandem refiner. Master's Thesis, Shandong: Ocean University of China, 2015, 41–42.
[37] Zhou B, Zhong H, Li ZJ, et al. New automatic rubber dosing and weighing control system for rubber dense refiner. Weighing Instrument 2013;042(001):6–17.
[38] Zhao GX. Calendering, extrusion, vulcanization, injection molding. Rubber and Plastic Technology and Equipment 2006;32(002):16–24.
[39] Farag H, Sisinni E, Gidlund M, et al. Priority-aware wireless fieldbus protocol for mixed-criticality industrial wireless sensor networks. IEEE Sensors Journal 2018;19 (7):2767–80.
[40] Wang WC. Overview of the production process and equipment of rubber seals. Special Rubber Products 2000;21(4):47–9.
[41] Wu HP. Structural performance and application of calender. Master's Thesis, Henan: Henan Science and Technology, 2018, (26): 49–52.
[42] China Machinery Industry Year book Editorial Committee, China Plastics Machinery Industry Association. China Plastics Machinery Industry Yearbook, 2009. p. 17–18.
[43] Ayyappan SK, Subramaniam RB. Analysis of in-site grinding process using new equipment for calendar roll machines. Materials and Manufacturing Processes 2019;34(10):1136–42.
[44] Eldridge MM. Extrusion processes. Applied plastics engineering handbook, New York: William Andrew Publishing, 2011.
[45] Zhang YG. The current situation of extrusion equipment and the study of adjusting product structure. Shanghai Plastics 2010;4:25–33.
[46] Han HC. Numerical simulation of the effect of screw structure on the extrusion flow field of screw extruder. Master's Thesis, Shanxi: Taiyuan University of Technology, 2020, 58–60
[47] Van HH, Noordermeer J, Heideman G, et al. Best practice for de-vulcanization of waste passenger car tire rubber granulate using 2-2′- dibenzamidodiphenyldisulfide as de-vulcanization agent in a twin-screw extruder. Polymers 2021;13(7):1139.
[48] Zhang LY. Research and application of fast intelligent extruder barrel temperature control algorithm. Master's Thesis, Shandong: Shandong University, 2005, 63–64.
[49] Zhao YJ. Research on the technology and method of intelligent CAD for rubber extruder. Master's Thesis, Beijing: Beijing University of Chemical Technology, 2005. 70–71.
[50] Liu HY. Application and research of fuzzy control in twin-screw extrusion and tablet press. Master's Thesis, Guangxi: Guilin University of Electronic Science and Technology, 2011, 61–62.
[51] Zhang YG. Re-talk about the independent innovation of plastic machinery. Shanghai Plastics 2010;3:20–5.
[52] Yu QX. Current status and outlook of tire curing press. China Rubber 2007;23 (22):8–19.
[53] Yu HL. The current situation and development trend of flat vulcanizing machine. Rubber and Plastic Technology and Equipment 1995;21(5):22–8.
[54] Yuan ZH. Design of intelligent monitoring system for flat vulcanizing machine. Master's Thesis, Shandong: Qingdao University of Science and Technology, 2021, 55–56.
[55] Zhang XC, Duan ZY, Shi WM. Current status and development trend of tire curing equipment. Rubber and Plastic Technology and Equipment 2010;36(9):18–23.
[56] Shang WL. Research on the mechanism and equipment of electromagnetic induction pulse heating for uniform vulcanization of thick rubber products. Master's Thesis, Beijing: Beijing University of Chemical Technology, 2016, 57–59.

[57] Yu QL, Feng QX. Application of PLC on the control system of flat vulcanizing machine. Science and Technology Information 2011;14:1.
[58] Qiu W, Wang J, Liu W, et al. Structural design and mechanical analysis of a new equipment for tire vulcanization. Mechanics Based Design of Structures and Machines 2021;1–17.
[59] Alimardani M, Abbassi-Sourki F. New and emerging applications of carboxylated styrene butadiene rubber latex in polymer composites and blends: review from structure to future prospective. Journal of Composite Materials 2014;49(10):1267–82.
[60] Sirisinha C, Sae-Oui P, Suchiva K, et al. Properties of tire tread compounds based on functionalized styrene butadiene rubber and functionalized natural rubber. Journal of Applied Polymer Science 2020;137(20):48696.
[61] Song P, Wu X, Wang S. Effect of styrene butadiene rubber on the light pyrolysis of the natural rubber. Polymer Degradation and Stability 2018;147:168–76.
[62] Chen B, Mu C, Bai Y, et al. 1,1-Diphenylhexyl lithium synthesis of star-shaped solvated butadiene rubber. Journal of Beijing University of Chemical Technology: Natural Science Edition 2011;131–7.
[63] Li B, Liu J. Research and development progress of synthetic rubber for high performance tires in China. Rubber Science and Technology Market 2011;09(012):11–14.
[64] Zhang LQ, Ye X, Xi MM, et al. Preparation method of silica-modified styrene-butadiene rubber for green tires and its products. China Rubber 2012;187–93.
[65] Heike K, Thomas G. Bimodal neodymium-catalyzed polybutadiene. US Patent, US 2013/0172489; 2013.
[66] Li Y, Hu YM, Shi ZH, et al. Rare-earth catalyzed system of star-branched polybutadiene and its preparation. CN Patent, CN2010106176201; 2010.
[67] Dai Q., Wen Z.Q., C. Hongguang, et al. Sulfonic acid rare earth catalysts for the preparation of polyisoprene and their preparation and application. CN Patent, CN200910217980X; 2009.
[68] Cai H.G., Zhang X., Bai C.X., et al. Naphthalenesulfonic acid rare earth catalysts, preparation and application. CN Patent, CN2009102179814; 2009.
[69] Ghoreishi A, Koosha M, Nasirizadeh N. Modification of bitumen by EPDM blended with hybrid nanoparticles: physical, thermal, and rheological properties. Journal of Thermoplastic Composite Materials 2018;33(3):343–56.
[70] Zhang M, Fan M, Peng S, et al. Synthesis and properties of EPDM-based oil-absorptive gels with different types of EPDM and styrene derivatives. RSC Advances 2021;11(3):1605–13.
[71] Na LH, Zhang XQ, Jiang LS, et al. Vanadium catalysts, vanadium catalyst compositions and methods for the preparation of ethylene propylene rubber. CN Patent, CN201210496187X; 2012.
[72] Liu C. Research progress on the third and fourth monomers of ethylene propylene rubber. Petrochemical Technology 2006;2:52–6.
[73] Na LH, Hao XF, Yu QZ, et al. Preparation of EPDM rubber from ethylene, propylene and LPB. National Conference on Polymer Academic Paper Presentation 2011.
[74] Guo WL, Wu YB, Li SX, et al. Direct production of halobutyl rubber from butyl rubber liquid by solution method. CN Patent, CN201110198624.5; 2011.
[75] Ismail H, Leong HC. Curing characteristics and mechanical properties of natural rubber/chloroprene rubber and epoxidized natural rubber/chloroprene rubber blends. chloroprene rubber blends. Polymer Testing 2001;20(5):509–16.
[76] Mochel WE, Nichols JB, Mighton CJ. The structure of neoprene. I. The molecular-weight distribution of neoprene type GN. Rubber Chemistry and Technology 1949;22(3):680–9.
[77] Zhang YL, Zhang JS. Special rubber and applications. Beijing: Chemical Industry Press; 2011. p. 107–14.

[78] Ren WT, Zhang Y, Zhang Y. Application technology and progress of special rubber materials in the automotive industry. Rubber Science and Technology 2006;8:100–4.
[79] Huang JM. Application and development of special rubber in rubber products in China. Rubber Industry 2000;47(2):94–8.
[80] Xie ZL. Some technological developments of special rubber. Rubber Industry 2000;3:145–54.
[81] Alessandro G. Polymers from renewable resources: a challenge for the future of macromolecular materials. Macromolecules 2008;41(24):9491–504.
[82] Ma C, Su X. Overview of biomass energy. World Forestry Research 2005;18 (6):32–8.
[83] Jong ED, Higson A, Walsh P, et al. Product developments in the bio-based chemicals arena. Biofuels Bioproducts & Biorefining 2012;6(6):606–24.
[84] Drumright RE, Gruber PR, Henton DE. Polylactic acid technology. Advanced Materials 2000;12(23):340–1.
[85] Marques AP, Reis RL, Hunt JA. The biocompatibility of novel starch-based polymers and composites: *in vitro* studies. Biomaterials 2002;23(6):1471–8.
[86] Sudesh K, Abe H, Doi Y. Synthesis, structure and properties of polyhydroxyalkanoates: biological polyesters. Progress in Polymer Science 2000;25(10):1503–55.
[87] Kurian JV. A new polymer platform for the future-sorona from corn derived 1,3-propanediol. Journal of Polymers & the Environment 2005;13(2):159–67.
[88] Witt U, Müller RJ, Augusta J, et al. Synthesis, properties and biodegradability of polyesters based on 1,3-propanediol. Macromolecular Chemistry & Physics 2010;195(2):793–802.
[89] Deng L. Synthesis and characterization of PBS and its copolymers. Master's Thesis, Sichuan: Sichuan University, 2004.
[90] Ji N, Qiao H, Wang Z, et al. Research progress of bio-based synthetic rubber. Journal of Materials Engineering 2019;47(12):1–9.
[91] Lipinsky ES. Chemicals from biomass: petrochemical substitution options. Science 1981;212:1465–71.
[92] Yao K, Tang C. Controlled polymerization of next-generation renewable monomers and beyond. Macromolecules 2013;46(5):1689–712.
[93] Ragauskas AJ, Williams CK, Davison BH, et al. The path forward for biofuels and biomaterials. Science 2006;311:484–9.
[94] Choi S, Song CW, Shin JH, et al. Biorefineries for the production of top building block chemicals and their derivatives. Metabolic Engineering 2015;28:223–39.
[95] Morais A, Dworakowska S, Reis A, et al. Chemical and biological-based isoprene production: green metrics. Catalysis Today 2015;239:38–43.
[96] Whited GM, Feher FJ, Benko DA, et al. Technology update: development of a gas-phase bioprocess for isoprene-monomer production using metabolic pathway engineering. Industrial Biotechnology 2010;6(3):152–63.
[97] Yang J, Zhao G, Sun Y, et al. Bio-isoprene production using exogenous MVA pathway and isoprene synthase in *Escherichia coli*. Bioresource Technology 2012;104:642–7.
[98] Fischer CR, Andrew AP, Jefferson WT. Production of C_3 hydrocarbons from biomass via hydrothermal carboxylate reforming. Industrial & Engineering Chemistry Research 2011;3:511–20.
[99] Ezeji TC, Qureshi N, Blaschek HP. Microbial production of biofuel (acetone-butanol-ethanol) in a continuous bioreactor: impact of bleed and simultaneous product removal. Bioprocess & Biosystems Engineering 2013;36(1):109–16.
[100] Daijiro T, Satoshi S, Masateru I, et al. Production of bio-based 1,3-butadiene by highly selective dehydration of 2,3-butanediol over SiO_2-supported cesium dihydrogen phosphate catalyst. Chemistry Letters 2016;45(7):831–3.

[101] Lieberei R. South American Leaf Blight of the Rubber Tree (Hevea spp.): new steps in plant domestication using physiological features and molecular markers. Annals of Botany 2007;100(6):1125–42.
[102] Martin WJ. The occurrence of South American leaf blight of hevea rubber trees in Mexico. Phytopathology 1948;38(2):157–8.
[103] Mo Y. Import and export of natural rubber in China in 2012. World Tropical Agriculture Information 2013;2:1.
[104] Keiper-Hrynko N.M., Hallahan D.L. Cis-prenyltransferases from the rubber-producing plants russian dandelion (taraxacum kok-saghyz) and sunflower (helianthus annus). AU Patent AU2003295489, 2003.
[105] Coates W, Ayerza R, Ravetta D. Guayule rubber and latex content-seasonal variations over time in Argentina. Industrial Crops & Products 2001;14(2):85–91.
[106] Du H, Xie B, Shao S. Research progress and development prospects of eucommia gum. Journal of Central South Forestry College 2003;04:96–100.
[107] Zhang L, Zhang J, Liao S. Natural rubber and bio-based elastomers. Beijing: Chemical Industry Press, 2014.
[108] Sun S, Zhang J, Zhang L. Research progress of bio-based rubber. In: "Linglong Tire Cup" The 17th China Tire Technology Symposium, 2012. p. 184–193.
[109] Rasutis D, Soratana K, McMahan C, et al. A sustainability review of domestic rubber from the guayule plant. Industrial Crops and Products 2015;70:383–94.
[110] Qian BZ. Accelerated breeding by SGB and Yulex to double the yield of silver gum rubber. Rubber Science and Technology 2013;11(12):47.
[111] Yang Y, Gan L, Qin B, et al. Progress of research on silver gum daisy. Tropical Agricultural Science 2017;37(7):75–9.
[112] Feng T. Yulex receives FDA approval for its silver chrysanthemum gum marketing program. Rubber Industry 2008;8:50.
[113] Zhu YK. Use of silver chrysanthemum latex as a dental material. Rubber Science and Technology 2013;6:30.
[114] Chen J. Goodyear bio-rubber technology concept tires debut, <https://www.pcauto.com.cn/drivers/yongpin/tire/knowledge/0912/1045932.html> 2023, (accessed 29.07.23).
[115] Qian B. Bridgestone successfully produces passenger car tires from 100% silver gum chrysanthemum-derived rubber. World Rubber Industry 2015;10:38.
[116] China Nature Rubber Association. Breakthrough in the silver rubber substitution project of Gupta Tire in the United States. Special Rubber Products 2015;05:82.
[117] Zhao M. Study completed by Goodpill proves that silver gum chrysanthemum plant can be a source of tire rubber. Tire Industry 2017;10:638.
[118] China Nature Rubber Association. Pirelli high-performance silver chrysanthemum rubber tires complete initial runway testing. Rubber Reference 2016;46:56.
[119] Pang T, Cheng R, Lun H, et al. Report of a trial study on the introduction of silver gum daisy. Tropical Agricultural Science 2007;27(1):1–11.
[120] Peng J, Peng P, Chen W, et al. Research on the introduction and cultivation of silver gum daisy. Anhui Agricultural Science 2007;16:4795–7.
[121] Yan N, Sa R, Sun S, et al. Dandelion rubber-a NR in urgent need of intensive research. Rubber Industry 2011;58(10):632–7.
[122] Zhang L, Zhang J, Wang F, et al. Global development trend of natural rubber and the road of diversification in China. China Rubber 2013;22:18–20.
[123] An Q. Gubo will try to use Russian Taraxacum to produce rubber. Rubber Technology Market 2009;7(3):21.
[124] Lin T, Xia X, Ruan J, et al. Genome analysis of Taraxacum kok-saghyz Rodin provides new insights into rubber biosynthesis. National Science Review 2018;5(1):78–87.

[125] Buranov AU, Elmuradov BJ. Extraction and characterization of latex and natural rubber from rubber-bearing plants. Journal of Agricultural & Food Chemistry 2010;58(2):734–43.
[126] Eskew RK, Edwards PW. Process for recovering rubber from fleshy plants, US Patent: US51345843A; 1946.
[127] Ho KC, Chan MK, Chen YM, et al. Treatment of rubber industry wastewater review: recent advances and future prospects. Journal of Water Process Engineering 2023;52:103559.
[128] Marianella HS, Michael dB, Santiago G. Routes to make natural rubber heal: a review. Polymer Reviews 2018;58(4):585–609.
[129] Wang F. The development of foreign bio-rubber resources. China Rubber 2012;28 (13):4–7.
[130] Ding F, Su Y, Du S, et al. Biological extraction of *Eucommia gum* strain screening and optimization of fermentation conditions. Journal of Northwest Forestry Academy 2012;27(2):149–54.
[131] Xie X, Lu BQ, Wang ZB, et al. Study on the extraction process of juniper gum from juniper leaf residue. Biomass Chemical Engineering 2013;5:39–43.
[132] Yuan YF, Cao YP. Ultrasonic-assisted extraction of juniper gum and juniper flavonoids from Eucommia bark. Journal of Food Science and Technology 2014;32 (2):67–71.
[133] Wang S, Urban MW. Self-healing polymers. Nature Reviews Materials 2020;5:562–83.
[134] Wang L, Kang H, Yang F, et al. Study on the performance of natural juniper/paraben rubber blends. Special Rubber Products 2017;4:13–17.
[135] Li ZG, Li C, Cai MD. Mechanism and effect of grafted juniper gum on the modification of dry rubber asphalt mixture. Journal of Southeast University (Natural Science Edition) 2014;44(4):845–8.
[136] Long H, Zhang FY, Feng YZ. Feasibility of eucalyptus gum as root canal filling material. Chinese Tissue Engineering Research 2015;19(16):2511–17.
[137] Xin ZC. Modern tire structure design. Beijing: Chemical Industry Press; 2011.
[138] Xiao Yi. Characteristics of Michelin C3M. Tire Industry 1997;1:50–1.
[139] Shi HX, Li QH, Jia CB, et al. Research and application of honeycomb seal assembly process technology. Chinese Mechanical Engineering Society 2012;18:45–9.
[140] Tu X. The first MIRS process tire Eufori@ shines. Tire Industry 2002;02:28.
[141] Gao M, Qiao L, Feng XJ, et al. Application of numerical simulation methods in the optimal design of tire structures. Tire Industry 2008;10:594–600.
[142] Anckaert D, Fontaine S, Besnoin E, et al. Reinventing the wheel innovations in the tyre industry. Atz Worldwide 2017;55–60.
[143] Ocheretianyi O. Spherical tyres for self-driving cars. Materials of conferences held in Zhytomyr Polytechnic 2017;1–2.
[144] Feng P. Research on technological innovation of intelligent tires of goodyear. Modern Rubber Technology 2019;2:1–8.
[145] Qian B. Tire manufacturing enters a new era of 3D printing. Modern Rubber Technology 2017;43(6):40–3.
[146] Goodyear's Eagle-360: a visionary tire concept for future autonomous vehicles, <https://corporate.goodyear.com/us/en/media/news/goodyear-reveals-concept-tires-for-autonomous-cars.html> 2023, [accessed 29.07.23].

CHAPTER 3

Polymer materials for fuel cell

Contents

3.1 Fuel cell proton exchange membrane materials

As a new power supply device with high energy density, high energy conversion, and environmental protection, the fuel cell has attracted extensive attention all over the world. There are many kinds of fuel cells. Currently, fuel cells are mainly divided into six categories [1]: alkaline fuel cell, phosphorous acid fuel cell, molten carbonate fuel cell, solid oxide fuel cell, direct methanol fuel cell, and proton exchange membrane fuel cell (PEMFC) (as shown in Fig. 3.1) [2]. PEMFC has a low working temperature (within 100°C) and good volume adaptability [3,4].

The main purpose of PEMFC research and development is to reduce the cost of the fuel cell by reducing the cost of membrane and electrolysis. Compared with other types of fuel cells, PEMFC using polymer proton exchange membrane (PEM) as an electrolyte has the advantages of low working temperature, fast starting speed, modular installation, and convenient operation, which is considered to be the best alternative power supply for electric vehicles, submarines, various movable power supplies, power supply grid, and fixed power supply.

New Polymeric Products
DOI: https://doi.org/10.1016/B978-0-443-19407-8.00004-X

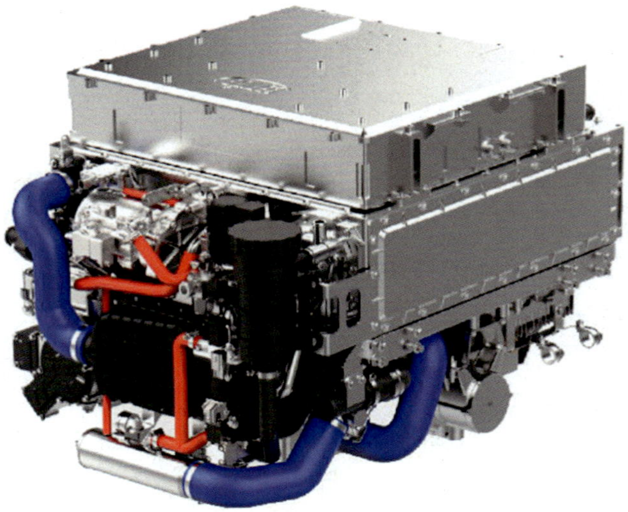

Figure 3.1 Example photo of a fuel cell [2].

Compared with other types of fuel cells, the advantages of PEMFC are as follows [5]: (1) Polymer films in solid form can reduce the volume and weight of fuel cells; (2) simple structure, convenient carrying and installation, can avoid problems such as leakage of liquid electrolyte; (3) under the work condition of catalyst, the oxidation–reduction reaction in the fuel cell occurs very quickly, which makes it possible to start quickly in practical use; (4) the suitable working temperature has a lower requirement and broader application range; and (5) the specific power and the theoretical energy conversion rate are high. Although the practical application often cannot reach the theoretical value, it is still at a high level.

3.1.1 Fuel cell structure and working principle

The fuel cell is a new type of energy conversion device. It directly generates current through the electrochemical oxidation of fuel (such as H_2) and electrochemical reduction of oxidant (such as O_2). Fuel cells are different from ordinary primary cells and batteries; instead of being stored in the battery, the electrode reactive substances are supplied externally. In theory, as long as fuel is continuously supplied from the outside, the fuel cell can continuously output electric energy. Fuel cell power generation does not go through the combustion process (a fuel cell is not a heat engine), nor is it limited by the Carnot cycle. It has high energy conversion efficiency (theoretically, it can be greater than 80%), and because the reaction product is water, the power generation process will not cause environmental pollution. Therefore, the fuel cell is recognized as the most promising clean energy in the future.

PEMFC comprises a membrane electrode assembly (MEA) and bipolar plate with a gas flow channel. The core component of the membrane electrode is made by hot pressing a polymer electrolyte membrane and two electrodes on both sides. The solid electrolyte membrane in the middle plays a dual role in ion transfer and separation of fuel and oxidant gas. The electrodes on both sides are the places for the electrochemical reaction of fuel and oxidant [6].

The working principle of PEMFC is shown in Fig. 3.2. PEMFC usually uses perfluorosulfonic acid PEM as an electrolyte, Pt/C or Pt Ru/C as an electrocatalyst, hydrogen or purified reforming gas as fuel, air and pure oxygen as oxidant, and graphite or surface modified metal plate with gas flow channel as bipolar plate [7]. When PEMFC works, fuel gas and oxidant gas reach the anode and cathode of the cell respectively through the gas channel on the bipolar plate, the reaction gas reaches the active reaction center of the electrode catalytic layer through the diffusion layer on the electrode, hydrogen dissociates into hydrogen ions (protons), and negatively charged electrons under the action of anode catalyst. Hydrogen ions migrate from one sulfonic acid group (—SO_3H) to another in the PEM in the form of hydrated proton H^+ (nH_2O, n is about 3–5) and finally reach the cathode to realize proton conductivity. At the same time, the oxygen molecules of the cathode react with electrons under the action of the catalyst to become oxygen ions, making the cathode become a positively charged terminal (positive pole), and a voltage is generated

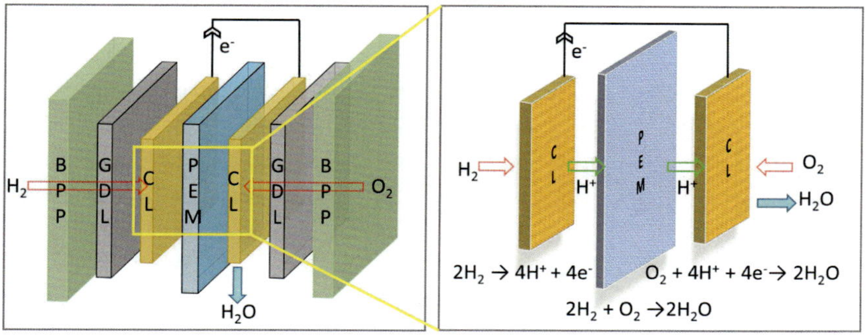

Figure 3.2 Mechanism of proton exchange membrane fuel cell.

between the negative terminal of the anode and the positive terminal of the cathode. If the two ends are connected through an external circuit, electrons will flow from the anode to the cathode through the circuit, resulting in a current. At the same time, hydrogen ions and oxygen react with electrons to form water.

$$\text{Anode: } 2H_2 \rightarrow 4H^+ + 4e^- \quad (3.1)$$

$$\text{Cathode: } O_2 + 4H^+ + 4e^- \rightarrow 2H_2O \quad (3.2)$$

$$\text{Total reaction: } 2H_2 + O_2 \rightarrow 2H_2O \quad (3.3)$$

Several key components of the PEM fuel cell are (1) PEM, (2) electrocatalyst, (3) electrode, and (4) bipolar plate.

The PEM is the core component of the fuel cell.

The fuel cell the PEM shall have the following characteristics: (1) High conductivity (highly selective ionic conductivity rather than electronic conductivity); (2) outstanding chemical stability (acid and alkali resistance and antioxidation reduction ability); (3) good thermal stability; (4) excellent mechanical properties; (5) low air permeability of reaction gas; (6) small electroosmosis coefficient of water; (7) it should be conducive to electrode reaction as a reaction medium; (8) low price.

Currently, the common PEMs mainly include fluorine-containing membranes, nonfluorine membranes, high-temperature membranes, and ceramic membranes [8].

Nafion membrane stands out at the current commercial development stage because of its high proton conductivity and good stability and has become the most commonly used PEM. However, the Nafion membrane still has some disadvantages, such as: (1) the synthesis and preparation of monomers are complex and difficult to operate. The high cost of Nafion limits further development and application. (2) It is very sensitive to temperature and humidity. The water content in the membrane has a fatal impact on proton conduction. The too-high temperature will make the membrane enter the water loss state, and the proton conductivity is very low. Therefore, the optimal working temperature of the Nafion membrane is 60°C–90°C. (3) Because Nafion contains fluorine atoms, it will pollute the environment.

To solve the development limitation caused by the defects of Nafion materials, researchers focus more on finding new polymer materials to prepare PEMs to replace Nafion or modify and strengthen Nafion. According to the research ideas of recent research articles, most studies still focus on sulfonated polymer materials, which are divided into perfluorosulfonic acid PEM, partially fluorinated PEM and fluorine-free PEM according to the content of fluorine [9,10].

3.1.2 Perfluorosulfonic acid membrane

Perfluorosulfonic acid ion exchange membrane comprises a main fluorocarbon chain and ether branch chain with the sulfonic acid group. It has high chemical stability. It is the most widely used fuel cell membrane material at present. Perfluorosulfonic acid membrane has the advantages of high mechanical strength, good chemical stability, high conductivity under high humidity, high current density, and low proton conduction resistance at low temperatures. However, perfluorosulfonic acid PEM also has some disadvantages, such as poor proton conductivity caused by temperature rise, chemical degradation of the membrane at high temperatures, difficulty monomer synthesis, high cost and high price, and methanol penetration is easy to occur when the membrane is used in the methanol fuel cell. Currently, the Nafion series perfluorosulfonic acid PEM developed by DuPont company of the United States is the most widely used. In addition to high fuel permeability, it has almost no shortcomings, but it only applies to the working environment of 60°C−90°C. In addition, due to the use of Pt-based catalysts, the cost is also relatively high. Fig. 3.3 [11] is the photo of commercial Nafion membrane.

Considering those mentioned above, different strategies were proposed to completely improve the performance or replace Nafion in PEMFCs [12]. For example, some aromatic polymers, such as sulfonated poly(ether ether ketone) (SPEEK) [13], sulfonated poly(sulfone) [14], or polybenzimidazole (PBI), are considered good PEM candidates due to their high thermal and oxidation stability and mechanical strength. Many studies related to membrane functionalization (through cross-linking or sulfonation) and the preparation of composite PEMs with (in)organic fillers were performed [15]. However, up to date, no polymer membrane with a better performance/durability ratio (especially in terms of proton conductivity/durability) than that for the Nafion membrane was obtained. In addition, even if some of these conductive membranes revealed a high conductivity value, their performance and efficiency over time and under real conditions of PEMFC use were not tested. Therefore, Nafion remains

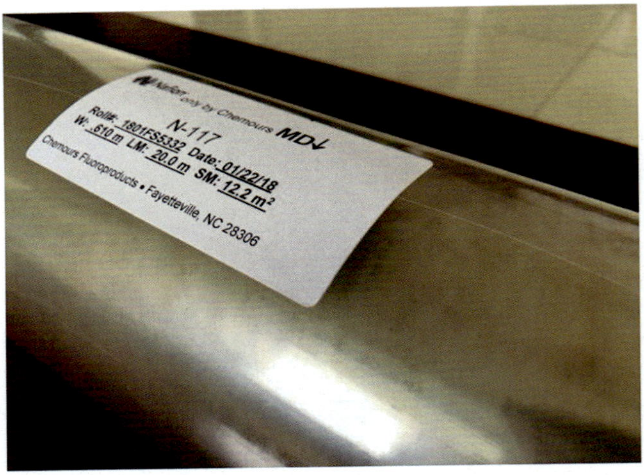

Figure 3.3 Nafion117 used in actual production [11].

the mainly used PEMFC membrane, and numerous recent studies aim to overcome its limitations by elaborating hybrid composite Nafion-based membranes [16].

The term "hybrid composite Nafion-based membrane" means that Nafion contains various (in)organic fillers [17,18] or is mixed with other aromatic [19] or aliphatic [20] polymers. Predicting how the filler nature will affect the Nafion performance is possible. While the ability of mineral particles to retain water may improve the proton conductivity, their possible elution from the organic matrix presents a technological limitation. The strategies for reducing their elution by cross-linking or grafting reactions are either insufficient or lead to a strong decrease in their hydric behavior and, so, to their conductivity decrease in time. Using inorganic fillers and composite membranes usually badly influences the proton conductivity but effectively improves Nafion thermomechanical parameters and reduces the fuel permeability.

On the contrary, applying ionic liquids (ILs) allows us to provide outstanding proton conductivity in anhydrous conditions but degrades Nafion's thermomechanical properties [21]. Therefore, a compromise should be found to minimize the loss of performances (i.e., proton conductivity, fuel permeability, oxidative stability, mechanical properties, and durability) of hybrid Nafion-based membranes. Recently, special attention has been paid to the hybrid Nafion membranes containing both inorganic and organic fillers.

It should be noted that usually (in)organic fillers are previously modified by grafting different proton conducting groups (—SO_3H or—PO_3H_2) [22]. This approach allows us to increase the filler compatibility with the Nafion matrix and decrease the conductivity loss in non-modified fillers. In addition, (in)organic fillers should be stable at high temperatures and pH values and, thus, should not impair the homogeneity of the Nafion membrane as it could significantly decrease transport properties, especially due to the destruction of ion-conducting channels.

It is assumed that ion-conducting channels in Nafion are formed between sulfonic groups depending on the membrane morphology. However, until now, no experimental study can accurately establish the form of these channels. Several theoretical assumptions (such as cluster-channel, core–shell, rod, sandwich, and water channel models) may be found in the literature to explain the Nafion membrane's transport properties [23]. For example, in the first proposed morphology model (i.e., cluster-channel), the Nafion sulfonic groups are considered to be in the form of clusters (diameter about 4–6 nm) that are interconnected through narrow channels (diameter about 1 nm) that provide the transport properties, while in more recent water channel model, sulfonic groups are supposed to be self-assembled into blocks of hydrophilic water channels (each about 2.5 nm in diameter) [24].

3.1.2.1 Inorganic fillers

Such fillers are often used in the Nafion membrane due to their excellent chemical and physical stability and commercial availability. Also, owing to the hydrophilic nature of inorganic fillers, the composite Nafion/inorganic filler membranes maintain water more easily and, as a result, demonstrate higher performance at high temperatures ($>50°C$) and low relative humidity (RH) ($\leq 50\%$).

The size of the inorganic filler is a key parameter as it increases the probability of obtaining a homogeneous membrane with a uniform distribution of particles across the membrane. Therefore, various methods of nanomaterial elaboration have been developed. By the size of the particles, these methods can be divided into two main groups:

1. dispersion methods—based on the grinding of raw materials. The minimum particle size ranges, in this case, from 20 to 30 nm;
2. condensation methods—based on the deposition of the obtained nanoparticles. The minimum particle size may be several nanometers.

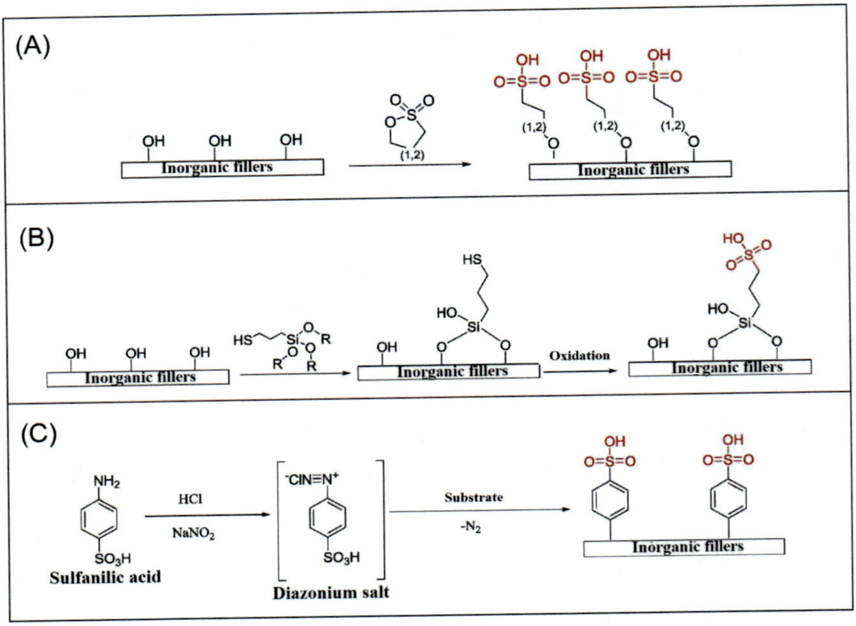

Figure 3.4 Different grafting reactions of sulfonic groups onto the surface of hydrophilic inorganic fillers using: (A) sultones, (B) silanes, (C) amines [17].

Also, the nonconducting behavior of inorganic fillers should be taken into account. This disadvantage of inorganic fillers is usually overcome by grafting the proton conducting groups onto their surface using different organic compounds, such as sultones (Fig. 3.4A), silanes (Fig. 3.4B), or amines with sulfonic groups (Fig. 3.4C).

3.1.2.2 Organic fillers

The grafting of similar lipophilic groups (sulfonic or phosphonic) can also be performed on the organic molecules. It was found that organic compounds with lipophilic groups allowed us to significantly improve the Nafion membrane performances at low RH (40%). For example, the sulfonic groups can be grafted using 1,3-propane sultone to obtain a hydrophilic zwitterion-functionalized compound (Z-COF) (50 nm). The Nafion/Z-COF (10 wt.%) membrane shows thermal stability up to 200°C, the same as the recast Nafion membrane. On the other hand, due to the high hydrophilic nature of the Z-COF nanofiller, the proton conductivity values of the hybrid Nafion membrane increase. Thus, at 80°C and 40% RH, the Nafion/Z-COF (10 wt.%) membrane shows a proton conductivity of 10.9 mS/cm, which is 7.3 times higher than that of the recast Nafion membrane (1.5 mS/cm). In addition, the Nafion/Z-COF (10 wt.%) membrane demonstrates the performance increase during the single fuel cell test (H_2/O_2, at RH of 100 and 50% and 80°C). The hybrid membrane shows the maximum power density of about 170 mW/cm^2–500 mA/cm^2 (at 100% RH) and about 65 mW/cm^2 (at 50% RH), respectively, while for recast Nafion demonstrates the maximum power density of about 100 mW/cm^2 (at 100% RH) and about 45 mW/cm^2 (at 50% RH) [25].

The organic compounds with phosphoric acid (PA) groups were also used for the Nafion modifications. Zhang et al. grafted the diphosphoric acid group to polydopamine (FDPA) [22]. Thus, hybrid Nafion/FDPA membranes demonstrate the increase in thermal stability: the decomposition temperature is shifted from 330°C for recast Nafion to 360°C and 400°C for the Nafion/FDPA (10 wt.%) and the Nafion/FDPA (15 wt.%) membranes, respectively. Also, FDPA positively affects the mechanical stability of the Nafion membrane. The hybrid Nafion/FDPA (10 wt.%) membrane reveals the increase of the tensile strength value up to (22.9 ± 1.0) MPa as well as the elongation at break of (336.5 ± 15.0)%, while for recast Nafion, the tensile strength and the elongation at break equal to (20.4 ± 1.2) MPa and (215.5 ± 9.1)%, respectively [22]. In addition, the hybrid Nafion/FDPA membranes show increased proton conductivity, especially at 80°C and 40% RH. The highest proton conductivity was obtained for the Nafion/FDPA (10 wt.%) membrane (31.4 mS/cm), which is one order of magnitude higher as compared with the recast Nafion membrane (2.1 mS/cm). Furthermore, the FDPA presence decreases the methanol permeability of Nafion due to the tortuosity increase of the methanol diffusion pathways. For example, at 25°C and 2 M methanol solution, the lowest methanol permeability was obtained for the Nafion/FDPA (10 wt.%) membrane (1.65×10^{-6} cm^2/s), while for recast Nafion, it was 3.82×10^{-6} cm^2/s [22]. Also, the single fuel cell test (H_2/O_2 fuel cell and direct methanol fuel cell at 60°C) was carried out for the Nafion/FDPA (10 wt.%) and recast Nafion membranes. The Nafion/FDPA (10 wt.%) membrane demonstrates the better performance of both fuel cell types. Thus, the maximum power density equal to 234.8 mW/cm^2 (H_2/O_2 fuel cell) was noticed, while for recast Nafion membrane equals ~150 mW/cm^2 (H_2/O_2 fuel cell).

Similar to the case of the inorganic fillers, it can be seen that the application of surface-modified organic compounds positively affects the

Nafion performance. In addition, the significant advantages of organic fillers as compared with inorganic fillers should be noted:

1. Desirable compatibility with the polymer matrix due to similar organic nature;
2. Negligible electron conductivity.

In addition, a higher amount of surface-modified organic fillers (up to 10 wt.%) can be used [22]. In contrast, the amount of inorganic fillers in the membrane is limited because of both the formation of agglomerates (i.e., a decrease of mechanical stability) [26] and the risk of the electron conductivity presence (especially in the case of the carbon nanofillers) [27].

Generally, special attention is paid to the Nafion membrane, which possesses the best technical parameters compared with other PEMs. Different types of hybrid composite Nafion-based membranes have already been developed for the PEMFC application, such as Nafion/inorganic fillers [28], Nafion/IL [18], blended Nafion, and modified Nafion membranes (i.e., NC700). The other issue that should be considered is PEM recycling, as it will allow easy PEMFC industrial implementation to be respectful to the environment [29]. Until now, not many works may be found concerning this problem, but the success of PEMFC use on the industrial scale depends on it strongly. In any case, the MEA market was US$ 603.18 million in 2022 and is forecast to a readjusted size of US$ 2032.96 million by 2028 with a compound annual growth rate of 22.45% [30]. This trend demonstrates that PEMFC will be a fast-growing green technology in the next decades.

However, improving the proton conductivity of the membranes is still a significant challenge for enabling the PEMFC development and making them economically reliable. Based on the scientific literature, we can say that one of the most effective ways of membrane improvement is incorporating (in)organic fillers into Nafion. However, to avoid the decrease of the membrane proton conductivity and to increase the compatibility of the Nafion matrix, the fillers should be modified by grafting proton conductivity groups onto their surface. Such hybrid composite Nafion-based membranes reveal increased water retention and, thus, improved fuel cell operation performances under severe conditions ($>80°C$ and $\leq 80\%$ RH).

However, the filler incorporation into the Nafion matrix cannot guarantee an improvement of all membrane characteristics. Based on the already obtained experimental results, one can say that the hybrid composite Nafion-based membrane requires some compromise between its durability, proton conductivity, and thermomechanical and oxidative stability. Depending on the filler type, different membrane characteristics can be improved or, vice versa, degraded. On the one hand, using inorganic fillers increases membrane thermomechanical properties and oxidative stability while decreasing the Nafion proton conductivity. On another side, using ionic organic fillers (i.e., ILs) contributes to the conductivity increase, especially at high temperatures, but deteriorates the membrane's thermomechanical stability. Thus, using both filler types seems promising to obtain a highly productive composite Nafion-based membrane due to the possibility of reaching the optimum compromise in the membrane characteristics depending on the inorganic/organic filler ratio. Indeed, as inorganic fillers offer good water retention ability and ionic organic fillers ensure conductive properties, combining both filler types should provide a conductive Nafion-based membrane at high temperatures and moderate humidity levels. However, the durability of these hybrid membranes, which is an essential property for their industrial use, mainly depends on the filler migration and elution during operation. Therefore, in the near future, it is necessary to develop a hybrid composite membrane through the filler stabilization inside the membrane. Some success has already been achieved in the case of the pure Nafion matrix.

At the same time, the search for a new proton conductive polymer membrane capable of operating at temperatures above 80°C should be continued, especially because the performance of hybrid composite Nafion-based membranes is still insufficient to ensure the long operational stability of PEMFC. In addition, these membranes remain too expensive. Therefore, the elaboration of loaded membranes consisting of Nafion and other polymers (such as sulfonated polymers and PBI) can be considered among future trends. The filler stabilization in these composite membranes may be achieved by creating interpenetrating networks, cross-linking reactions between polymer chains and fillers, or confinement effect in a structured multilayer. In addition to the expected improved conductivity/durability performance, these hybrid membranes' large-scale production and recycling will be essential for their industrial development, which will enormously increase in the future.

3.1.3 Fluorinated free proton exchange membrane

The sulfonated hydrocarbon polymers (SHPs) have been intensively studied due to their outstanding thermal stability, high mechanical strength, and low fuel crossover [31,32]. In addition, it is possible to develop PEMs at low cost due to the relatively convenient synthetic process compared with the perfluorosulfonic acid (PFSA) ionomers. Moreover, various SHPs with different ion exchange capacities (IECs) can be obtained by incorporating diverse monomers and/or postmodifying intermediates [33]. Representative SHP structures include sulfonated poly(arylene ether sulfone) [34,35], SPEEK [36,37], sulfonated poly(phenylene oxide) (SPPO), and sulfonated polyimide [38,39]. Although PEMs with these polymers have been studied as alternatives to the PFSA ionomer-based PEMs due to the advantages described, the proton conductivity of the SHP-based PEMs is generally lower than that of the PFSA ionomer-based PEMs, because the interconnected hydrophilic channels are not as well developed as PFSA ionomer-based PEMs [40]. Typically, SHPs with high sulfonation (DS) can form large hydrophilic domains, resulting in high proton conductivity. However, when the DS of SHPs is high enough to reach a comparable proton conductivity as that of PFSA ionomer-based PEMs, they do not maintain the necessary high physicochemical stability for PEMFC operation [41,42]. To improve the proton conductivity of SHP-based PEMs without deteriorating physicochemical stability, structural engineering of the SHPs has been studied to form distinct phase-separated structures of the hydrophilic and hydrophobic domains, similar to those of PFSA ionomers [43,44]. It is generally known that the preparation of block can achieve control of hydrophilic and hydrophobic segments within SHPs, graft/comb-shaped and densely sulfonated copolymers [45,46]. Therefore, this study reports on recent research trends related to the development of SHP-based PEMs showing high performances in PEMFCs by pursuing rational design strategies for copolymer architectures.

3.1.3.1 Block copolymer-based PEM

In general, SHPs synthesized via the nucleophilic aromatic substitution reaction between dihalo monomers with or without sulfonic acid groups and difunctional monomers with nucleophiles (e.g., dihydroxy and dithiol) are composed of randomly distributed hydrophilic and hydrophobic moieties due to the random distribution of hydrophilic sulfonic acid groups (Fig. 3.5). Therefore, the PEMs prepared by random copolymers usually exhibit lower proton conductivity than the PFSA-based PEMs, especially at low RH conditions, due to the low hydrophilic/hydrophobic phase separation behavior, which forms the small ion-conducting channels [47–49]. Therefore, structural engineering of SHP-based polymer

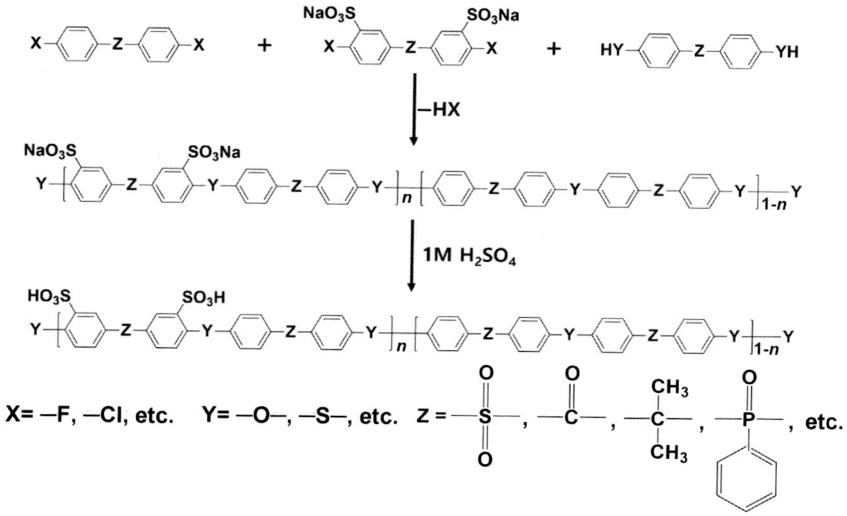

Figure 3.5 Synthetic procedures of sulfonated hydrocarbon-based random copolymers [47].

beginning with the synthetic process is highly required to control the nano-phase structures of the resulting PEMs.

It is well known that block copolymers synthesized by assembling the hydrophilic and hydrophobic oligomers forming diblock, triblock, and multiblocks can effectively control the nano-phase structure and facilitate the distinct phase separation characteristics of the hydrophilic/hydrophobic moieties [49]. Therefore, the block copolymer-based PEMs can exhibit outstanding proton conductivity even under low RH conditions and greatly reduce the conductivity dependence on temperature and humidity changes. In addition, due to the well-defined phase-separated structure, the dimensional and chemical stabilities of the block copolymer-based PEMs can be improved compared with those of the random copolymer-based PEMs having similar IECs.

3.1.3.2 Graft/comb-shaped copolymer-based PEMs

The graft or comb-shaped copolymers used in PEMFCs are usually composed of physically and chemically stable hydrophobic main chains and proton-conducting hydrophilic side chains. By geometric separation of the structural units of the copolymers that are responsible for stability and ion conductivity, the PEMs with graft copolymers were found to have a distinct phase-separated structure between hydrophilic/hydrophobic domains and reveal better proton conductivity, oxidative, and dimensional stabilities compared with their linear counterparts of similar compositions [50]. Notably, some studies have reported that comb-shaped copolymer-based membranes show better PEM properties, including proton conductivity and lower water absorption behavior, resulting in smaller dimensional changes than block copolymer-based PEMs with similar compositions [48].

Jianfu Ding et al. reported graft copolymers composed of partially fluorinated poly (arylene ether) main chains as a rigid hydrophobic main backbone and oligomeric sulfonated polystyrene as flexible hydrophilic side chains, as shown in Fig. 3.6A [51]. In this study, weight fractions (wt.%) of the side to the main chains in the graft copolymers were 19, 25, and 38% and designated as 1a, 2a, and 3a, respectively. Based on cross-sectional TEM images, as the hydrophilic side chain moieties increased, the size of the ionic clusters inducing the distinct phase-separated structure became larger. Accordingly, the 3a membrane with the largest sulfonated polystyrene side chains showed better proton conductivities than Nafion 117 measured in water under the range of temperatures from 20°C to 80°C. It was also demonstrated that the 2a membrane showed a smaller dimensional change value than Nafion 117, although it had a larger IEC than Nafion 117.

Moreover, the 1a membrane with a similar IEC of Nafion 117 showed half the dimensional change in the longitudinal direction compared with Nafion 117, significantly lower than that of previously reported hydrocarbon-based PEMs [52]. Overall, it is demonstrated that membranes composed of graft copolymer with poly (arylene ether) main chains and oligomeric sulfonated polystyrene side chains effectively controlled the water absorption behavior without deterioration of proton conductivity. However, preliminary cell performance tests using the MEAs with these membranes revealed a rapid drop in cell voltages with increases in the current density due to chemical degradation of the highly reactive benzylic position of the polystyrene [53]. Therefore, in addition to the geometric separation of hydrophilic and hydrophobic regions of the graft copolymer, the authors emphasized that one of the most important aspects in the future development of graft copolymer-based PEMs was that the chemical composition of the side chains should not contain chemically vulnerable bonds. This will be necessary to ensure long-term durability for practical fuel cell operation [54].

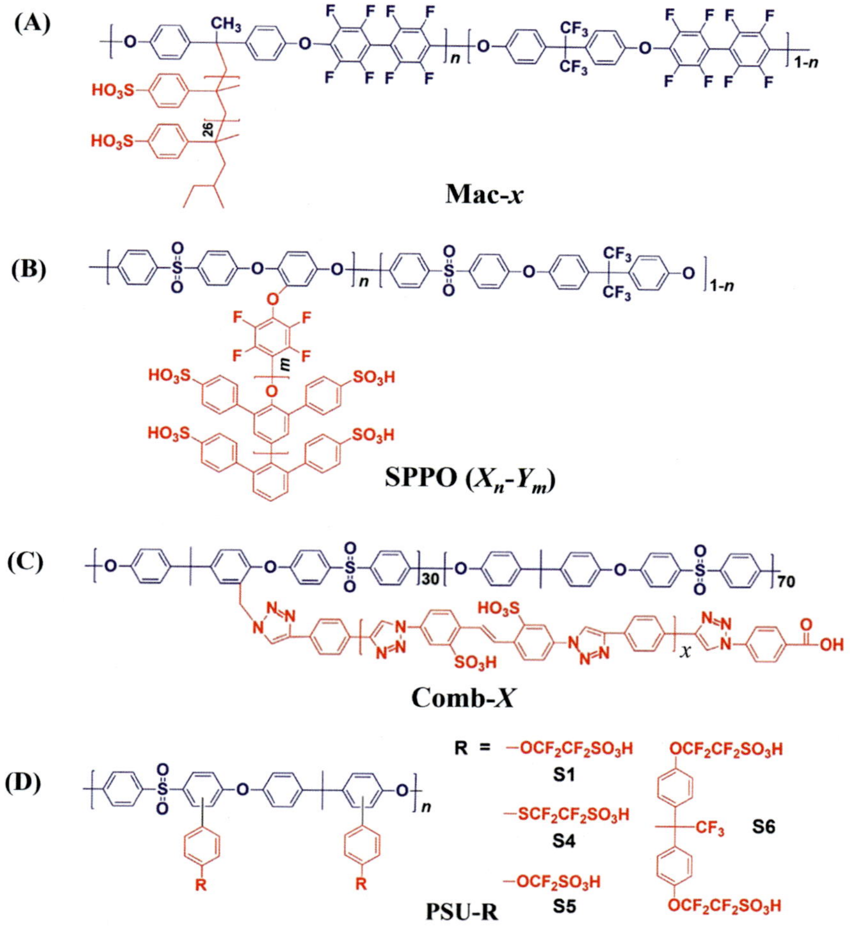

Figure 3.6 Representative chemical structures of sulfonated hydrocarbon-based graft/comb-shaped copolymers. (A) Mac-*x*,(B) SPPO $(X_n - Y_m)$, (C) Comb-*X*, and (D) PSU-R [47].

3.1.3.3 Densely sulfonated copolymer-based PEMs

One of the effective strategies for developing PEMs with outstanding proton conductivity under low RH conditions is the preparation of densely sulfonated copolymers with many sulfonic acid groups located at a specific portion of copolymers [55,56]. As the sulfonic acid groups are concentrated locally, well-defined phase-separated structures inducing comparable or better proton conductivity than PFSA-based ionomer PEMs have been reported with densely sulfonated copolymer-based PEMs [57,58].

Meanwhile, the degree of sulfonation and the bulkiness of the sulfonic acid moieties should be considered during structure engineering because the water absorption behavior, including water uptake and dimensional change in the corresponding PEMs, is highly affected by their sulfonated portion of copolymers [59].

Allan S. Hay et al. reported densely sulfonated copolymers containing six sulfonic acid groups at the end group of the main chain as shown in Fig. 3.7A [60]. As the main chain was composed of poly (sulfide ketone) (PSK)-based rigid aromatic polymer, the resulting PEMs had low methanol permeability. Furthermore, due to the electron-withdrawing effect of the ketone groups, the copolymer PEMs showed outstanding proton conductivity, possibly from the increase in the acidity of the sulfonic acid groups, in addition to the phased-separated structure between hydrophilic and hydrophobic moieties. Although the densely sulfonated copolymer membranes (termed 6b—SO_3H and 6c—SO_3H according to the structure of Ar1 and Ar2 in the main chain of PSK) revealed relatively low IEC values compared with other SHP-based PEMs, the proton conductivities of these membranes were superior to other reported PEMs with similar IECs due to the well-defined phase-separated structures. However, the proton conductivities of the 6b—SO_3H and 6c—SO_3H are lower than those of the PFSA-based PEMs due to the two times-lower IEC values of the copolymer PEMs. To improve the IEC and proton conductivity of these sulfonated PEMs, follow-up studies were conducted by the same group. Fig. 3.7B shows branched poly (ether ketone) (PEK) copolymers with three ends having multiple sulfonic acid groups [60]. The end-capping reagents of the PEK are hexaphenylbenzene and 3,6-trityl-9H-carbazole, respectively; thus, the maximum numbers of sulfonic acid groups that can be introduced in these branched PEKs are 18 and 24, respectively. The larger IEC and proton conductivity values observed in the PEM with the 3,6-trityl-9H-carbazole were clearly due to the higher sulfonic acid group content. For example, the maximum proton conductivity of the PEM containing PEK with 3,6-trityl-9H-carbazole was 95 mS/cm (at 30°C and 100% RH), while those with hexaphenylbenzene achieved 91 mS/cm. Although the size of ionic clusters observed in both membranes by TEM was smaller (2−3 nm) than that of Nafion 117 (5−8 nm), distinct phase-separated structures similar to that of Nafion 117 could be observed. Therefore, the better proton conductivities of the PEK-based PEMs than Nafion 117 are obtained due to the well-developed phase-separated structures, in addition to their larger IECs.

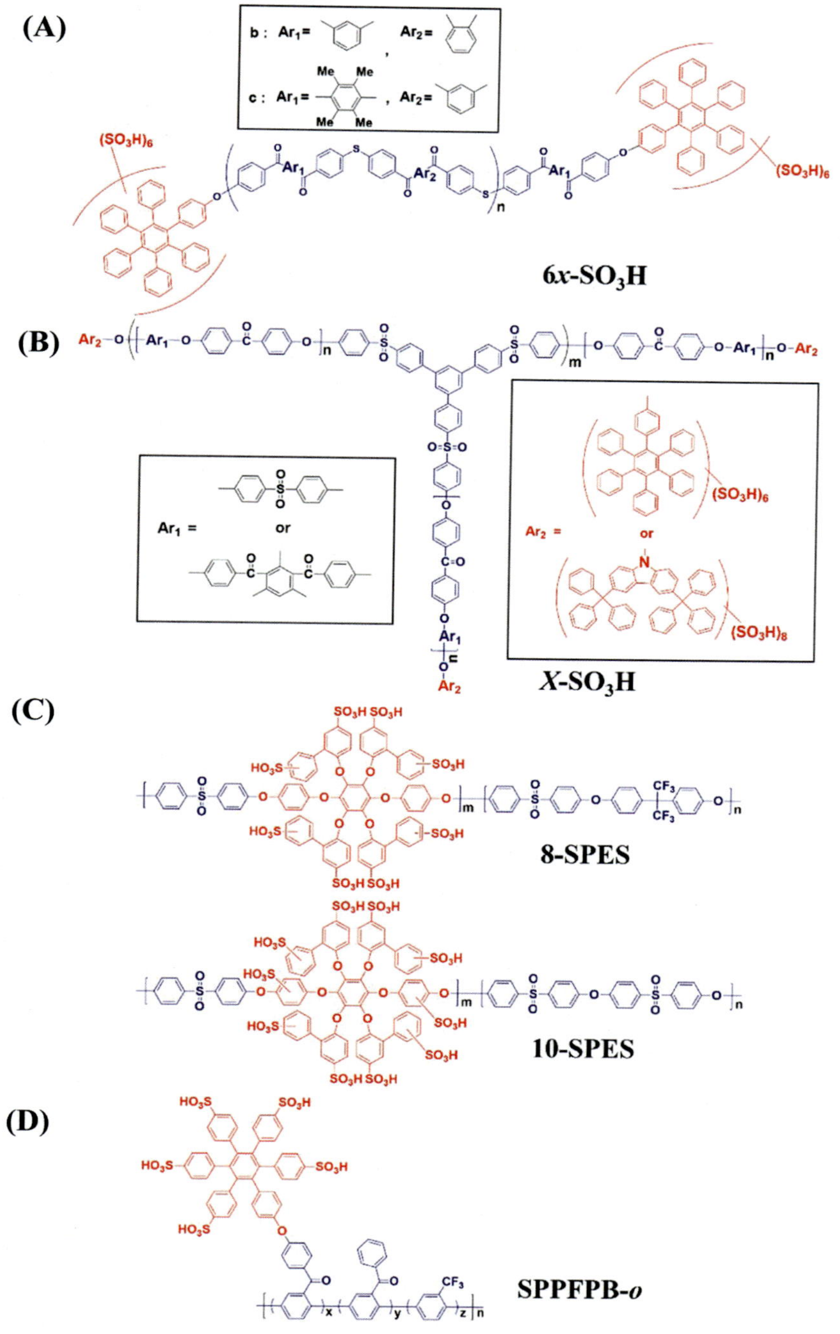

Figure 3.7 Representative chemical structures of densely sulfonated hydrocarbon-based copolymers. (A) 6*x*-SO_3H, (B) *X*-SO_3H, (C) 8-SPES, 10-SPES, and (D) SPPFPB-*o* [47].

In summary, the development of SHP-based PEMs has been actively pursued to overcome the inherent drawbacks of the PFSA ionomer-based PEMs. SHP-based PEMs composed of well-known aromatic random copolymers have shown relatively low proton-conducting behavior and physicochemical stability compared with PFSA ionomer-based PEMs due to the difficulty in forming the well-defined hydrophilic/hydrophobic phase-separated structures. To improve the proton conductivity of SHP-based PEMs without deterioration in physicochemical stability, the polymeric architecture must be controlled to form distinct phase-separated structures between the hydrophilic and hydrophobic domains. Therefore, this review has focused on the synthetic procedures that underlie structure-engineered copolymers, such as block, graft/comb-shaped, and densely sulfonated copolymers and their PEM properties. Studies on SHP-based PEMs using block copolymers indicated that the length and structure of each hydrophilic and hydrophobic block could affect the size of ionic clusters and the phase-separation architecture, thereby affecting the PEM properties of the corresponding membranes. For graft/comb-shaped copolymers, the degree, length, and structure of side chains control the PEM properties, such as proton conductivity and physical stability of the corresponding membranes. Finally, studies on densely sulfonated copolymers indicate that the degree of sulfonation of the hydrophilic reagents and the location of the densely sulfonated moieties significantly affect nano-phase structures and the proton-conducting behavior of the resulting PEMs. Based on this review of existing literature, we strongly believe that, with rational designs of polymeric architecture beginning with the synthetic process, it will be possible to develop high-performance SHP-based PEMs surpassing the PFSA ionomer-based PEMs operating in harsh conditions of low RH. In addition, the exclusion of chemically vulnerable heteroatom linkages by the novel synthetic processes should be further considered to develop next-generation SHP-based PEMs for practical PEMFC applications.

3.1.4 High-temperature membrane

3.1.4.1 Modification of Nafion membranes

One of the promising ways of tackling the problems of Nafion membranes encountered at elevated temperatures is to incorporate NPs [61], such as TiO_2 [62,63], SiO_2 [64,65], ZrO_2 [66], and graphene oxide (GO) [67] into Nafion polymer matrix to form heterogeneous membranes.

The morphology properties, such as ionic clustering and the distribution of ionogenic groups in Nafion membranes, have an important influence on Nafion membranes' mechanical and transport properties [68]. It is crucial to minimize the destruction of the Nafion membrane morphology during their modification. Xu et al. [69] prepared silica/Nafion composite membranes by a nondestructive filling of silica NPs into a Nafion matrix using a swelling-filling strategy. It was found that the proton conductivity of a composite membrane is enhanced by 30% compared with the pristine Nafion due to the nondestructive nanophase-separation morphology of Nafion and the high water retention capacity of silica NPs. The proton conductivity of the composite membrane is 0.033 S/cm at 110°C under 60% RH. What is more important, Xu et al. [65] modified the silica particles by in-situ sulfonation process to prepare the sulfonated silica/Nafion nanocomposite membrane to improve the proton conductivity at high temperatures and low RH. The prepared membrane showed a high proton conductivity of 0.07 S/cm at 110°C under 60% RH due to the synergistic effect of the silica and the efficient SO_3H-based proton conductive channel. Oh et al. [70] also prepared sulfonated silica/Nafion composite membranes and investigated their physicochemical properties, proton conductivity, and fuel cell performance. It was found that the thermal stability, mechanical properties, and proton conductivity were improved owing to the interaction between sulfonated silica and sulfonic acid groups of Nafion. The proton conductivity of composite membranes containing 1 wt.% modified silica particles is 0.088 S/cm which is 1.7 times higher than the value of the proton conductivity of pristine Nafion membranes at 80°C under 20% RH.

Incorporating NPs into Nafion membranes can limit water evaporation and improve thermal stability and mechanical properties at temperatures up to 140°C. However, the modified Nafion membranes exhibit high fuel cell performance only under humidified conditions. The humidification is crucial for modified Nafion membranes to obtain high fuel cell performance. Therefore, modifying Nafion membranes with NPs is insufficient for HT-PEMFC. The replacement of water with nonvolatile solvents such as PA and ILs should be considered to prepare more suitable membranes for HT-PEMFC.

3.1.4.2 HT-PEMs with phosphoric acid

PA is a favorable water substitute for PEMFC at high temperatures owing to its low volatility, greater stability compared with other acids, and independence from the RH [71]. To obtain the PA-containing PEMs with high proton conductivity and performance, the membranes must absorb an optimum amount of PA because of a trade-off between mechanical properties and proton conductivity. The rise of PA content in the membrane increases the proton conductivity but simultaneously decreases the mechanical strength of the membranes. Moreover, the membrane material should keep the maximum amount of PA under long-term operating conditions.

The other widely tested membranes for HT-PEMFC are the ones based on the RH-independent PBI and aromatic amides activated with acid. With PA as the electrolyte, PBI-based membranes with PA are the most preferential candidates for HT-PEMFC owing to their high proton conductivities under low RH. Moreover, PBI has good thermal, chemical, and mechanical stabilities, which ensure their performance at high temperatures [21,72]. The PA content determines the proton conduction rate of PBI/PA membranes under anhydrous and high-temperature operating conditions. It is crucial to improve the PA content and retention ability without sacrificing the stability of PBI membranes to improve the proton conductivity and fuel cell performance. Blending another material into PBI to form blend membranes is an effective and simple way to achieve such a purpose. Taherkhani et al. [73] prepared PBI-poly (acrylic acid) (PAA) blend membranes with different number-average molecular weights and molar ratios of PAA. They found that increasing the PAA molecular weight from 2 to 100 KD enhances tensile strength from 45 to 82 MPa and elongation at break from 2.4% to 7.9% due to the entanglement and hydrogen bonding between PBI and PAA. It also results in the pore structure of the membranes forming new channels for proton transport and facilitates an efficient proton delivery by hopping and vehicle mechanisms [74,75]. PBI: PAA100KD (1:4) (membrane with PAA of 100,000 g/mol and PBI: PAA molar ratio of 1:4) blend membrane showed the highest proton conductivities of 0.005 S/cm at 150°C in the anhydrous state and 0.5 S/cm at 150°C with increasing PA content.

PA leakage is a challenge for PA-doped PEMs. PA leaches out from the membrane during the process and undergoes auto dehydration above 140°C, which is a serious obstacle because of forming pyrophosphoric oligomers of lower conductivity [76]. Moreover, membrane materials containing the liquid acid component possess high anisotropic swelling due to the large uptake of the acid, which can significantly impair the mechanical properties of the membranes [77]. Nevertheless, the efforts to obtain composite polymer membrane materials with desirable conductive properties of the membranes for HT-PEMFC application have required research approaches. It is important to mitigate the PA leakage and keep high PA loading to achieve high proton conductivity during the long period of operation [78]. Many researchers have been improving PA retention ability, thus preventing PA leaching problems. Özdemir et al. cross-linked PBI membranes with different cross-linkers, including bisphenol A diglycidyl ether (BADGE), ethylene glycol diglycidyl ether, α-α′-dibromo-p-xylene (DBpX), and terephthalaldehyde [78]. It was found that the cross-linked membranes showed good thermal stability at high temperatures (above 200°C), and the acid leaching was mitigated effectively owing to a 14% reduction in PA leaching level. Among these membranes, the PBI/DBpX membrane showed the highest proton conductivity value, 0.151 S/cm at 180°C, and PBI/BADGE membranes containing an acid uptake level of 15 molecules of H_3PO_4/repeating unit of PBI presented the highest acid retention properties.

The nonfluorinated polymer membranes with PA are also suitable for the applications of HT-PEMFC due to their good thermal and chemical stabilities, high mechanical strength, and high proton conductivity at the anhydrous state or low RH. However, the proton conductivity of such membranes is strongly dependent on the acid content. The blending and cross-linking can simultaneously improve the acid content and mechanical strength of membranes with PA, which enhances the proton conductivity and power density of HT-PEMFC. The PA leaching leads to the loss of the proton conductivity of PEMs and severe corrosion problems. Cross-linking is one of the most commonly used methods to improve the acid retention ability and eliminate the adverse effects caused by PA leaching.

3.1.4.3 HT-PEMs with inorganic particles

The introduction of additives such as inorganic particles, apart from polymer blending and cross-linking, has been demonstrated as an effective approach to altering the physical and chemical properties of pristine polymer membranes. Such an approach to the elaboration of PEMs also improved the proton conductivity, acid content, and retention ability. As a result, the preparation of composite membranes for HT-PEMFC has drawn a vast deal of attention in recent years. Hence, the effects of different additives on the physical and chemical properties of PEMs and their performance in fuel cells have been reviewed by many researchers [79]. In this section, the influence of pristine and functionalized fillers on the membranes with PA is reviewed and discussed to understand how adding inorganic particles into polymer membranes improves proton conductivity and fuel cell performance. In general, incorporating inorganic particles into polymer membranes effectively alters the physical and chemical properties, such as the proton conductivity and the mechanical stability of the membranes Pan et al. [80]. Kuo and Lin [81] dispersed homogeneously different amounts of MCM-41 mesoporous silicates into a polymer solution to prepare PBI-based composite membranes. Compared with the pure PBI membranes, those with MCM-41 (10 wt.%) exhibited a significantly enhanced PA doping level and enhanced retention ability, referring to the 22% reduction of PA leaching level and a slight increase in Young's modulus from 5.78 to 7.82 MPa, leading to the improved HT-PEMFC performance (maximum power density of 310 mW/cm^2) at 160°C under nonhumidified condition. The improved performance of membranes was due to increased hydrogen bonds between the PA molecules and the -OH groups of the filler particles. However, the aggregation of particles may happen if an excess of filler particles is added. Moradi et al. [82] prepared PA-doped PBI-Fe_2TiO_5 nanocomposite membranes by dispersion of various Fe_2TiO_5 NPs in a PBI polymer matrix. The composite membrane containing 4 wt.% of Fe_2TiO_5 NPs showed the highest proton conductivity of 0.078 S/cm at 180°C and the highest value of PA doping level of 12%. Bai et al. [83] synthesized graphite-like carbon nitride (CN) nanosheets and dispersed them in poly(ether sulfones)-poly (vinyl pyrrolidone) (PES-PVP) polymer matrix to prepare a composite membrane. It was found that the proton conductivity of a composite membrane was improved up to 0.104 S/cm at 160°C when 0.5 wt.% of CN nanosheets were introduced due to the 25% higher PA doping level and faster proton dissociation in PA. At the same time, the tensile strength of the composite membrane increased from 40.4 to 34.7 MPa.

They directly incorporated inorganic particles into a polymer matrix that exports desirable properties to PEMs. For instance, the PA content and acid retention ability were enhanced due to the interaction between inorganic particles and PA molecules. The mechanical strength was improved due to the interaction between inorganic particles and polymer chains. However, the compatibility between inorganic particles and polymer matrix limits adding inorganic particles in a polymer matrix. Namely, agglomeration of the particles may appear when an excess of particles is added. The inorganic filler functionalization plays an important role in the compatibility improvement between the inorganic particles and the polymer matrix.

Moreover, the functional groups in modified particles can improve composite membranes' proton conductivity, acid retention ability, and mechanical strength. For instance, composite membranes incorporated with the ILs-modified inorganic fillers require less PA content while keeping high proton conductivity. This is because the ILs and inorganic filler provide a proton transport route. Adding ILs into PEMs helps avoid acid leaching and materials corrosion.

3.1.4.4 HT-PEMs with ionic liquids

ILs are characterized by good thermal stability, high ionic conductivity, negligible vapor pressure [84], nonflammability [85], and electrochemical stability [86]. Therefore, ILs have been widely investigated as promising and efficient additives for PEMs, making it possible to change and tailor the properties of membranes [87,88].

PILs are an ILs subclass that constitutes the combination of Brønsted's acids (HA) and bases (B) and, thus, they possess the ability to transfer protons from acid to the base [Eq. (3.1)] creating the proton donor and acceptor sites as well as building a hydrogen-bonded network [89,90].

$$HA + B \leftrightarrow BH^{+} + A^{-} \tag{3.4}$$

In general, the proton transfer in PEMs occurs owing to the presence of a proton conductor, and two mechanisms drive it: the vehicular and the Grotthuss one [91,92]. Nevertheless, Yaghini et al. [90] discovered how the structural and dynamic properties of PILs constituents determine the proton mobility in imidazole-based PILs. Based on the results of the computational and experimental studies of 1-ethylimidazolium bis (trifluoromethanesulfonyl)imide (C_2HImTFSI) PIL and imidazole, it was pointed out that at a low imidazole content ($X < 0.2$), imidazole behaves as a base that pulls the proton on the imidazolium cation and so creates a very stable Hebonds network and, thus, promotes proton transfer via the Grotthuss mechanism. On the other hand, the high imidazole content ($X > 0.5$) inhibits the vehicular mechanism owing to the presence of highly dissociated and fast diffusing ions in PIL [90].

Over the years, ILs have been tested as additives in PEMs based on various polymers, including SPEEK [91] and PBI [62,92]. Fig. 3.8 illustrates the timeline for PEMs modified with different types of ILs over the last 12 years. The discussed research comprises various approaches to PEMs functionalization based on the direct addition of ILs or the incorporation of NPs containing immobilized ILs into polymer membranes.

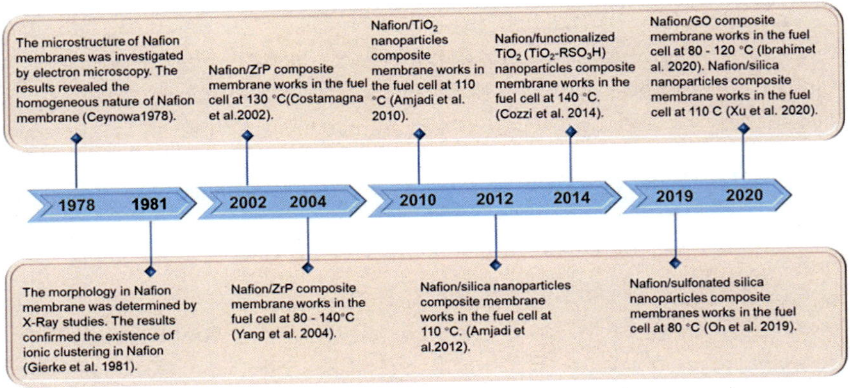

Figure 3.8 Timeline indicating the developments of Nafion-based membranes, modifications, and milestones in the fuel cell performance [62,67,68,93–95].

3.1.5 Preparation process of proton exchange membrane

The membrane-forming process affects the overall cost of the membrane and the microstructure of the membrane. The proton transfer mechanism and many test results show that the performance of a PEM is inevitably related to its microstructure morphology, such as the phase separation mechanism. Currently, many studies are based on the solvent casting method [96,97] to prepare PEMs of new materials. Most of them are prepared by this method in commercialization.

The process of tape casting (casting) preparation refers to coating the solvent with tools and evaporating the solvent on a flat surface or making the solvent flow. It extends spontaneously while evaporating on a flat surface, which can obtain films with good uniformity and thin thickness.

The traditional preparation process has many disadvantages, such as complex preparation process, high cost of materials and equipment, and high energy consumption. In recent years, the vigorous rise of nanomaterials has been recognized by more and more people. Because nanofibers have a high specific surface area, good mechanical properties, and a strong ability to penetrate with other substances, researchers also began to develop nanocomposite polymer PEMs through nanocomposite methods based on perfluorinated sulfonic acid resin or sulfonated polymer as proton transfer materials. It provides more possibilities for the popularization and application of fuel cells.

In recent years, solution jet spinning has been a new technology for preparing nanofibers [98]. The principle of preparing fibers is to use the

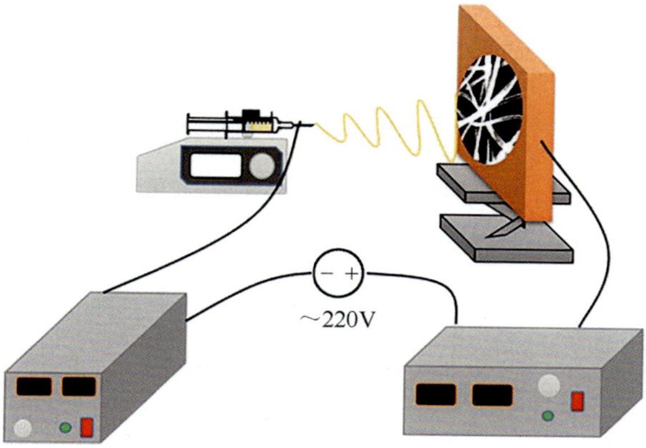

Figure 3.9 Schematic diagram of electrospinning [99].

tensile force of high-speed airflow on the solution. The solution trickle is stretched by the airflow, accompanied by solvent evaporation, trickle solidification and other processes to form nanofibers. The equipment is shown in Fig. 3.9. It is characterized by a simple process, easy operation, and high production efficiency [100]. Zhang et al. prepared polyacrylonitrile nanofibers by solution jet spinning technology [101], then treated them to obtain carbon nanofibers (CNF) and activated carbon nanofibers (ACNF) felt, immersed them in SPEEK solution to prepare a dense composite PEM. The results of a series of tests show that introducing CNF and ACNF into the composite membrane can significantly improve the mechanical properties and proton conductivity.

Due to its high production efficiency, solution jet spinning technology can be industrialized to prepare nanofibers, which can be widely used in many fields. It will become an important method for preparing nanofibers.

Electrospinning technology is also one of the most important methods in preparing nanofibers. American Professor Formalas first proposed it in 1934. After decades of development, the application scope has gradually expanded to various fields. The specific principle of electrospinning is similar to that of solution jet spinning. The difference is that a high-voltage electrostatic field replaces the high-speed airflow, and the droplet at the tip of the solution is stretched by electric field force to form a conical shape, that is, Taylor cone, which is also accompanied by solvent volatilization and trickle solidification. The advantage of electrospinning is that the electric field force drives the charged groups in the polymer to move and aggregate directionally, forming a long-distance proton transport channel. Zhao et al. prepared PFSA/PVP fiber membranes by electrospinning [102]. The experimental results show that the nanofiber structure has a high specific surface area and can expose more functional groups, making it more conducive to proton transport in the PEM.

Our research group has been researching the application of electrospinning technology (Fig. 3.9) for many years and has independently built various spinning devices, such as centrifugal electrospinning and melt electrospinning. In recent years, electrospinning technology has been expanded in fuel cells, especially in preparing the fuel cell catalytic layer. Zhang Shaopeng [103] et al. used it on the surface of a commercial Nafion PEM. Electrospinning technology was used to electrospun Pt containing catalytic layer and then hot pressed to obtain a nanofiber MEA. Compared with the membrane electrode prepared by electrospray with the same Pt load in the working environment of the fuel cell, it is found that the electrochemical active surface area (ECSA) of the catalytic layer prepared by electrospinning is larger under certain conditions. Its performance is better than that of the catalytic layer prepared by electrospray [99].

Solution electrospinning technology is simple. The diameter of nanofibers prepared by electrospinning can be controlled by adjusting the propulsion pump's voltage and propulsion rate. It can also coblend other compounds with polymers to realize the preparation of multifunctional nanofiber materials. Electrospinning technology has become increasingly mature in recent years, and the experimental utilization rate of all aspects is higher and higher. It has become a hot spot in the field of nanomaterial science.

3.2 Fuel cell fibrous catalyst layer

Fuel cell electrocatalysts need good conductivity, electrochemical stability, catalytic performance, and other characteristics. The surface of nanomaterials is very easy to adsorb other molecules and produce chemical reactions. The progress of nanotechnology directly promotes the progress of electrocatalysts. The structure of NPs, especially their surface composition, strongly affects the catalytic effect. Because the adsorption of reaction molecules on the catalyst surface is the prerequisite for catalysis, Pt materials have always been the focus of catalyst research because of their excellent molecular adsorption and dissociation behavior. However, the problems of expensive Pt and shortage of resources are becoming increasingly prominent. The research on reducing Pt load and exploring nonplatinum catalysts is still in depth; from the perspective of cathode and anode reactions, although there is anode polarization when methanol is used as fuel, cathode oxygen reduction reaction (ORR) is always the dominant polarization. Therefore, as the research focus of the reaction, ORR is the key to improving the catalytic performance. It can be seen from the above that the progress of electrocatalysts is a process of continuous breakthrough and extension in the research of particle size, structure, and morphology in the nano field. It is also a sublimation process of the optimization of existing Pt catalysts, the exploration of nonplatinum catalysts and the understanding of ORR reaction, the frontier intersection and synthesis of nanoscience, material structure, electrochemical reaction, and other disciplines are the focus of catalyst research [104].

At present, the catalyst used in PEMFC is mainly precious metal platinum, which has a high cost. Therefore, how to further reduce the use of platinum in the catalyst layer (CL) and improve the performance and durability of the CL has become the focus of PEMFC research [101,105]. At present, finding catalysts with higher performance and durability through material research and development, improving the active area of the CL, and improving the material transport in the CL through structural optimization are the two main methods to improve the performance and durability of the CL [106–108].

Based on the excellent catalytic effect of nanomaterials, the fuel cell CL was prepared by electrospinning technology in our laboratory [109], which combined the advantages of nanomaterials and a platinum catalyst to reduce the cost and improve the catalytic effect.

3.2.1 Evolution of catalyst layer preparation technology and structure

As shown in Fig. 3.10, the preparation methods of the PEMFC CL can be divided into the bonding method [110,111], deposition method [112,113], cylindrical array preparation method [114,115], and electrospinning preparation method [103,116,117]. The bonding method is to mix the catalyst, polytetrafluoroethylene (PTFE) emulsion or Nafion solution with alcohol solvent to form a catalyst slurry and to coat it on the gas diffusion layer to form an electrode. However, the CL prepared by the method is thicker, and the catalyst utilization rate is low. The deposition method directly deposits the catalyst slurry containing Pt and carbon carrier on the Nafion film. The deposition methods include chemical vapor deposition, physical or thermal vapor deposition, spraying, electrodeposition, etc. This method is

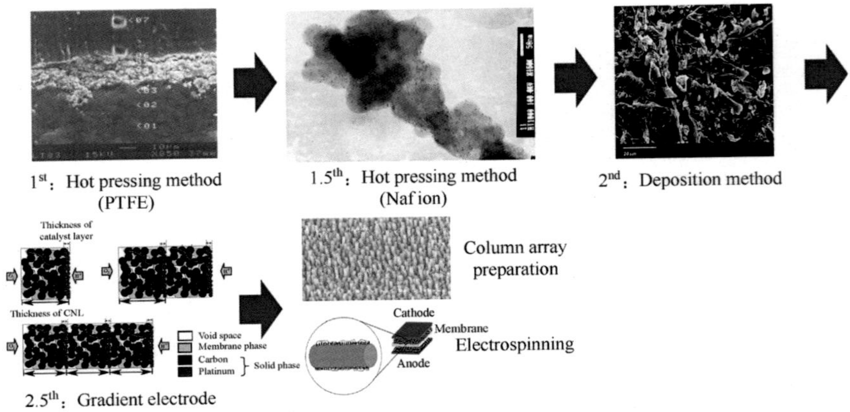

Figure 3.10 Evolution of catalyst layer fabrication methods [110].

the current mainstream commercial preparation method. The deposition method can reduce the thickness of the CL and has made remarkable progress in reducing the platinum load, but its durability still needs to be improved. The columnar array preparation method is to grow nanowires, nanorods, or nanotubes containing catalysts directionally or supports to form ordered electron channels [118], proton channels, or gas transport channels. The thickness of the CL prepared by the columnar array preparation method is generally 1 μM or less [119]. The amount of platinum in the CL can be significantly reduced, at the same time, generally, because the catalyst structure in the CL is not NPs, it can reduce the dissolution and agglomeration of platinum catalyst, reduce the occurrence of carbon corrosion [120], and effectively improve the durability of the CL [118]. The electrospinning preparation method forms nanofibers from the slurry through electrospinning technology and then makes a CL. Nanofiber structure can improve the specific surface area of the catalyst in the CL [121] and then improve the three-phase interface of the CL.

By further classifying the structure of the CL, the current CL can be divided into three categories [109], as shown in Fig. 3.11 [122].

The first type is the particle stack (PS) type, mainly composed of NPs catalyst stacking. Currently, the CLs prepared by commercial mainstream deposition methods belong to this type. The second type is the column array (CA) type, which is mainly composed of nanowires, or nanotubes,

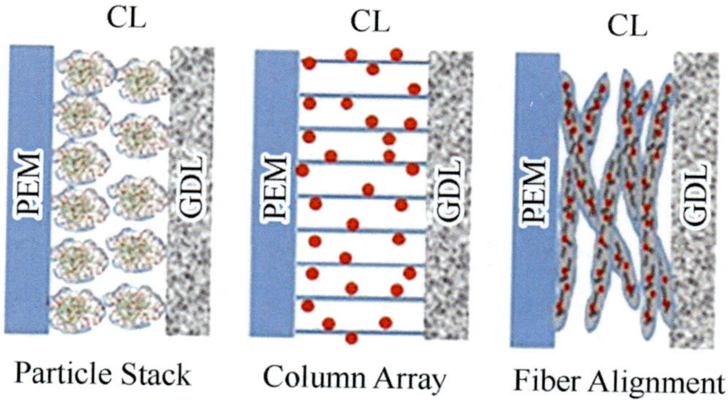

Figure 3.11 Types of catalyst layer structure [122].

or nanorods with one-dimensional structure, the NSTF CL of 3 M company [123,124] and the array nanotube CL prepared by Dalian Institute of Chemical Physics, Chinese Academy of Sciences [113] belong to this type. The third type is the fiber alignment type, which is mainly composed of nanofibers (NF) or nanotubes (NT) prepared by electrospinning, the CL prepared by Vanderbilt University, Tsinghua University [103]. Dalian Institute of Chemical Physics, Chinese Academy of Sciences [125–127] belongs to this type, and the prepared CLs have a high electrochemical active area.

PEMFC CL can be divided into particle stacking type, columnar array type, and fiber arrangement type according to its structure. The PS CL is the mainstream CL used in commercial PEMFC. In terms of performance, the particle-stacked CL's triple-phase boundary (TPB) is small because the structure between NPs is greatly affected by the preparation conditions. At the same time, its three-phase channel (TPC) is bent, and the material transmission resistance is large, so the particle-stacked CL performance is poor. In terms of durability, due to carbon carriers, carbon corrosion is easy to occur under high potential, and the durability of the CL still needs to be improved.

The columnar array CL can form ordered electron, proton, or gas transport channels by forming columnar array catalysts with ordered structures. Therefore, its outstanding advantages include (1) efficient three-phase transmission, (2) high Pt utilization, and (3) improved durability. However, the columnar array CL is generally thin, the water capacity is small, and water management is difficult.

Compared with the particle-stacked CL, the fiber-arranged CL increases the utilization of the catalyst through nano fibrosis in the structure. In terms of performance, the published research results show that it has a higher three-phase interface than the particle-stacked CL. It has also been proved that the fiber-arranged CL's proton and electron transport channels are orderly. At the same time, the research results in the published literature show that the fiber-arranged CL has higher durability than the particle-stacked CL. At the same time, compared with the columnar array CL, the fiber-arranged CL has outstanding characteristics of mass production. With the rapid development of the electrospinning industry, a variety of electrospinning fiber batch manufacturing equipment—no nozzle batch manufacturing equipment and multinozzle equipment have appeared, which can be conducive to the large-scale and rapid production of fiber arrangement CL. However, how to control the preparation mode of fiber arranged CL and further improve its three-phase transmission channel is an important problem to be solved in its development [122].

Columnar array CL and fiber arrangement CL contain more possibilities, which is an important development direction in the field of PEM fuel cell CL in the future and still needs further research by researchers.

3.2.2 Classification and current status of ORR catalysts

During the past decade, many highly promising ORR catalysts have been developed. Broadly, these catalysts can be categorized as (1) Pt/C, (2) Pt and Pt alloy/dealloy, (3) core – shell, (4) nonprecious metal catalysts, (5) shape-controlled nanocrystals, and (6) nanoframes. No specific dates are provided as this would be rather presumptuous and will depend largely on where current and future research efforts are focused [128].

Notably, these catalysts are usually supported by carbon and other nanomaterials. As this perspective is not focused on support material, readers are referred to other review papers published on this topic [129]. Additionally, the nanostructured thin film (NSTF) catalyst developed by 3 M can be considered a hybrid of several of the above catalysts and is also highly promising [119]. However, this catalyst is a free-standing structure, typically "ionomer-free" when used as a CL. Thus, this particular catalyst will not be covered in this perspective, and for further understanding of the NSTF, readers are referred to an overview published by 3 M [114].

As mentioned, the "current status" and development timeline are subjective and, among other factors, depend largely on what applications are being considered. For example, from a purely automotive perspective, the nonprecious metal catalysts (NPMCs) would likely be considered the furthest from commercialization, as they currently do not meet performance, durability (performance loss during voltage cycling), stability (performance loss during potentiostatic/galvanostatic experiments [130]), or power density requirements. However, NPMCs are now close to meeting the requirements for portable power applications. In addition to differences in the development timeline of each family of catalysts, each one offers unique advantages and disadvantages from an industrial or commercialization perspective. At a high level, these differences are summarized in Figs. 3.12 and 3.13 and will be discussed in more depth in the following section.

Pt/C. Commercial PEMFC products have relied heavily on Pt/C catalysts for the past decade. When first introduced, these catalysts offered significant advantages over unsupported Pt black because of the much smaller NPs achievable with supported catalysts. The simplicity of these catalysts is both a benefit and a drawback. From a synthetic perspective,

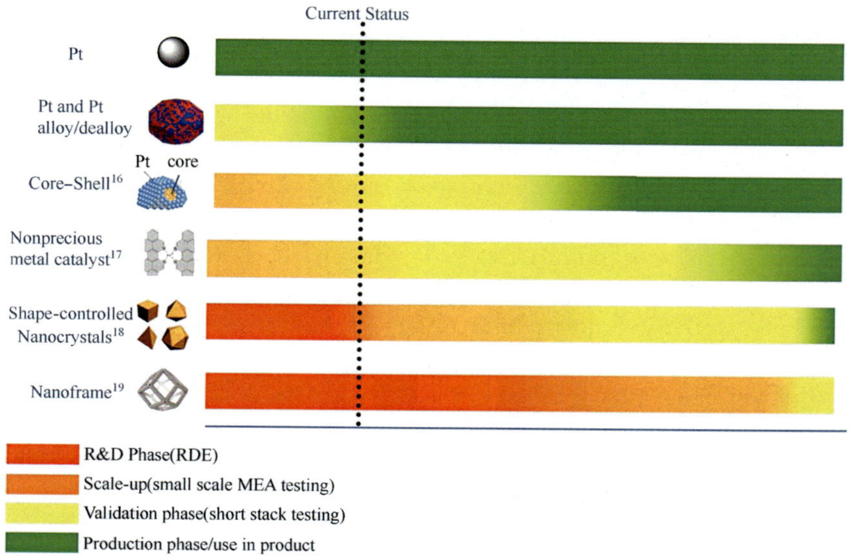

Figure 3.12 Development timelines for Pt, Pt alloy/dealloy, core – shell, nonprecious metal, shape-controlled, and nanoframe ORR electrocatalysts [129].

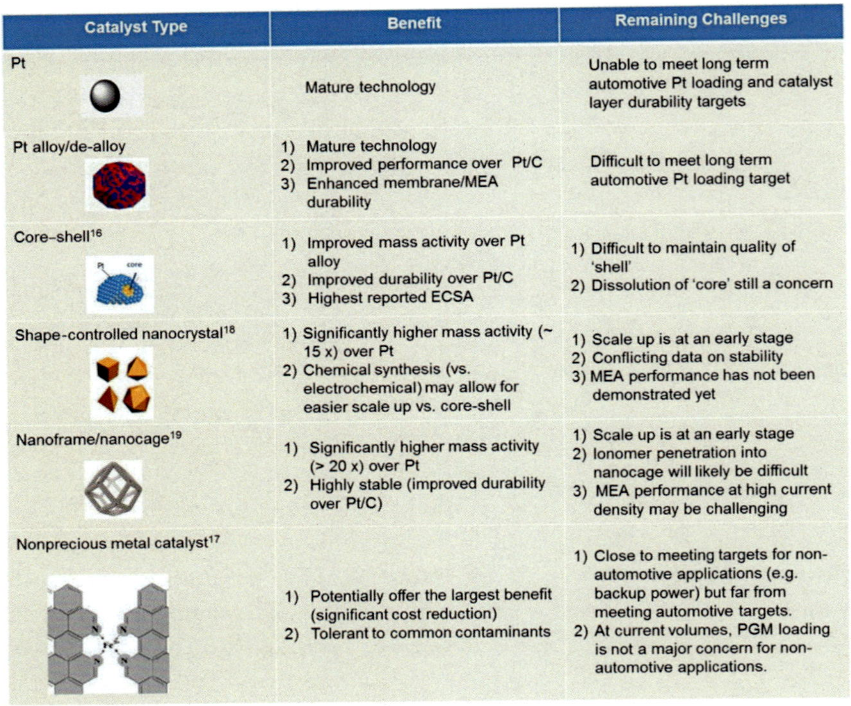

Catalyst Type	Benefit	Remaining Challenges
Pt	Mature technology	Unable to meet long term automotive Pt loading and catalyst layer durability targets
Pt alloy/de-alloy	1) Mature technology 2) Improved performance over Pt/C 3) Enhanced membrane/MEA durability	Difficult to meet long term automotive Pt loading target
Core-shell[16]	1) Improved mass activity over Pt alloy 2) Improved durability over Pt/C 3) Highest reported ECSA	1) Difficult to maintain quality of 'shell' 2) Dissolution of 'core' still a concern
Shape-controlled nanocrystal[18]	1) Significantly higher mass activity (~ 15 x) over Pt 2) Chemical synthesis (vs. electrochemical) may allow for easier scale up vs. core-shell	1) Scale up is at an early stage 2) Conflicting data on stability 3) MEA performance has not been demonstrated yet
Nanoframe/nanocage[19]	1) Significantly higher mass activity (> 20 x) over Pt 2) Highly stable (improved durability over Pt/C)	1) Scale up is at an early stage 2) Ionomer penetration into nanocage will likely be difficult 3) MEA performance at high current density may be challenging
Nonprecious metal catalyst[17]	1) Potentially offer the largest benefit (significant cost reduction) 2) Tolerant to common contaminants	1) Close to meeting targets for non-automotive applications (e.g. backup power) but far from meeting automotive targets. 2) At current volumes, PGM loading is not a major concern for non-automotive applications.

Figure 3.13 Benefits and remaining challenges for each of the primary categories of electrocatalysts [129].

there is little room to tailor activity and durability when limiting the design to a single element (Pt). Further improvements in activity and durability with conventional Pt/C now rely on advances in catalyst supports resulting in "catalyst – support" interactions [131], which have been reported to enhance both activity and durability of PGM-based ORR catalysts. While promising, these approaches are unlikely to meet long-term mass activity requirements using conventional Pt NPs. From a manufacturing perspective, simpler systems are advantageous.

Pt-Alloy. For these reasons, Pt-alloy (e.g., PtCo, PtNi) catalysts are becoming the new baseline catalyst at the commercial level as they can achieve high mass activities while demonstrating similar or better durability compared with Pt/C. Toyota's recent announcement that a PtCo-alloy is currently used in the Mirai highlights the technological maturity of these catalysts. The improved electrocatalytic activity of Pt-alloys (such as PtCo, PtNi, PtFe, PtCr, PtV, PtTi, PtW, PtAl, and PtAg) has been attributed to

(1) the smaller Pt—Pt bond distances resulting in more favorable sites that enhance the dissociative adsorption of oxygen and (2) the structure-sensitive inhibiting effect of OHads. Considerable work has been carried out over the past decades on carbon-supported binary or ternary alloys that demonstrate 2–3 times higher mass activity than Pt/C [132].

Core – sell. In recent years, significant progress has been made on the highly promising "core – shell" family of ORR catalysts. The core – shell concept relies on having the active ORR catalyst (Pt) located only on the NP's surface, with another metal (most typically Pd) making up the bulk. Theoretically, this unique design and concept can allow for the highest possible Pt utilization (surface Pt to bulk). Thus, from a cost perspective, it is highly attractive (provided less expensive cores can be developed).

In addition to having improved activity and ECSA versus conventional PGM catalysts, core shells have demonstrated improved durability during voltage cycling [133,134]. One reason for this is the "self-healing" mechanism that Dr Adzic has identified at BNL. It is known that these core – shell catalysts often have imperfections in the shell, and the less noble core is prone to dissolution during voltage cycling. However, as this occurs, the Pt shell experiences lattice contraction, leading to a higher specific activity and dissolution potential, thus reducing overall Pt dissolution [135]. Additionally, it was shown that using a Pd – Au alloy core can further enhance the stability of these catalysts as the Au preferentially diffuses to any defects in the Pt shell, thus preventing further dissolution of the core.

Shape-controlled nanocrystals. While at an earlier stage of development than core – shell catalysts, shape-controlled catalysts appear to be a highly promising class of ORR catalysts because of their extremely high mass activities. As described above, mass activity depends on specific activity and Pt utilization (Pt dispersion). Core – shell catalysts exemplify the "Pt utilization" strategy to generate high-mass activities. Conversely, shape-controlled catalysts rely primarily on achieving high specific activities to generate high-mass activities. This way, these catalysts are closely tied to the fundamental single-crystal studies in this perspective. In principle, these catalysts attempt to recreate the ideal crystal structure identified by single-crystal studies [135], but at the nanometer scale. An excellent example is the 9 nm Pt2.5Ni octahedra developed at the Georgia Institute of Technology. However, due to the relatively poor Pt utilization (Pt dispersion) afforded by the large 9 nm particles in this study, a mass activity of 3.3 A/mg was achieved (more than sevenfold higher than the 2020 Department of Energy (DOE) MEA target, albeit at the RDE level). This was accomplished by maintaining the ideal Pt2.5Ni (111) crystal structure (which has >50 higher specific activity versus commercial Pt/C) at the nanometer scale.

Nanoframe. The most recent family of PGM ORR catalysts is the "nano-frame" [136]. These catalysts consistently show significantly higher-mass activity than commercial Pt/C (up to 20-fold higher than commercial Pt/C [137] based on RDE studies). However, the largest advantage of this catalyst type is its superb stability and durability during voltage cycling [136]. The unique design of nano-frame catalysts allows them to benefit from the high activity and stability typically associated with extended platinum surfaces [138] while still achieving excellent ECSAs (> 50 m^2/g) due to their relatively thin frames (<2–3 nm). These catalysts are typically terminated by highly uniform crystal surfaces, which is advantageous for durability as it is known that defect sites in conventional NPs are more susceptible to dissolution [133]. Additionally, having such uniform structures helps to reduce the Ostwald ripening process typically observed for conventional NPs, whereby large particles get larger and small particles get smaller. Thus, conceptually it is very reasonable to expect high durability from these catalysts (unlike some shape-controlled catalysts), and there appears to be no disagreement in the literature on this topic.

3.2.3 Preparation of nanofiber catalyst layer by electrospinning

The research content of the electrospinning nanofiber CL can be divided into two parts: the exploration of the preparation process and the characterization of the nanofiber CL.

In terms of characterization, the research on nanofiber CL in the reported literature can be divided into physical characterization, electrochemical performance, and durability characterization. Fig. 3.14 is the SEM image of the nanofiber CL. In terms of physical properties, it is mainly studied from the fiber diameter and distribution of the nanofiber CL, pore characteristics of the CL and so on. The fiber diameter of the electrospinning nanofiber CL is generally 100–1800 nm. The pore size distribution of the electrospun nanofiber CL was measured by BET (Brunauer Emmett Teller, bet). The results show that the electrospun nanofiber CL has a higher pore volume than the particle stacking CL [103,116,139,140]. Electrochemical performance is mainly analyzed by testing the polarization curve, ECSA, and electrochemical impedance

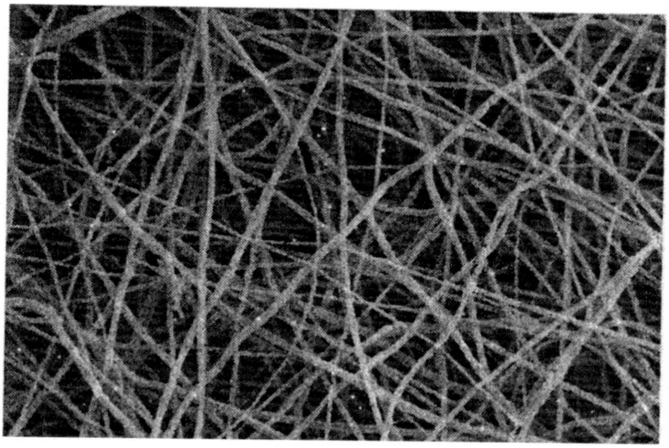

Figure 3.14 Nanofiber catalyst layer [138].

spectroscopy of the CL. From 2009 to now, the electrospinning nanofiber CL has significantly improved in reducing platinum load and improving performance. Regarding durability characterization, the research shows that under the same test conditions and CL formula, the maximum power loss of the electrospun nanofiber CL is less than that of particle-stacked CL [141].

3.2.3.1 Preparation of electrospinning nanofiber catalyst layer

Electrospinning technology mainly uses a high-voltage electrostatic field to charge and deform polymer solution or melt and form suspended conical droplets (Taylor cone) at the nozzle or end; when the charge repulsion force on the droplet surface exceeds its surface tension, a small liquid flow will be ejected at high speed on the droplet surface, which is referred to as "jet" [142]; these jets undergo high-speed stretching of electric field force, solvent volatilization, and fiber solidification in a short distance and finally deposit on the receiving device to form nanofibers.

As shown in Fig. 3.15, it is the schematic diagram of an electrospinning device. The utility model mainly comprises a high-pressure generator, a propulsion pump, a syringe, and a receiving device.

The CL prepared by electrospinning technology can be divided into electrospun nanofiber CL and electrospun nanotube (ESNT) CL based on different structures. Electrospun nanofiber CL can be divided into an electrospinning nanofiber CL based on carbon NP and an electrospinning nanofiber CL based on CNF. The former is obtained by preparing the catalyst with carbon NPs as the carrier into a slurry and then preparing it

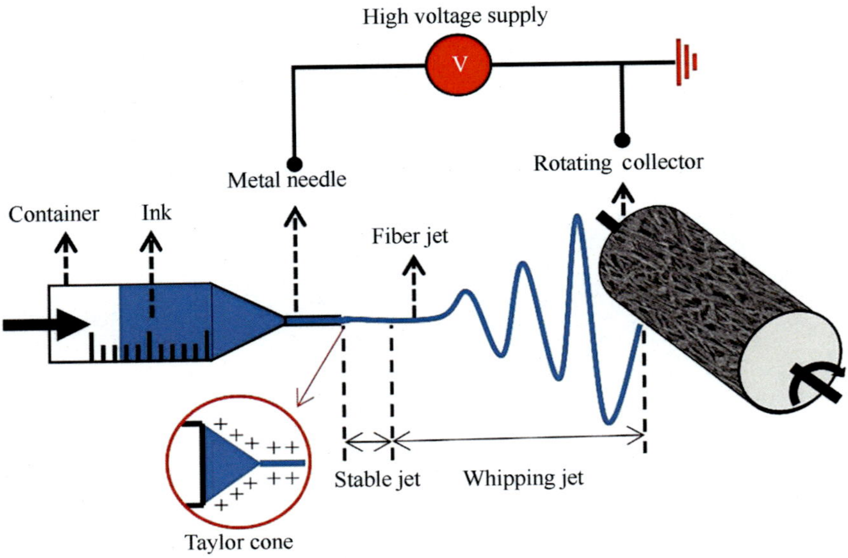

Figure 3.15 Schematic diagram of electrospinning [109].

on the receiving substrate, such as a PEM or aluminum foil by electrospinning technology. The latter can be called the electrospinning CNF CL. Firstly, the solution containing polymer organic matter is prepared, the nanofibers are prepared by the electrospinning method, and the CNFs are prepared after sintering. Finally, Pt is deposited on the CNFs to form the CNF CL [116,143,144].

The preparation process of ESNT CL is as follows: Firstly, the precursor solution containing high molecular organic matter and catalyst carrier is prepared, then the nanofibers are prepared by electrospinning, then the nanotubes are prepared by sintering, and finally, Pt is deposited on the nanotubes to obtain the nanotube fiber-dimensional CL.

3.2.3.2 Characterization of the nanofiber catalyst layer

Determining the CL's physical properties can reveal the internal structure and characteristics of the CL. The CL's physical properties mainly include material and structural properties. Its material characteristics mainly include hydrophilicity, hydrophobicity, component content, and distribution. The hydrophilicity of the CL has an important impact on water generation, transportation, and distribution in the CL. Regulating the hydrophilicity of the CL is an important method for the optimization of water management of fuel cells. The components in the CL generally include a catalyst, ionomer, and support. The proportion and distribution of components in the CL are the main factors affecting the CL's three-phase channel and three-phase interface. The structural characteristics of the CL mainly include fiber diameter, CL thickness, and pore characteristics of the CL. The fiber diameter influences the specific surface area and material transport in the CL. The CL's thickness and pore size distribution are the key parameters of gas and liquid transmission in the CL. It is an important adjustment method for high-current, subzero start and other working conditions.

The functional characteristics of electrospun nanofiber CL mainly include ECSA, exchange current density, and proton conductivity. ECSA is not only an important index to reflect the catalyst activity in the CL but also an important parameter to evaluate the three-phase interface of the CL. It is mainly measured by cyclic voltammetry. The current exchange density is an important index to reflect the reaction speed in the CL, mainly measured by the polarization curve or electrochemical impedance spectrum. Proton conductivity can reflect the proton conductivity of the CL. The measurements include electrochemical impedance spectroscopy and the DC method [109].

3.2.4 Overview of subzero start

The subzero starting capacity of PEM fuel cells is the biggest challenge for fuel cell vehicles in winter. It is also the main obstacle to promoting fuel cell vehicles in warm and cold regions. According to the indicators released by the U.S. DOE in 2017, the fuel cell stack must be successfully started from minus 20°C within 30 seconds in 2025. At the same time, the fuel cell stack needs to be successfully started from minus 30°C within 30 seconds by 2025. Japan's New Energy and Industrial Technology Development Organization (NEDO) formulated relevant technical indicators for subzero startups in 2017. Its technical indicators pointed out that fuel cell stacks need to be started from minus 20°C and 40°C in 2020 and 2040, respectively.

Performance degradation often occurs in the process of subzero startup of fuel cells. This is mainly due to the fuel cell's repeated freezing/melting phenomenon during the subzero startup process, which will damage the structures of the PEM, CL, and gas diffusion layer. Generally, the performance attenuation caused by a subzero start is irreversible, especially when the water production is relatively large.

Because the water inside the PEM usually exists in the state of bound water and is not easy to freeze, the degradation of the PEM mainly occurs at the interface between the PEM and the CL [145]. Usually, during the subzero startup process, the surface icing of the PEM will cause many problems. It is mainly as follows [145,146]: (1) The expansion of the PEM causes mechanical damage to the membrane. (2) Ohmic impedance increases. (3) The PEM cannot absorb the water produced by the electrochemical reaction in the cathode CL. (4) It will cause the separation of the PEM and CL at the interface.

The degradation of the CL is mainly caused by repeated freezing/melting [147]. The main aging mechanisms are as follows [148]: (1) The PEM and CL are separated at the interface. (2) Repeated freezing/melting causes the collapse and densification of micropores in the CL. (3) The freezing/melting cycle also leads to the agglomeration of platinum particles in the CL. (4) Platinum particles dissolve in PEM. (5) Phase change occurs in the CL, resulting in performance degradation.

The aging reason for the gas diffusion layer is relatively simple, mainly because: (1) the porous structure is damaged [149]. (2) Water contact angle decreases [150].

As shown in Fig. 3.16, during the subzero startup of the PEM fuel cell, the water generated by the ORR of the cathode will freeze. The ice will block the transmission channel of reaction gas, thus preventing the reaction gas from reaching the reaction site, thereby preventing the electrochemical reaction and leading to the failure of the subzero startup.

For the subzero starting condition, academia and industry have done a lot of research and made important progress in improving the subzero starting ability of fuel cell vehicles.

Although significant automobile companies have made breakthroughs in the subzero starting capacity of fuel cell vehicles, they are all aimed at PEM fuel cell passenger vehicles based on metal bipolar plates. Compared with fuel cell passenger vehicles with a stacked life of 5000 hours, commercial fuel cell vehicles require a stacked life of 20,000 hours and need to use PEM fuel cells based on graphite bipolar plates with better durability. Because the processing performance of graphite-based bipolar plate is not as good as that of metal bipolar plate, and the thickness of the bipolar plate is high, it will double the heat capacity of the stack, which will bring greater difficulties in solving the problem of the subzero startup.

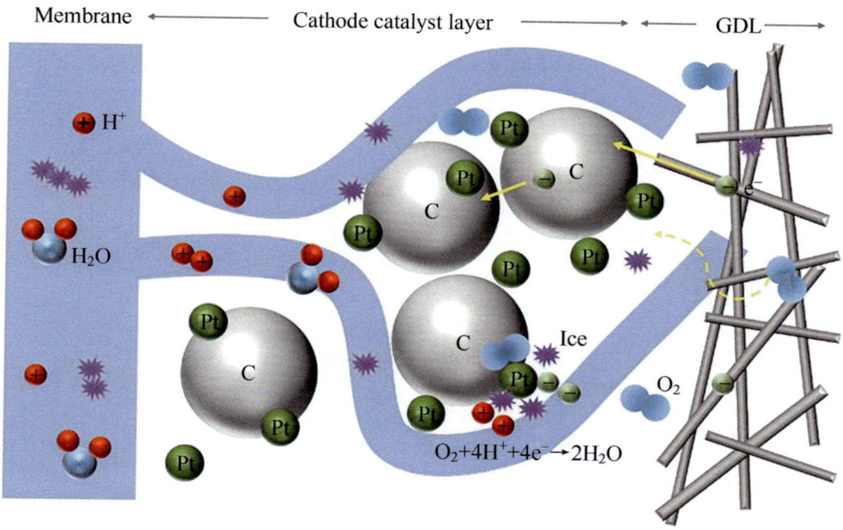

Figure 3.16 The distribution of ice during subzero startup.

Currently, the proton exchange fuel cell commercial vehicle based on graphite bipolar plate cannot operate subzero self-start. Hence, it needs external preheating in advance to ensure its use in a subzero environment.

During the subzero startup of the fuel cell, the essence is the competition between the utilization of water storage space and the heat generation rate inside the cell. The key to the successful subzero startup of the fuel cell is to raise the internal temperature above 0°C before the cell fails. As shown in Fig. 3.17, when the fuel cell stops working, there is usually more water. After purging, the water content in the cell can be reduced to a lower level, so a certain water storage space can be increased to store the water/ice generated during subzero startup. In the process of subzero startup, the internal electrochemical reaction will generate heat to heat the battery. At the same time, the reduction reaction of the cathode will produce water and occupy the water storage space. In the subzero environment, the water produced by the fuel cell reaction is easy to freeze, blocking the gas transmission channel and resulting in the failure of the subzero startup. If the temperature inside the fuel cell is raised above zero by the heat generated by the electrochemical reaction before the space inside the fuel cell is completely occupied by water/ice, the subzero startup is successful. Conversely, if the temperature inside the fuel cell is

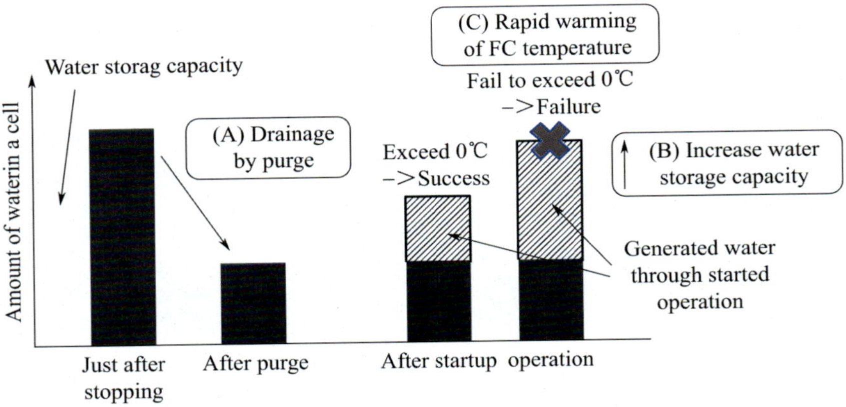

Figure 3.17 Schematic diagram of subzero startup restriction factors [122].

still below zero after the water storage space inside the fuel cell is completely occupied by water/ice, the subzero start fails.

The industry and academia have also done much work on the research of subzero start strategy. There are three types of subzero startup strategies: the first is to improve the subzero startup ability of fuel cells by optimizing the material and structure of fuel cells. Second, the rapid heat generation in the subzero start is realized by controlling different electrical parameters in the subzero start. Third, the fuel cell's subzero starting ability is improved by auxiliary starting [151]. This chapter mainly discusses the first strategy.

As shown in Fig. 3.18, the subzero starting ability of fuel cells can be improved by optimizing the design of fuel cell materials and structures. For the PEM, the ultimate goal of optimization is to improve the water absorption capacity of the PEM during subzero startup to help the CL open up more water/ice storage space. At present, there are two main methods: the first is to increase the concentration gradient of water on both sides of the PEM by reducing the thickness of the PEM to promote the absorption of water on the cathode side of the PEM and then improve the subzero starting ability of the fuel cell. The second is to improve the water absorption capacity of traditional PEMs by adding inorganic materials [152]. Optimizing the CL can be divided into two primary purposes: improving the water absorption capacity. The second is to increase the water storage space of the CL. Accordingly, the optimization design of the CL mainly includes two aspects: the material, including

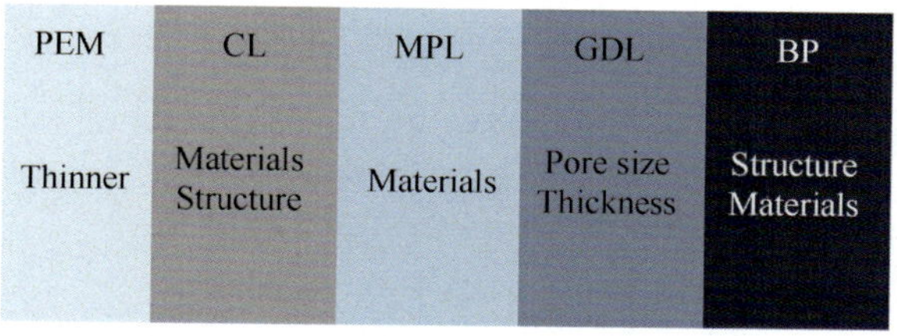

Figure 3.18 Improving subzero startup capability of fuel cells by optimizing material and structure design.

increasing the ionomer content and adding inorganic matter. The second is the structure, mainly the thickness of the CL, for example, by reducing the ratio of Pt/C. The second is the structure, mainly the thickness of the CL, for example, by reducing the ratio of Pt/C. By increasing the ionomer content in the CL, the water generated in the subzero startup process can be effectively absorbed, which will increase the storage space of water/ice to improve the subzero startup capacity of fuel cells [151]. By adding specific inorganic substances to the cathode CL of the fuel cell, the water absorption capacity can be enhanced to improve the fuel cell's subzero starting capacity.

Similarly, under the condition of the same platinum load, reducing the proportion of Pt/C will increase the thickness of the CL, thus increasing the water storage space of the CL so as to improve the subzero starting capacity of the fuel cell [153]. The microporous layer is a transition layer connecting the catalyst and gas diffusion layers. It can increase the water storage space in the cell during the subzero startup process to improve the fuel cell's subzero startup capacity. The temperature range at which the gas diffusion layer works during the subzero startup process is −10°C to 0°C [154]. There are two main optimization ideas for the gas diffusion layer: aperture and thickness. At −5°C, smaller pore size is conducive to subzero startup [148]. The gas diffusion layer with a small thickness is conducive to accelerating ice melting and the subzero startup of fuel cells [155]. For the bipolar plate, the optimization methods include hydrophilic and hydrophobic treatment, the selection of bipolar plate materials, the thickness of the bipolar plate, and the design of the flow field. The surface hydrophobicity of bipolar plate has a great influence on the aggregation and freezing of liquid water. Different materials have a different heat capacities and thermal conductivity, so they have different effects on heat during subzero startup, which also has a great impact on subzero startup.

The thinner bipolar plate will absorb less heat generated by electrochemical reaction during subzero startup, which is conducive to the rise of internal temperature of fuel cell and the improvement of subzero startup capacity. Through the design of bipolar plate flow field, water transport can be promoted, to improve the subzero starting ability.

3.2.5 Development trend of fuel cell catalyst layer

According to the structure classification, the CL can be divided into particle stacking, cylindrical array, and fiber arrangement types.

The particle stacking CL is the mainstream CL used in commercial PEMFC, and the carrier of the CL is carbon with a NP structure. In the preparation process, the structure of NPs is greatly affected by the preparation conditions, the utilization rate of the catalyst is low, and the TPB is small. Secondly, the TPC is bent due to the particle stacking structure, and the material transmission resistance is large. Finally, due to the use of carbon carriers, carbon corrosion is easy to occur at high potential, and the durability of the CL still needs to be improved.

The columnar array CL can form an ordered electron channel, proton channel, or gas transmission channel by forming a columnar array catalyst with an ordered structure. Therefore, its outstanding advantages include (1) efficient three-phase transmission, (2) high Pt utilization, and (3) improved durability. However, the columnar array CL also has obvious disadvantages. Firstly, the synthesis of ordered carriers is difficult. Transferring the prepared CL to the PEM may destroy the structure of the ordered CL. Therefore, the mass production of the cylindrical array CL is challenging.

Compared with the particle-stacked CL, the fiber-arranged CL increases the utilization of the catalyst through nano fibrosis in the structure. The research results in the literature show that it has a higher three-phase interface than the particle-stacking CL. And its proton and electron channels are orderly. At the same time, the research results in the literature show that it has higher durability than the particle-stacked CL. Compared with the columnar array CL, the fiber-arranged CL has outstanding characteristics of mass production. With the rapid development of the electrospinning industry, various electrospinning fiber batch manufacturing equipment—no-nozzle batch manufacturing equipment and multinozzle equipment have emerged, which can realize large-scale and rapid production. However, the in-plane arrangement of nanofibers prepared by electrospinning generally has disorder characteristics, and the material transport needs to be further improved. Therefore, controlling the preparation method of fiber-arranged CL to realize the ordered structure of the fiber is an important problem to be solved in its development.

Overall, many theoretical and research foundations exist for the fuel cell CL. Still, the important reason limiting its commercialization is how to design an efficient and stable fuel cell cathode catalyst. It cannot effectively reduce the use of precious metal platinum.

At present, the common catalyst of PEMFCs is platinum (PT) carbon catalyst. However, there are still some problems in the practical application of this catalyst, such as: (1) in the cathodic ORR, the binding ability between Pt and oxygen-containing intermediates is too strong, the oxygen-containing intermediates are not easy to desorb from the Pt surface, resulting in slow reaction kinetics, which reduces the performance of the battery [156]. (2) PT is a precious metal with scarce reserves in the earth's crust and a high price, resulting in high battery costs [157]. (3) Under the harsh working conditions of PEMFCs, the catalyst is prone to metal dissolution, particle agglomeration, and falling off, resulting in the rapid attenuation of catalyst activity. Therefore, developing catalysts with low cost, high activity, and long life is the key to promoting the commercialization of PEMFCs.

In addition, a series of key problems in membrane electrodes caused by low Pt need scientific understanding and rational response. First, low Pt brings more obvious cathode oxygen mass transfer resistance, significantly reducing the battery performance in the high-current region. The cathode oxygen mass transfer resistance mainly comes from the mass transfer in the catalytic layer. The reduction of Pt loading will reduce the active catalytic sites and greatly improve the local mass transfer resistance. Therefore, it is important to clarify the local mass transfer mechanism and improve it [158]. The pore structure in the catalytic layer also affects the bulk mass transfer. The experiments show that the pore utilization rate in the porous catalytic layer commonly used at present is too low. If the appropriate microstructure of the catalytic layer can be designed and manufactured, the cathode oxygen mass transfer can also be significantly improved to compensate for the concentration polarization. The mass transfer resistance of the catalytic layer caused by low Pt will also accelerate the attenuation of the catalyst. The consequences are multifaceted and need to be dealt with together. Future research can further explore the local and bulk oxygen mass transfer mechanism in the catalytic layer and optimize the structure according to the performance and service life of the battery.

Because of some shortcomings of the platinum catalyst itself, Pt is mainly alloyed with other transition metals (m), such as Fe, Co, Ni, and other late transition metals, to improve the ORR activity of Pt catalyst [159]. These methods can improve the stability and activity by doping or atomic arrangement. There have been some studies on using nonnoble metal catalysts, such as transition metal chelates, transition metal oxygen/nitrogen compounds, transition metal nitrogen carbon compounds, and nonmetallic catalysts. Among them, transition metal nitrogen carbon (M-N-C, M = Fe, CO) materials are considered the most promising nonnoble metal catalyst materials because of their low price and high activity advantages. However, although much progress has been made in the research on nonnoble metal catalysts, most of the research still focuses on the ORR activity of the catalyst itself, and the electrochemical test characterization is mostly based on the RDE test in an acidic solution. There are few reports on the research on the membrane electrode of nonnoble metal catalysts.

On the other hand, fast decay is also difficult when applying nonnoble metal catalysts. On the one hand, there is a great controversy over the reasons for the rapid decay of the nonnoble metal catalyst. On the other hand, many processes are involved in the membrane electrode CL, including the transport of oxygen in the pore, the conduction of the proton on the electrolyte ionic polymer, and the reaction on the surface of the active site. Many factors may cause the performance degradation of membrane electrodes. Therefore, it is necessary to explore further the reasons for attenuating membrane electrode performance of nonnoble metal catalysts.

References

[1] Sopian K, Wan, Daud WR. Challenges and future developments in proton exchange membrane fuel cells. Renewable Energy 2006;31(5):719–27.

[2] 63kW Fuel Cell system https://www.refire.com/products/prisma-6/ 2023, [accessed 28.07.23].

[3] Gang L, Zhang H, Hu J, et al. Studies of performance degradation of a high temperature PEMFC based on H_3PO_4-doped PBI. Journal of Power Sources 2006;162(1):547–52.

[4] Kumar GG, Kim AR, Nahm KS, et al. Nafion membranes modified with silica sulfuric acid for the elevated temperature and lower humidity operation of PEMFC—ScienceDirect. International Journal of Hydrogen Energy 2009;34(24):9788–94.

[5] Zhang L, Chae SR, Hendren Z, et al. Recent advances in proton exchange membranes for fuel cell applications. Engineering 2012;204–206:87–97.
[6] Olabi AG, Wilberforce T, Abdelkareem MA. Fuel cell application in the automotive industry and future perspective. Energy 2021;214:118955.
[7] Wang Y, Diaz DFR, Chen KS, et al. Materials, technological status, and fundamentals of PEM fuel cells—a review. Materials Today 2020;32:178–203.
[8] Liu ZX, Qian W, Guo JW, et al. Materials for proton exchange membrane fuel cells. Progress in Chemistry 2011;23(Z1):487–500.
[9] Xiang Z, Zhao X, Junjie GE, et al. Effect of sulfonation degree on performance of proton exchange membranes for direct methanol fuel cells. Chemical Research in Chinese Universities 2016;2:5.
[10] Pourmahdian S, Salarizadeh P, Javanbakht M, et al. Fabrication and physicochemical properties of iron titanate nanoparticles based sulfonated poly (ether ether ketone) membrane for proton exchange membrane fuel cell application. Solid State Ionics 2015;281(15):12–20.
[11] Coovee. <https://www.coovee.com/detail44728063.html>; 2017 [accessed 16.01.22].
[12] Wong CY, Wong WY, Ramya K, et al. Additives in proton exchange membranes for low- and high-temperature fuel cell applications: a review. International Journal of Hydrogen Energy 2019;44(12):6116–35.
[13] Parnian MJ, Rowshanzamir S, Prasad AK, et al. High durability sulfonated poly (ether ether ketone)-ceria nanocomposite membranes for proton exchange membrane fuel cell applications. Journal of Membrane Science 2018;556:12–22.
[14] Yang S, Kim D. Antioxidant proton conductive toughening agent for the hydrocarbon based proton exchange polymer membrane for enhanced cell performance and durability in fuel cell. Journal of Power Sources 2018;393:11–18.
[15] Kim DH, Park IK, Lee DH. Fluorinated sulfonated poly (arylene ether)s bearing semi-crystalline structures for highly conducting and stable proton exchange membranes. International Journal of Hydrogen Energy 2020;45(43):23469–79.
[16] Prykhodko Y, Fatyeyeva K, Hespel L, et al. Progress in hybrid composite Nafion®-based membranes for proton exchange fuel cell application. Chemical Engineering Journal 2021;409:127329.
[17] Li Y, Li Z, Li Y, et al. Preparation and electrochemical characterization of organic–inorganic hybrid poly(vinylidene fluoride)-sio2 cation-exchange membranes by the sol-gel method using 3-mercapto-propyl-triethoxyl-silane. Materials 2019;12(19):3265.
[18] Danyliv O, Martinelli A. Nafion/protic ionic liquid blends: nanoscale organization and transport properties. The Journal of Physical Chemistry C 2019;123(23):14813–24.
[19] Lin CC, Lien WF, Wang YZ, et al. Preparation and performance of sulfonated polyimide/Nafion multilayer membrane for proton exchange membrane fuel cell. Journal of Power Sources 2012;200:1–7.
[20] Shao ZG, Wang X, Hsing IM. Composite Nafion/polyvinyl alcohol membranes for the direct methanol fuel cell. Journal of Membrane Science 2002;210(1):147–53.
[21] Sun X, Simonsen SC, Norby T, et al. Composite membranes for high temperature pem fuel cells and electrolysers: a critical review. Membranes 2019;9(7):83.
[22] Zhang H, Hu Q, Zheng X, et al. Incorporating phosphoric acid-functionalized polydopamine into Nafion polymer by *in situ* sol-gel method for enhanced proton conductivity. Journal of Membrane Science 2019;570–571:236–44.
[23] Mauritz KA, Moore RB. State of understanding of Nafion. Chemical Reviews 2004;104(10):4535–86.
[24] Parker SF, Shah S. Characterisation of hydration water in Nafion membrane. RSC Advances 2021;11(16):9381–5.

[25] Li Y, Wu H, Yin Y, et al. Fabrication of Nafion/zwitterion-functionalized covalent organic framework composite membranes with improved proton conductivity. Journal of Membrane Science 2018;568:1–9.

[26] Cozzi D, De Bonis C, D'epifanio A, et al. Organically functionalized titanium oxide/Nafion composite proton exchange membranes for fuel cells applications. Journal of Power Sources 2014;248:1127–32.

[27] Sasikala S, Selvaganesh SV, Sahu AK, et al. Block co-polymer templated mesoporous carbon–Nafion hybrid membranes for polymer electrolyte fuel cells under reduced relative humidity. Journal of Membrane Science 2016;499:503–14.

[28] Patel HA, Mansor N, Gadipelli S, et al. Superacidity in Nafion/MOF hybrid membranes retains water at low humidity to enhance proton conduction for fuel cells. ACS Applied Materials & Interfaces 2016;8(45):30687–91.

[29] Xu F, Mu S, Pan M. Recycling of membrane electrode assembly of PEMFC by acid processing. International Journal of Hydrogen Energy 2010;35(7):2976–9.

[30] Marketresearch. <https://www.marketresearch.com/QYResearch-Group-v3531/Global-Membrane-Electrode-Assemblies-MEA-32543235/>; Octo, 2022 [accessed 16.11.22].

[31] Kim J, Kim K, Han J, et al. End-group cross-linked membranes based on highly sulfonated poly(arylene ether sulfone) with vinyl functionalized graphene oxide as a cross-linker and a filler for proton exchange membrane fuel cell application. Journal of Polymer Science 2020;58(24):3456–66.

[32] Byun GH, Kim JA, Kim NY, et al. Molecular engineering of hydrocarbon membrane to substitute perfluorinated sulfonic acid membrane for proton exchange membrane fuel cell operation. Materials Today Energy 2020;17:100483.

[33] Oh K, Ketpang K, Kim H, et al. Synthesis of sulfonated poly(arylene ether ketone) block copolymers for proton exchange membrane fuel cells. Journal of Membrane Science 2016;507:135–42.

[34] Han J, Kim K, Kim J, et al. Cross-linked highly sulfonated poly(arylene ether sulfone) membranes prepared by in-situ casting and thiol-ene click reaction for fuel cell application. Journal of Membrane Science 2019;579:70–8.

[35] Park JE, Kim J, Han J, et al. High-performance proton-exchange membrane water electrolysis using a sulfonated poly(arylene ether sulfone) membrane and ionomer. Journal of Membrane Science 2021;620:118871.

[36] Sun H, Tang B, Wu P. Two-dimensional zeolitic imidazolate framework/carbon nanotube hybrid networks modified proton exchange membranes for improving transport properties. ACS Applied Materials & Interfaces 2017;9(40):35075–85.

[37] Erce Ş, Erdener H, Akay RG, et al. Effects of sulfonated polyether-etherketone (SPEEK) and composite membranes on the proton exchange membrane fuel cell (PEMFC) performance. International Journal of Hydrogen Energy 2009;34(10):4645–52.

[38] Asano N, Aoki M, Suzuki S, et al. Aliphatic/aromatic polyimide ionomers as a proton conductive membrane for fuel cell applications. Journal of the American Chemical Society 2006;128(5):1762–9.

[39] Zhang H, Shen PK. Advances in the high performance polymer electrolyte membranes for fuel cells. Chemical Society Reviews 2012;41(6):2382–94.

[40] Ko T, Kim K, Kim SK, et al. Organic/inorganic composite membranes comprising of sulfonated Poly(arylene ether sulfone) and core–shell silica particles having acidic and basic polymer shells. Polymer 2015;71:70–81.

[41] Kim K, Jung BK, Ko T, et al. Comb-shaped polysulfones containing sulfonated polytriazole side chains for proton exchange membranes. Journal of Membrane Science 2018;554:232–43.

[42] Robertson GP, Mikhailenko SD, Wang K, et al. Casting solvent interactions with sulfonated poly(ether ether ketone) during proton exchange membrane fabrication. Journal of Membrane Science 2003;219(1):113–21.

[43] Lee HS, Lane O, Mcgrath JE. Development of multiblock copolymers with novel hydroquinone-based hydrophilic blocks for proton exchange membrane (PEM) applications. Journal of Power Sources 2010;195(7):1772–8.
[44] Liu D, Xie Y, Cui N, et al. Structure and properties of sulfonated poly(arylene ether)s with densely sulfonated segments containing mono-, di- and tri-tetraphenylmethane as proton exchange membrane. Journal of Membrane Science 2021;620:118856.
[45] Lee KH, Chu JY, Kim AR, et al. Enhanced performance of a sulfonated poly(arylene ether ketone) block copolymer bearing pendant sulfonic acid groups for polymer electrolyte membrane fuel cells operating at 80% relative humidity. ACS Applied Materials & Interfaces 2018;10(24):20835–44.
[46] Kim M, Ko H, Nam SY, et al. Study on control of polymeric architecture of sulfonated hydrocarbon-based polymers for high-performance polymer electrolyte membranes in fuel cell applications. Polymers 2021;13(20):3520.
[47] Higashihara T, Matsumoto K, Ueda M. Sulfonated aromatic hydrocarbon polymers as proton exchange membranes for fuel cells. Polymer 2009;50(23):5341–57.
[48] Tsang EMW, Zhang Z, Shi Z, et al. Considerations of macromolecular structure in the design of proton conducting polymer membranes: graft versus diblock polyelectrolytes. Journal of the American Chemical Society 2007;129 (49):15106–7.
[49] Dimitrov I, Takamuku S, Jankova K, et al. Polysulfone functionalized with phosphonated poly(pentafluorostyrene) grafts for potential fuel cell applications. Macromolecular Rapid Communications 2012;33(16):1368–74.
[50] Dimitrov I, Takamuku S, Jankova K, et al. Proton conducting graft copolymers with tunable length and density of phosphonated side chains for fuel cell membranes. Journal of Membrane Science 2014;450:362–8.
[51] Norsten TB, Guiver MD, Murphy J, et al. Highly fluorinated comb-shaped copolymers as proton exchange membranes (pems): improving pem properties through rational design. Advanced Functional Materials 2006;16(14):1814–22.
[52] Xing P, Robertson GP, Guiver MD, et al. Sulfonated poly(aryl ether ketone)s containing the hexafluoroisopropylidene diphenyl moiety prepared by direct copolymerization, as proton exchange membranes for fuel cell application. Macromolecules 2004;37(21):7960–7.
[53] Qiao X, Wang X, Liu S, et al. The alkaline stability and fuel cell performance of poly(N-spirocyclic quaternary ammonium) ionenes as anion exchange membrane. Journal of Membrane Science 2021;630:119325.
[54] Park CH, Lee CH, Guiver MD, et al. Sulfonated hydrocarbon membranes for medium-temperature and low-humidity proton exchange membrane fuel cells (PEMFCs). Progress in Polymer Science 2011;36(11):1443–98.
[55] Liu YL. Developments of highly proton-conductive sulfonated polymers for proton exchange membrane fuel cells. Polymer Chemistry 2012;3(6):1373–83.
[56] Wang C, Li N, Shin DW, et al. Fluorene-based poly(arylene ether sulfone)s containing clustered flexible pendant sulfonic acids as proton exchange membranes. Macromolecules 2011;44(18):7296–306.
[57] Han X, Pang J, Liu D, et al. Novel branched sulfonated poly(arylene ether)s based on carbazole derivative for proton exchange membrane. International Journal of Hydrogen Energy 2020;45(7):4644–52.
[58] Long Z, Miyake J, Miyatake K. Proton exchange membranes containing densely sulfonated quinquephenylene groups for high performance and durable fuel cells. Journal of Materials Chemistry A 2020;8(24):12134–40.
[59] Li N, Hwang DS, Lee SY, et al. Densely sulfophenylated segmented copoly(arylene ether sulfone) proton exchange membranes. Macromolecules 2011;44(12):4901–10.

[60] Matsumura S, Hlil AR, Lepiller C, et al. Ionomers for proton exchange membrane fuel cells with sulfonic acid groups on the end groups: novel linear aromatic poly (sulfide − ketone)s. Macromolecules 2008;41(2):277−80.
[61] Li G, Kujawski W, Rynkowska E. Advancements in proton exchange membranes for high-performance high-temperature proton exchange membrane fuel cells (HT-PEMFC). Reviews in Chemical Engineering 2022;38(3):327−46.
[62] Amjadi M, Rowshanzamir S, Peighambardoust SJ, et al. Investigation of physical properties and cell performance of Nafion/TiO_2 nanocomposite membranes for high temperature PEM fuel cells. International Journal of Hydrogen Energy 2010;35 (17):9252−60.
[63] Saccà A, Carbone A, Gatto I, et al. Composites nafion-titania membranes for polymer electrolyte fuel cell (PEFC) applications at low relative humidity levels: chemical physical properties and electrochemical performance. Polymer Testing 2016;56:10−18.
[64] Wang H, Tang C, Zhuang X, et al. Novel structure design of composite proton exchange membranes with continuous and through-membrane proton-conducting channels. Journal of Power Sources 2017;365:92−7.
[65] Xu G, Wei Z, Li S, et al. In-situ sulfonation of targeted silica-filled Nafion for high-temperature PEM fuel cell application. International Journal of Hydrogen Energy 2019;44(56):29711−16.
[66] Saccà A, Gatto I, Carbone A, et al. ZrO_2−Nafion composite membranes for polymer electrolyte fuel cells (PEFCs) at intermediate temperature. Journal of Power Sources 2006;163(1):47−51.
[67] Ibrahim A, Hossain O, Chaggar J, et al. GO-nafion composite membrane development for enabling intermediate temperature operation of polymer electrolyte fuel cell. International Journal of Hydrogen Energy 2020;45(8):5526−34.
[68] Ceynowa J. Electron microscopy investigation of ion exchange membranes. Polymer 1978;19(1):73−6.
[69] Xu G, Wu Z, Wei Z, et al. Non-destructive fabrication of Nafion/silica composite membrane via swelling-filling modification strategy for high temperature and low humidity PEM fuel cell. Renewable Energy 2020;153:935−9.
[70] Oh K, Kwon O, Son B, et al. Nafion-sulfonated silica composite membrane for proton exchange membrane fuel cells under operating low humidity condition. Journal of Membrane Science 2019;583:103−9.
[71] Haque MA, Sulong AB, Loh KS, et al. Acid doped polybenzimidazoles based membrane electrode assembly for high temperature proton exchange membrane fuel cell: a review. International Journal of Hydrogen Energy 2017;42(14):9156−79.
[72] Araya SS, Zhou F, Liso V, et al. A comprehensive review of PBI-based high temperature PEM fuel cells. International Journal of Hydrogen Energy 2016;41 (46):21310−44.
[73] Taherkhani Z, Abdollahi M, Sharif A. Proton conducting porous membranes based on poly(benzimidazole) and poly(acrylic acid) blends for high temperature proton exchange membranes. Solid State Ionics 2019;337:122−31.
[74] Hempelmann R, Karmonik C. Proton diffusion in proton conducting oxides. Phase Transitions 1996;58(1−3):175−84.
[75] Tai CC, Chen CL, Liu CW. Computer simulation to investigate proton transport and conductivity in perfluorosulfonate ionomeric membrane. International Journal of Hydrogen Energy 2017;42(7):3981−6.
[76] Eguizábal A, Lemus J, Pina MP. On the incorporation of protic ionic liquids imbibed in large pore zeolites to polybenzimidazole membranes for high temperature proton exchange membrane fuel cells. Journal of Power Sources 2013;222:483−92.
[77] Díaz M, Ortiz A, Ortiz I. Progress in the use of ionic liquids as electrolyte membranes in fuel cells. Journal of Membrane Science 2014;469:379−96.

[78] Özdemir Y, Özkan N, Devrim Y. Fabrication and characterization of cross-linked polybenzimidazole based membranes for high temperature pem fuel cells. Electrochimica Acta 2017;245:1−13.
[79] Bose S, Kuila T, Nguyen TXH, et al. Polymer membranes for high temperature proton exchange membrane fuel cell: recent advances and challenges. Progress in Polymer Science 2011;36(6):813−43.
[80] Pan J, Wu B, Wu L, et al. Proton exchange membrane from tetrazole-based poly (phthalazinone ether sulfone ketone) for high-temperature fuel cells. International Journal of Hydrogen Energy 2016;41(28):12337−46.
[81] Lin B, Yuan W, Xu F, et al. Protic ionic liquid/functionalized graphene oxide hybrid membranes for high temperature proton exchange membrane fuel cell applications. Applied Surface Science 2018;455:295−301.
[82] Moradi M, Moheb A, Javanbakht M, et al. Experimental study and modeling of proton conductivity of phosphoric acid doped PBI-Fe_2TiO_5 nanocomposite membranes for using in high temperature proton exchange membrane fuel cell (HT-PEMFC). International Journal of Hydrogen Energy 2016;41(4):2896−910.
[83] Bai H, Wang H, Zhang J, et al. Simultaneously enhancing ionic conduction and mechanical strength of poly(ether sulfones)-poly(vinyl pyrrolidone) membrane by introducing graphitic carbon nitride nanosheets for high temperature proton exchange membrane fuel cell application. Journal of Membrane Science 2018;558:26−33.
[84] Yoon J, Lee HJ, Stafford CM. Thermoplastic elastomers based on ionic liquid and poly(vinyl alcohol). Macromolecules 2011;44(7):2170−8.
[85] Liew CW, Ramesh S, Arof AK. Good prospect of ionic liquid based-poly(vinyl alcohol) polymer electrolytes for supercapacitors with excellent electrical, electrochemical and thermal properties. International Journal of Hydrogen Energy 2014;39 (6):2953−63.
[86] Liew CW, Ramesh S, Arof AK. A novel approach on ionic liquid-based poly(vinyl alcohol) proton conductive polymer electrolytes for fuel cell applications. International Journal of Hydrogen Energy 2014;39(6):2917−28.
[87] Pereiro AB, Araújo JMM, Martinho S, et al. Fluorinated ionic liquids: properties and applications. ACS Sustainable Chemistry & Engineering 2013;1(4):427−39.
[88] Rynkowska E, Fatyeyeva K, Kujawski W. Application of polymer-based membranes containing ionic liquids in membrane separation processes: a critical review. Reviews in Chemical Engineering 2018;34(3):341−63.
[89] Chen K, Wang Y, Yao J, et al. Equilibrium in protic ionic liquids: the degree of proton transfer and thermodynamic properties. The Journal of Physical Chemistry B 2018;122(1):309−15.
[90] Yaghini N, Gómez-González V, Varela LM, et al. Structural origin of proton mobility in a protic ionic liquid/imidazole mixture: insights from computational and experimental results. Physical Chemistry Chemical Physics 2016;18(33):23195−206.
[91] Jothi PR, Dharmalingam S. An efficient proton conducting electrolyte membrane for high temperature fuel cell in aqueous-free medium. Journal of Membrane Science 2014;450:389−96.
[92] Bao X, Zhang F, Liu Q. Sulfonated poly(2,5-benzimidazole) (ABPBI)/ MMT/ ionic liquids composite membranes for high temperature PEM applications. International Journal of Hydrogen Energy 2015;40(46):16767−74.
[93] Amjadi M, Rowshanzamir S, Peighambardoust SJ, et al. Preparation, characterization and cell performance of durable nafion/SiO_2 hybrid membrane for high-temperature polymeric fuel cells. Journal of Power Sources 2012;210:350−7.
[94] Costamagna P, Yang C, Bocarsly AB, et al. Nafion® 115/zirconium phosphate composite membranes for operation of PEMFCs above 100°C. Electrochimica acta 2002;47(7):1023−33.

[95] Yang C, Srinivasan S, Bocarsly AB, et al. A comparison of physical properties and fuel cell performance of Nafion and zirconium phosphate/Nafion composite membranes. Journal of Membrane Science 2004;237(1):145–61.
[96] Kopitzke RW, Linkous CA, Anderson HR, et al. Conductivity and water uptake of aromatic-based proton exchange membrane electrolytes. Journal of the Electrochemical Society 2000;147(5):1677–81.
[97] Guo Q, Pintauro PN, Hao T, et al. Sulfonated and crosslinked polyphosphazene-based proton-exchange membranes. Journal of Membrane Science 1999;154(2):175–81.
[98] Li L, Kang W, Zhuang X, et al. A comparative study of alumina fibers prepared by electro-blown spinning (EBS) and solution blowing spinning (SBS). Materials Letters 2015;160(DEC.1):533–6.
[99] Hu H.W. Preparation and properties of novel nanofiber composite proton exchange membrane. Master's thesis, Beijing: Beijing University of Chemical Technology, 2020.
[100] Zhuang X, Jia K, Cheng B, et al. Solution blowing of continuous carbon nanofiber yarn and its electrochemical performance for supercapacitors. Chemical Engineering Journal 2014;237:308–11.
[101] Bo Z, Zhuang XP, Cheng B, et al. Carbonaceous nanofiber-supported sulfonated poly(ether ether ketone) membranes for fuel cell applications. Materials Letters 2014;115:248–51.
[102] Zhao J, Wang ZY, Xu A, et al. Perfluorinated sulfonic acid ionomer/poly(N-vinylpyrrolidone) nanofiber membranes: electrospinning fabrication, water stability, and metal ion removal applications. Reactive and Functional Polymers 2011;71(11):1102–9.
[103] Si D, Zhang S, Huang J, et al. Electrochemical characterization of pre-conditioning process of electrospun nanofiber electrodes in polymer electrolyte fuel cells. Fuel Cells 2018;18(5):576–85.
[104] Nie M, Zhang LY, Li Q, et al. Research status of proton exchange membrane fuel cell. Surface Technology 2012;41(3):3.
[105] Ohma A, Mashio T, Sato K, et al. Analysis of proton exchange membrane fuel cell catalyst layers for reduction of platinum loading at Nissan. Electrochimica Acta 2011;56(28):10832–41.
[106] Cho YH, Park HS, Cho YH, et al. Effect of platinum amount in carbon supported platinum catalyst on performance of polymer electrolyte membrane fuel cell. Journal of Power Sources 2007;172(1):89–93.
[107] Litster S, Mclean S. PEM fuel cell electrodes. Science 2004;130(1–2):61–76.
[108] Jiang SF, Yi BL. Progress in ordered membrane electrodes. Journal of Electrochemistry 2016;22(3):6.
[109] Liu Y, Ding H, Si DC, et al. Review on preparation of catalyst layer for proton exchange membrane fuel cell by electrospinning. Journal of Electrochemistry 2018;24(6):16.
[110] Ticianelli E, Derouin C, Redondo A, et al. Methods to advance technology of proton exchange membrane fuel cells. Mass Transfer 1988;135(9):2209.
[111] Paganin V, Ticianelli E, Gonzalez E. Development and electrochemical studies of gas diffusion electrodes for polymer electrolyte fuel cells. Journal of Applied Electrochemistry 1996;26(3):297–304.
[112] O'Hayre R, Lee SJ, Cha SW, et al. A sharp peak in the performance of sputtered platinum fuel cells at ultra-low platinum loading. Journal of Power Sources 2002;109(2):483–93.
[113] Fofana D, Hamelin J, Bénard E. Modelling and experimental validation of high performance low platinum multilayer cathode for polymer electrolyte membrane fuel cells (PEMFCs). International Journal of Hydrogen Energy 2013;38(24):10050–62.

[114] Debe T. Nanostructured thin film electrocatalysts for PEM fuel cells-a tutorial on the fundamental characteristics and practical properties of NSTF catalysts. Journal of the Electrochemical Society 2012;45(2):47.
[115] Tian ZQ, Lim SH, Poh CK, et al. A highly order-structured membrane electrode assembly with vertically aligned carbon nanotubes for ultra-low Pt loading PEM fuel cells. Advanced Energy Materials 2011;1(6):1205–14.
[116] Park JH, Ju YW, Park SH, et al. Effects of electrospun polyacrylonitrile-based carbon nanofibers as catalyst support in PEMFC. Journal of Applied Electrochemistry 2009;39(8):1229–36.
[117] Zhang W, Brodt MW, Pintauro T. Nanofiber cathodes for low and high humidity hydrogen fuel cell operation. ECS Transactions 2011;41(1):891.
[118] Lu Y, Du S, Steinberger-Wilckens E. One-dimensional nanostructured electrocatalysts for polymer electrolyte membrane fuel cells-a review. Applied Catalysis B: Environmental 2016;292–314.
[119] Vliet D, Chao W, Tripkovic D, et al. Mesostructured thin films as electrocatalysts with tunable composition and surface morphology. Nature Materials 2012;11 (12):1051–8.
[120] Zhang S, Yuan XZ, Hin J, et al. A review of platinum-based catalyst layer degradation in proton exchange membrane fuel cells. Journal of Power Sources 2009;194 (2):588–600.
[121] Brodt M, Wycisk R, Pintauro S. Nanofiber electrodes with low platinum loading for high power hydrogen/air PEM fuel cells. Journal of the Electrochemical Society 2013;160(8):F744.
[122] Ding H. Study on preparation and performance of fiber arrangement catalyst layer for fuel cell. Master's thesis, Beijing: Beijing University of Chemical Technology, 2019.
[123] Debe MK, Schmoeckel A, Hendricks S, et al. Durability aspects of nanostructured thin film catalysts for PEM fuel cells. ECS Transactions 2006;1(8):51.
[124] Park YC, Tokiwa H, Kakinuma K, et al. Effects of carbon supports on Pt distribution, ionomer coverage and cathode performance for polymer electrolyte fuel cells. Journal of Power Sources 2016;315:179–91.
[125] Hong S, Hou M, Xiao Y, et al. Investigation of a high-performance nanofiber cathode with ultralow platinum loading for proton exchange membrane fuel cells. Energy Technology 2017;5(8):1457–63.
[126] Hong S, Hou M, Zeng Y, et al. High-performance low-platinum electrode for proton exchange membrane fuel cells: pulse electrodeposition of Pt on Pd/C nanofiber mat. ChemElectroChem 2017;4(5):1007–10.
[127] Hong S, Hou M, Zhang H, et al. A high-performance PEM fuel cell with ultralow platinum electrode via electrospinning and underpotential deposition. Electrochimica Acta 2017;245:403–9.
[128] Banham D, Ye S. Current status and future development of catalyst materials and catalyst layers for proton exchange membrane fuel cells: an industrial perspective. ACS Energy Letters 2017;2(3):629–38.
[129] Yu X, Ye S. Recent advances in activity and durability enhancement of Pt/C catalytic cathode in PEMFC: part II: degradation mechanism and durability enhancement of carbon supported platinum catalyst. Journal of Power Sources 2007;172(1):145–54.
[130] Banham D, Ye S, Pei K, et al. A review of the stability and durability of non-precious metal catalysts for the oxygen reduction reaction in proton exchange membrane fuel cells. Journal of Power Sources 2015;285:334–48.
[131] Mukerjee S, Srinivasan S, Soriaga MP, et al. Role of structural and electronic properties of Pt and Pt alloys on electrocatalysis of oxygen reduction: an *in situ* XANES and EXAFS investigation. Journal of the Electrochemical Society 1995;142(5):1409.

[132] Prachayawarakorn S. Drying technologies for foods: fundamentals and applications. Drying Technology 2019;37(6):801.
[133] Adzic RR. Platinum monolayer electrocatalysts: tunable activity, stability, and self-healing properties. Electrocatalysis 2012;3(3):163–9.
[134] Stamenkovic VR, Fowler B, Mun BS, et al. Improved oxygen reduction activity on $Pt_3Ni(111)$ via increased surface site availability. Science 2007;315(5811):493–7.
[135] Choi SI, Xie S, Shao M, et al. Synthesis and characterization of 9 nm Pt-Ni octahedra with a record high activity of 3.3 A/mg Pt for the oxygen reduction reaction. Nano Letters 2013;13(7):3420–5.
[136] Park J, Wang H, Vara M, et al. Platinum cubic nanoframes with enhanced catalytic activity and durability toward oxygen reduction. ChemSusChem 2016;9 (19):2855–61.
[137] Zhu J, Xiao M, Li K, et al. Active Pt_3Ni (111) Surface of Pt_3Ni icosahedron for oxygen reduction. ACS Applied Materials & Interfaces 2016;8(44):30066–71.
[138] Debe MK. Electrocatalyst approaches and challenges for automotive fuel cells. Nature 2012;486(7401):43–51.
[139] Yu J.R., Wang Y., Hu Z.M., et al. A preparation method of composite nanofiber non-woven fabric supported with catalyst, CN Patent, 201711138965.7, 2018.
[140] Nabil Y, Cavaliere S, Harkness I, et al. Novel niobium carbide/carbon porous nanotube electrocatalyst supports for proton exchange membrane fuel cell cathodes. Journal of Power Sources 2017;363:20–6.
[141] Brodt M, Han T, Dale N, et al. Fabrication, in-situ performance, and durability of nanofiber fuel cell electrodes. Journal of the Electrochemical Society 2014;162(1): F84.
[142] Reneker DH, Yarin AL, Fong H, et al. Bending instability of electrically charged liquid jets of polymer solutions in electrospinning. Journal of Applied Physics 2000;87(9):4531–47.
[143] Chan S, Jankovic J, Susac D, et al. Electrospun carbon nanofiber catalyst layers for polymer electrolyte membrane fuel cells: structure and performance. Journal of Power Sources 2018;392:239–50.
[144] Chan S, Jankovic J, Susac D, et al. Electrospun carbon nanofiber catalyst layers for polymer electrolyte membrane fuel cells: fabrication and optimization. Journal of Materials Science 2018;53(16):11633–47.
[145] Alink R, Gerteisen D, Oszcipok S. Degradation effects in polymer electrolyte membrane fuel cell stacks by sub-zero operation-An *in situ* and *ex situ* analysis. Journal of Power Sources 2008;182(1):175–87.
[146] Mehta V, Cooper S. Review and analysis of PEM fuel cell design and manufacturing. Journal of Power Sources 2003;114(1):32–53.
[147] Mukundan R, Kim YS, Garzon F, et al. Freeze/thaw effects in PEM fuel cells. ECS Transactions 2006;1(8):403.
[148] Hwang GS, Kim H, Lujan R, et al. Phase-change-related degradation of catalyst layers in proton-exchange-membrane fuel cells. Electrochimica acta 2013;95:29–37.
[149] Lee C, Merida S. Gas diffusion layer durability under steady-state and freezing conditions. Journal of Power Sources 2007;164(1):141–53.
[150] Oszcipok M, Riemann D, Kronenwett U, et al. Statistic analysis of operational influences on the cold start behaviour of PEM fuel cells. Journal of Power Sources 2005;145(2):407–15.
[151] Luo Y, Jiao E, Science C. Cold start of proton exchange membrane fuel cell. Progress in Energy and Combustion Science 2018;64:29–61.
[152] Nicotera I, Coppola L, Rossi CO, et al. NMR investigation of the dynamics of confined water in Nafion-based electrolyte membranes at subfreezing temperatures. The Journal of Physical Chemistry B 2009;113(42):13935–41.

[153] Ko J, Ju E. Effects of cathode catalyst layer design parameters on cold start behavior of polymer electrolyte fuel cells (PEFCs). International Journal of Hydrogen Energy 2013;38(1):682–91.
[154] Hirakata S, Mochizuki T, Uchida M, et al. Investigation of the effect of pore diameter of gas diffusion layers on cold start behavior and cell performance of polymer electrolyte membrane fuel cells. Electrochimica acta 2013;108:304–12.
[155] Dursch T, Trigub G, Liu J, et al. Non-isothermal melting of ice in the gas-diffusion layer of a proton-exchange-membrane fuel cell. International Journal of Heat and Mass Transfer 2013;67:896–901.
[156] Ma Z, Cano ZP, Yu A, et al. Enhancing oxygen reduction activity of pt-based electrocatalysts: from theoretical mechanisms to practical methods. Angewandte Chemie International Edition 2020;59(42):18334–48.
[157] Ma YB. Progress in morphology control and stability of platinum alloy nanocrystalline catalysts for fuel cells. Chinese Science Bulletin 2017;62(25):2905–18.
[158] Cheng XJ, Shen SY, Wang C, et al. Analysis and prospect of material transport in proton exchange membrane fuel cells during ultra-low platinum conversion. Chinese Science Bulletin 2016;1–16.
[159] Liang JS, Liu X. Principles, strategies and methods for improving the stability of platinum based catalysts for fuel cells. Journal of Physical Chemistry 2020;37(9) 2010072–2010070.

CHAPTER 4

High performance fiber materials and applications

Contents

New Polymeric Products
DOI: https://doi.org/10.1016/B978-0-443-19407-8.00005-1

With the continuous development of science and technology, the cutting-edge technology industry has put forward higher requirements for fiber materials. Fibers with high strength, modulus, and high-temperature resistance have become new materials that urgently need to be developed. High-performance fiber is a unique fiber that has been overgrown in fiber materials recently, and its application is extensive [1].

High-performance fibers generally refer to fibers with a strength greater than 17.6 cN/dtex and an elastic modulus greater than 440 cN/dtex. High-performance fibers are a new generation of synthetic fibers with high strength, modulus, and high-temperature resistance developed by fiber science and engineering circles. High-performance fibers have unique properties that ordinary fibers do not have and are mainly used in various fields of military and high-tech industries [2].

4.1 Carbon fiber

Carbon fiber (CF) is a fiber whose chemical composition contains carbon more than 90% of its total mass. The industrial production of polyacrylonitrile (PAN) fibers as CF precursors began in the 1960s. Currently, most CF precursors used in industrial production are PAN fibers. CF can be obtained by PAN fiber after preoxidation, carbonization or further graphitization and posttreatment. High-performance CF is a new material with higher strength compared with steel and lower specific gravity compared with aluminum, and its specific gravity is between 115 and 210 g/cm^3 [3].

CF has a graphite-like chemical structure; its work temperature reaches 2000°C. CF has high tensile strength. Its tensile strength is twice that of steel and six times that of aluminum. The modulus of CF is seven times that of steel and eight times that of aluminum. CF has stable dimensions, good rigidity, self-lubricity, low friction coefficient, and good wear resistance. CF has outstanding heat resistance when not exposed to air or in an oxidizing atmosphere. The thermal expansion coefficient of CF is small, the thermal conductivity is high, and it decreases with the temperature increase. At 1500°C, its thermal conductivity is 15%—30% of the average temperature [3].

CF is composed of carbon elements, does not burn, has good chemical stability, and has good chemical resistance. It can be combined with resin, rubber, ceramics, glass, and metal to make various structural materials and thermal insulation materials, which are applied in high-tech fields such as rockets, satellites, missiles, airplanes, automotive, machinery, chemical, sports, medical, and other industries. The application is also expanding day by day [3].

4.1.1 Development status of carbon fiber

In 2002, the world's production capacity of high-performance CF was 3.11 million tons, and small-tow CF was about 2.13 million tons, accounting for 3/4. Large-tow CF is about 1.18 million tons, accounting for 1/4. The production capacity of high-performance small-tow CF of Japan's Toray, Toho, and Mitsubishi (MRC) is 11.75 million tons, accounting for 3/4 of the world's total production capacity of small-tow CF. The company controls the world's high-performance small-tow CF production. The world's total PAN-based large-tow CF production capacity is 8145 t. PORTAFIL, SGL, ZOLTEK, and ALDILA companies monopolize the production of PAN-based large-tow CFs worldwide [4].

T2300 CF is the main variety of small-tow CF. So far, T2300 CF has accumulated more than 1.15 million tons and has been widely used in the aerospace industry. But from the development point of view, T2300 will be gradually replaced by T2700S because the tensile strength of T2700S reaches 4900 MPa, the performance is about 40% higher than T2300, and the price only increased by less than 10% [5].

Large-tow CFs are more significant than 48, while 1K, 3K, 6K, 12K, and 24K are called small-tow CFs. Large-tow CF has many advantages compared with small-tow. Commercialized civilian PAN filaments are the primary filaments for preparing large-tow CFs with low prices and high performance. Therefore, most manufacturers now value large-tow CFs, and the production technology has made significant progress, becoming the most promising variety of CF series. Its output will soon exceed that of aerospace-grade small-tow CF [6].

The new technology of CF preparation is the research focus of countries worldwide, especially the low-cost CF preparation technology. The state-of-the-art technologies for preparing CFs can be summarized in three aspects: [6]

1) Research and development of cheap raw yarn. In the cost of high-performance CF, the proportion of the raw yarn is 40%–60%, so the research and development of cheap raw yarn are essential in reducing the cost of CF. All countries try to reduce the cost of raw silk from two aspects. One is to try to use materials other than PAN as the precursor of high-performance CF, and the other is to improve the process technology to achieve the purpose of reducing costs;
2) Development of new preoxidation technology. The current new idea in this area is to use plasma technology;
3) Development of new carbonization and graphitization technologies, such as microwave technology, has achieved good results.

4.1.2 Application of carbon fiber

As shown in Table 4.1 [7], the world's high-performance CF was first applied in the aerospace industry, then promoted and applied to sports and leisure products. In recent years, it has been used in the industry.

In addition, many new applications are being developed. For example, Hyunkee Hong et al. [8] proposed a new method for manufacturing carbon fiber–reinforced plastics (CFRP) to improve its thermal conductivity, conductivity, and mechanical properties. PAN and pitch-based CFs are used as suture materials to provide conductive paths through the CFRP body in the through-plane direction. Coaxial fibrous scaffolds integrating with CF prepared by Meng et al. [9] can promote cardiac tissue regeneration postmyocardial infarction. CF and coaxial fibrous mesh were integrated, which combined the high modulus and excellent electrical conductivity of CFs and the fibrous and porous structures of the electrospun fiber. The scaffold is manufactured by simply adjusting the coaxial electrospun fiber and CF through a freeze–drying procedure. This scaffold can promote myocardial cell maturation and

Table 4.1 Applications of high-performance carbon fiber [7].

Application	Aerospace industry	Ports and leisure products	Industrial applications
Detailed examples	Commonly used in aircraft, rockets and spacecraft parts, such as speed plates and brakes for military airplanes and large civil airliners, thermal shields for optical instruments in the control cabin of Apollo spacecraft, and internal combustion engine pistons, etc.	Fishing rods, golf clubs, tennis rackets, badminton rackets, sleds, sailing boats, racing cars, high vaulting poles, airplane models, etc.	Load-bearing components include leaf springs, coil springs, and drive shafts of cars and trucks.
Advantage	The mechanical performance is improved, the quality level increases, and the application effect is good.	It can significantly increase the strength and rigidity of the product.	The weight loss effect is obvious.

angiogenesis, establish cross talk between the myocardial and vascular endothelial cells, and promote myocardial regeneration and functional recovery.

4.1.3 Preparation of carbon fiber

In principle, any polymer with a carbon skeleton can be used as the precursor of CF. According to existing literature, there are many precursors used to prepare CFs, such as PAN, viscose, asphalt, polyethylene oxide (PEO), and polyvinyl alcohol (PVA) [10]. PAN has become the essential precursor due to its high spinnability and carbon content, and its output accounts for more than 90% of the world's total CF. However, PAN is more expensive and has limited reserves, and the solvent used has a certain degree of toxicity, which limits its future application in the CF field [11]. Therefore, choosing better raw materials to prepare CF has attracted wide attention from researchers. Lignin is the second largest renewable resource after cellulose and is the only biomass polymer containing many aromatic rings. The carbon content is as high as 60%. It is an ideal raw material for preparing CF.

a. Melt spinning

The melt-spinning process is simple, efficient, and has a low production cost. It is currently a common method for preparing general-purpose lignin-based CFs. Wang et al. [10] used lignin/polylactic acid (PLA) blend materials to prepare a series of continuously wound columnar lignin/PLA-based CFs through melt spinning, thermal stabilization, and carbonization processes. As shown in Fig. 4.1, the mechanical properties of lignin/PLA-based CF were tested by a fiber tensile testing machine, and it was found that the tensile modulus of

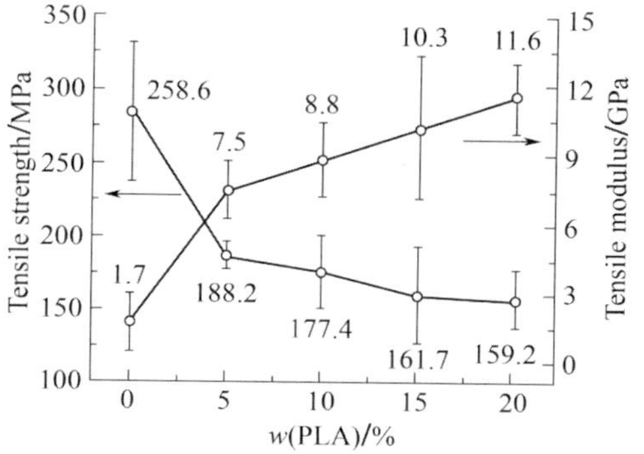

Figure 4.1 Tensile strength and tensile modulus as a function of PLA contents for lignin/PLA-based carbon fiber [12].

lignin/PLA (80/20)-based CF was as high as 11.6 GPa, which was significantly higher than that of pure CF. However, with lignin (1.7 GPa), due to the volatilization of PLA during the preoxidation and carbonization process, voids appear on the surface and cross section of the fiber, and the tensile strength of CF is reduced.

b. Wet spinning

To improve the spinnability of lignin, researchers have used excellent thermoplastic polymers to blend with lignin. Jiang et al. [13] used wet spinning (the general process of wet spinning is shown in Fig. 4.2) [14] to prepare CF with wheat straw lignin and PAN, but the CF prepared by this method has a large diameter. Tensile strength is only 300–500 MPa, the tensile modulus is less than 100 GPa, and the mechanical properties need further improvement. To improve this defect, Ouyang [15] and Xia [16] used lignosulfonate (LS) and acrylonitrile (AN) as raw materials. They improved spinnability and thermal properties by two-step modification of esterification and free radical copolymerization. A LS-AN copolymer with good stability was prepared. With wet spinning, LS-AN copolymer was successfully changed into CF with almost no apparent pores after the stabilization and carbonization. The average tensile strength of LS-AN copolymer-based CF is about twice that of LS/PAN blend-based CF. This study provides a new idea for the practical application of lignin-based CF.

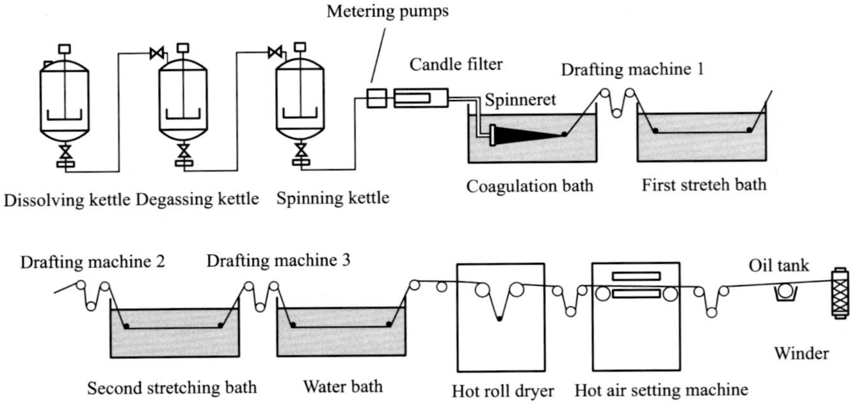

Figure 4.2 Wet spinning process diagram [14].

4.2 Aramid fiber

4.2.1 Overview

Aramid fiber, or aromatic polyamide fiber, is a synthetic fiber made by polycondensation and spinning aromatic compounds as raw materials. The main varieties are poly-p-phenylene terephthalamide fiber and poly-m-phenylene isophthalamide fiber. Aramid fiber is the most versatile reinforcing fiber, second only to CF in high-performance composite materials [17]. It has high strength, modulus, high-temperature resistance, and low specific gravity. Its strength is more than three times higher than general organic fibers, and its modulus is 10 times that of nylon and nine times that of polyester. Its relative strength is 6–7 times that of steel wire, its modulus is about 2–3 times that of steel wire and glass fiber, and its specific gravity is only about 1/5 of that of steel wire. Aramid fiber has good impact resistance and fatigue resistance, suitable dielectric and chemical stability, resistance to organic solvents, fuels, organic acids, and dilute concentrations of strong acids and alkalis, and good inflection resistance and processing performance. It can be woven into a fabric with an ordinary loom, and its strength after weaving is not less than 90% of the strength of the original fiber [17].

Aramid fiber was first introduced to the market in 1965 by DuPont of the United States. This metaoriented aromatic polyamide fiber is called Nomex. In the early 1970s, DuPont developed the second product, the paraaramid fiber Kevlar. It has occupied the primary position of aramid since 1986 until Twaron of Akzo in the Netherlands in 1986, and Technora of Teijin in 1987 started to produce the same fiber. And the emergence of the Soviet ARMOC fiber made the Kevlar monopoly system collapse [18].

Aramid fiber has low lateral strength due to its structure and is prone to fracture under compression and shear forces. Therefore, to give full play to the excellent mechanical properties of aramid fiber, it is necessary to modify the surface of aramid fiber. At present, the surface modification technology of aramid fiber mainly focuses on improving the composition and structure of the fiber surface by chemical reaction or improving the wettability between aramid fiber and matrix resin through physical action. Specific methods include surface coating, chemical modification, plasma surface modification, gamma radiation, and ultrasonic immersion modification [17]. Aramid fiber can form a good interface layer with the resin matrix only after surface treatment to maximize the comprehensive performance of the aramid-reinforced composite material. Surface modification technology has gradually developed from intermittent chemical modification to continuous, multiangle online processing. Batch continuous processing and processing methods that are easy to achieve industrialization characteristics are the main trends in the future research and development of surface modification technologies [19].

4.2.2 Application of aramid fiber

Aramid fiber is widely used for its excellent performance, as shown in Table 4.2. In the aerospace field, aramid fiber—reinforced resin-based composites are applied as structural materials for aerospace, rockets, and airplanes, which can reduce weight, increase the effective load, and save fuel. Kevlar49, the main variety of aramid fiber from DuPont, has been successfully used in the shells, interior trims, and seats of Boeing 757 and 767 aircraft, reducing weight by 30%. Aramid composite materials can be used to manufacture components such as elevated instrument panels, sidewall panels, and high-span shelves on the L1011 transport aircraft of Lockheed Aircraft Corporation of the United States [17].

Aramid fiber has high strength, good toughness, and weaving properties. The impact force of the bullet hitting the aramid braid can be absorbed and dispersed on each fiber. Therefore, aramid is widely used to make bulletproof vests and knee pads. The bulletproof capacity is increased by 23%, making it comfortable. Aramid, titanium, aluminum, and ceramic composites can be used for tank armor, and the bulletproof level can be increased by 50% [17].

The application of aramid fiber is expanding. Ding et al. [20] used a simple blade coating method to manufacture ultrathin and ultrastrong Kevlar aramid nanofiber membranes with interconnected three-dimensional (3D) nanofluid channels for highly stable osmotic energy conversion. The negatively charged 3D nano channels show typical surface charge—controlled nanofluid ion transport and excellent cation selectivity. Thi Tuong Vi Tran et al. [21] prepared high thermal conductivity nanopaper based on aramid nanofibers and ultralong copper nanowires, which have excellent electromagnetic interference effects and mechanical properties.

4.2.3 Preparation of aramid fiber

DuPont's first preparation method of aramid fiber in the United States was condensation polymerization of terephthaloyl chloride and

Table 4.2 Application of aramid fiber [17].

Application field	Aerospace industry	Bulletproof material	Industrial application	Sports Equipment
Detailed examples	Aramid fiber–reinforced resin-based composite materials are used as structural materials for aerospace, rockets and airplanes.	Bulletproof vests and knee pads, tank armor, etc.	Automobile tires, high-strength parachutes, thermal shields, protective clothing, gloves, sealing materials, etc.	Rowing boats, surfboards, hockey sticks, golf clubs, javelins, archery bows, sleds, safety protective racing suits, etc.
Advantage	It can reduce weight, increase payload and save fuel.	High strength, good toughness and weaving significantly improve the level of bulletproof.	Lightweight, wear resistance, good self-lubricity, heat resistance and toughness, long life.	Lightweight and good impact resistance.

p-phenylenediamine in a low-temperature solution. Later, the Teijin company in Japan also used this method to prepare Conex aramid fiber. The research on this method is relatively mature and currently used in industrial production.

Kong and others [22] used calcium chloride N-methyl pyrrolidone (NMP) as a solvent, added *n*-hexane (immiscible with NMP), used p-phenylenediamine and terephthaloyl chloride as reaction monomers, and prepared PPTA by a low-temperature solution polycondensation reaction. The research results show that the introduction of *n*-hexane can delay the gel formation time during PPTA polymerization, which is beneficial to the progress of the polymerization reaction. When the volume ratio of *n*-hexane to NMP is 1:2, a higher molecular weight can be obtained. Adding *n*-hexane can make the specific surface area of PPTA resin particles larger but has little effect on its structure and performance.

Zhang et al. used lithium chloride NMP as the solvent, p-phenylenediamine and terephthaloyl chloride as the reactive monomers, and added 2-(4-aminophenyl)-5-aminobenzimidazole (DAPBI) as the solvent. The third monomer is prepared by a low-temperature solution polycondensation method to obtain a modified PPTA copolymer. The results show that the molecular weight of the regular copolymer prepared by this method is larger than that of the random copolymer; the introduction of DAPBI makes the modified PPTA an amorphous polymer, and the regular copolymer is more orderly than the random copolymer. And the crystallization performance is good; the initial decomposition temperature of the regular copolymer is slightly lower than that of PPTA, but the solubility and mechanical properties have been improved [23].

Liang uses lithium chloride dimethylacetamide as the solvent, p-phenylenediamine and terephthaloyl chloride as the reactive monomers while adding 4,4-diamino diphenyl sulfone as the third monomer, using low temperature. The modified PPTA copolymer was prepared by the solution polycondensation method. Research results show that copolymers with high viscosity, good solubility, and good thermal stability can be obtained after copolycondensation [24].

Liu et al. used dimethylacetamide as a solvent and metaphenylenediamine and isophthaloyl chloride as reaction monomers to prepare PMTA. The polymerization reaction is carried out in a microchannel reactor, the ratio of the reactants is 1.005–1.000, the reaction temperature is controlled from 0°C to 10°C, the water content in the solvent is below 300×10^{-6} g/g, and two monomers. The feed rate is 5:4–5:3, there is no need to add acid-binding agent during the preparation process, the prepared polymer has a viscosity of 1.9−2.1 and a stable molecular weight distribution [25].

Other synthetic methods of aramid fiber also include the interfacial polycondensation method [26], direct polycondensation method [27], gas-phase polymerization method [28], and so on. However, these methods are limited to theoretical research and have not been industrialized due to their respective limitations.

4.3 Ultrahigh-molecular-weight polyethylene fiber

4.3.1 UHMWPE development status

UHMWPE fiber, also known as straight-chain polyethylene fiber or high-strength and high-modulus polyethylene (HSHMPE) fiber, is spun from polyethylene with an average molecular weight of more than 1 million. It has high strength, modulus, good chemical and weather resistance, high energy absorption, low conductivity, can pass through X-rays, and has certain waterproof properties. Table 4.3 lists the performance of UHMWPE fiber manufacturers and their products. It has a broad application prospect in the military, aerospace, navigation engineering, high-performance lightweight composite materials, sports equipment, biological materials, and other fields [29].

In 1978, Pennings Smith, a senior consultant of the Dutch DSM company, invented the HSHMPE fiber and applied for a patent. It uses high-density polyethylene with a molecular weight of more than 1×10^6, gel spinning with its dilute solution, and stretching 30 times. In 1984, DSM

Table 4.3 Performance of UHMWPE fiber manufacturers at home and abroad and their products [29]

Manufacturer	Brand	The tensile strength /GPa	The tensile modulus /GPa	Elongation at break /%
Ningbo Dacheng	Strong nylon	3.1−3.7	79−129	3−7
ZhongFang Investment	Futai	2.8−3.0	83−149	2.8−3.0
DSM Netherlands	DyneemaSK66	3.0	95	3.7
	DyneemaSK77	3.7	132	3.75
American Honeywell	Spectra1000	3.0	116	2.9
	Spectra2000	3.0	119	2.9
Japan Mitsui Petrification	TEKMILON series	3.2	93	4.0

of the Netherlands and Toyota of Japan jointly established a pilot plant with an output of 50 tons/year, and the fiber product was named dynema. The American Honeywell Company purchased the patent and carried out commercial production. Its fibers are called Spectra900 and Spectra1000, and it began market development and application in 1988. The UHMWPE fiber produced by Japan's Mitsui Petrochemical Company is called Tenilon. China started the research on UHMWPE fiber in 1985, achieved a series of breakthroughs at the end of the last century, and invested in pilot tests and industrial development. Its fiber performance has reached the international high and middle levels. For example, the Futai brand UHMWPE fiber is produced by China Textile Investment Development Co., Ltd. The Beijing Tongyizhong New Material Technology Corporation has a production capacity of 250 t/year [30].

Currently, there is only gel spinning to produce high-strength and high-modulus UHMWPE fiber. This method has a long production cycle, complex equipment, expensive solvents, low yield, and high cost. People try to find other high-efficiency production methods, such as the surface-crystal growth method, single-crystal superstretching method, single-crystal precipitation method, melt extrusion stretching method, and plasticized melt stretching method, but the overall effect is not ideal. The research of the UHMWPE fiber spinning method mainly focuses on the selection of solvents used in gel spinning and the influence of draw ratio on fiber structure and performance [30].

The disadvantages of UHMWPE fiber are poor surface adhesion, high creep, low melting point, and low compressive strength, which limit its application in composite materials and pressure vessels. Therefore, another current research direction of UHMWPE fiber is to modify its shortcomings. Treatment to improve the surface adhesion of UHMWPE fibers includes physical methods, chemical methods, plasma, radiation cross-linking, corona, photooxidation, photo-cross-linking, etc., so that the surface of UHMWPE fibers will be etched after treatment, forming convexities and concaves. The uneven surface increases the contact area and bonding force between the fiber and the resin. The primary method to improve the creep properties of UHMWPE fibers is to cross-link them or mix them with other fibers. Cross-linking is mainly done by radiation treatment of UHMWPE fiber, and it can also be mixed with CF and aramid fiber to achieve creep resistance. Currently, the creep performance of UHMWPE fiber has not been well resolved, so improving the creep resistance of UHMWPE fiber is still one of the main tasks and directions of future research [31,32].

4.3.2 Application of UHMWPE fiber

UHMWPE fiber is lightweight and has high strength, long service life, abrasion resistance, moisture resistance, and large elongation at break. It is commonly used in opposing force ropes, heavy-duty ropes, salvage ropes, towing ropes, sailing ropes, and fishing lines. The breaking length of UHMWPE fiber rope under its weight is eight times that of steel rope and twice that of aramid fiber. Due to its good impact resistance and excellent energy absorption, it can be made into protective clothing, helmets, and bulletproof materials in the military, such as tank shields, radar shields, missile shields, and body armor. Among them, the application of body armor is the most popular. In sporting goods, it is used in helmets, snowboards, windsurfing boards, fishing rods, rackets and bicycles, gliding boards, ultra-lightweight aircraft parts, etc. Its performance is better than traditional materials. Because UHMWPE fiber composite material has a high specific strength, specific modulus, good toughness, and damage tolerance. The sports equipment is durable and can produce good results [29].

4.3.3 Preparation of UHMWPE fiber

a) Gel spinning

Early research shows that gel spinning is the best method to reduce the secondary bond between UHMWPE molecules, reduce the entanglement of macromolecules, and improve their fluidity. In the process of gel spinning, the quality of the spinning solution is the decisive factor in obtaining high-quality UHMWPE gel yarn. It is also a prerequisite for superstretching and obtaining high-strength, high-modulus UHMWPE fibers. The quality of the spinning solution is closely related to factors such as concentration, molecular weight, solvent, dissolution temperature, and dissolution form. At present, most UHMWPE gel wet spinning enterprises use UHMWPE suspension swelling liquid, twin-screw extruder dissolution, gel spinning, gel fiber cooling, solvent removal (extraction), extractant removal (Drying), superstretching process (as shown in Fig. 4.3) [33].

b) Melt spinning

Compared with gel wet and dry spinning, UHMWPE melt spinning has apparent advantages such as a simple process, high production efficiency, low energy consumption, and low environmental pollution. Therefore, its technological development has attracted long-term attention from academic and industrial circles. The strong secondary bond

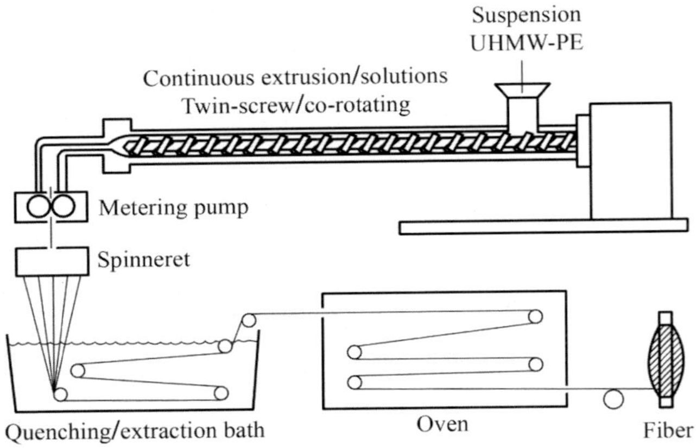

Figure 4.3 Schematic diagram of typical UHMWPE gel spinning process [33].

and the entanglement of macromolecules make HMWPE almost no melt fluidity. Therefore, the key to realizing the melt spinning of UHMWPE is to overcome the intermolecular interaction, reduce the entanglement of macromolecules, and improve its melt fluidity [33].

External plasticization is the most commonly used method to improve UHMWPE's melt fluidity. Adding a certain proportion of additives with good fluidity to UHMWPE can improve its melt fluidity. Considering compatibility, many researchers use low-density polyethylene, linear low-density polyethylene, high-density polyethylene, or paraffin wax as additives to spin UHMWPE fiber after melt blending [34]. The general process of melt spinning is shown in Fig. 4.4.

4.4 Nanofiber materials

4.4.1 Concept and characteristics of nanofibers

Nanofibers have a diameter ranging from 1 to 1000 nm. The polymers that produce nanofibers, including natural and synthetic materials, need high molecular weight, few and small side chains, and can be easily and completely dissolved in specific solutions and conditions. Nanofibers have been used in various applications, such as air filtration, tissue engineering, and drug delivery, as microsensors, wastewater treatment, fuel cells, photocatalysis, lithium-ion batteries, and nanostructured piezoelectric materials. They can be customized with physical and chemical properties, such as a

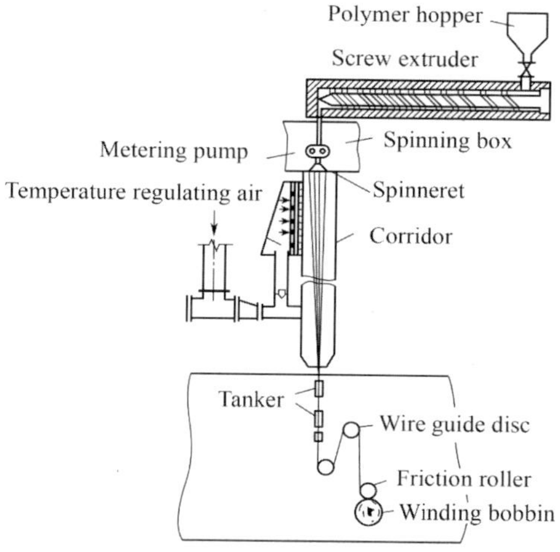

Figure 4.4 Melt spinning process of the polymer [35].

large surface-area-to-volume ratio (about 10^3 higher than microfiber), flexibility, lightweight, small diameter (300−1000 nm), and small pore size (<3 μm) [36].

4.4.2 Preparation method

There are many methods for preparing nanofibers, such as stretching, template synthesis, self-assembly, microphase separation, and electrospinning. Among them, the electrospinning method is widely used for its advantages of simple operation, wide application, and relatively high production efficiency.

1) Stretching method

   The stretching method to prepare nanoscale fibers is similar to the dry spinning in the fiber industry. While the stretching method can produce very long single nanofiber filaments, only those viscoelastic materials that can withstand sufficient stress and deformation can be stretched into nanofibers [37].

2) Template synthesis method

   Template synthesis uses nanoporous membranes as templates to prepare nanofibers or hollow nanofibers. The main feature of this method is that it can spin different raw materials, such as conductive polymers, metals, semiconductors, carbon nanotubes and fibrils. Lin et al. used an anodized aluminum oxide membrane as a template to prepare PAN nanofibers [38,39].

3) Phase separation method

The phase separation method includes the processes of dissolution, gelation, extraction with different solvents, condensation, drying, etc. Finally, a three-dimensional interconnected fibrous network with a fiber diameter ranging from 50 to 500 nm is obtained. The disadvantage of this method is that it takes a considerable amount of time to convert a solid polymer into a nanoporous fibrous network [40].

4) Self-assembly method

The self-assembly method is a process of spontaneously assembling existing components into a desired pattern and function. Liu et al. synthesized cylindrical fibers tens of micrometers long and about 100 nm across from a diblock copolymer. Similar to the phase separation method, the self-assembly process is very time-consuming [41].

5) Electrospinning method

Electrospinning is when a charged solution or melt flows or deforms in an electrostatic field and obtains fibrous substances through solvent evaporation or melt cooling and solidification, referred to as electrospinning. According to the state of the spinning material, it can be divided into solution electrospinning and melt electrospinning. Electrospinning is different from traditional spinning technology. The traditional spinning method is difficult to spin fibers with a diameter of less than 500 nm. In comparison, electrospinning even can spin ultrafine fibers with a minimum diameter of 1 nm [42,43]. At the same time, the device and principle of electrospinning are relatively simple. The schematic diagram of a typical electrospinning device is shown in Fig. 4.5, which is mainly composed of a high-voltage power supply, metering pump, spinning solution container, spinneret, collector, and other components.

Electrospinning can produce a variety of nanofibers and submicron fiber continuously, which is environmentally friendly and low-consumption. The range of raw materials used for electrospinning is broad, covering two major material types: organic and inorganic. Nanofiber materials with different components, functions, and morphologies can be prepared by designing and controlling the composition of the material [44].

Polymer nanofibers: In electrospinning, polymers are the most common raw material for filament formation. The filament formation properties of polymers directly affect the preparation and performance of nanofibers. Studying different raw materials for electrospinning can lay a good foundation for improving the performance and application of nanofibers. At present, electrospinning can be used for the

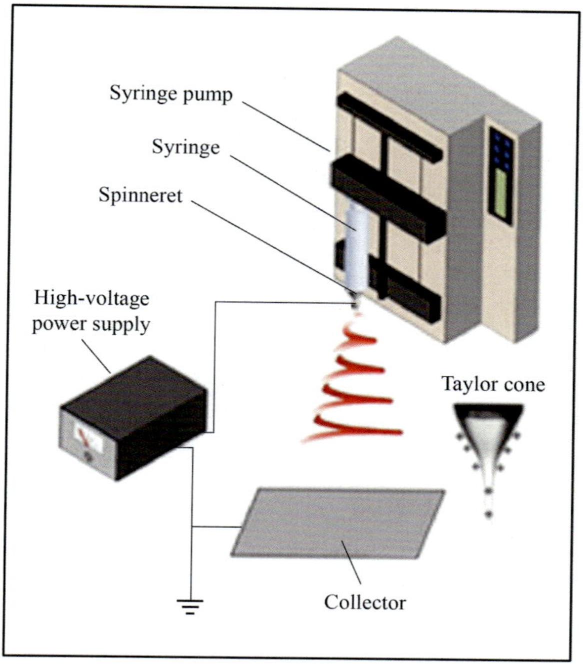

Figure 4.5 Schematic diagram of electrospinning device [44].

preparation of most polymer nanofibers, including conventional semi-crystalline polymer materials, such as nylon (PA), polyethylene terephthalate (PET), and PVA; liquid crystal rigid polymer materials, such as poly(p-phenylenediamine terephthalate) and polyaniline; and elastomeric polyurethane (PU) [44]. Table 4.4 [45–52] lists examples of polymers used in electrospinning to prepare nanofibers.

Inorganic nanofibers: Initially, the preparation of nanofibers by electrospinning was mainly concentrated in polymer solution spinning. In recent years, many studies have been formulated into solutions as precursors by the sol–gel method, organic–inorganic hybrid method, and precursor method, which can also meet the process requirements of electrospinning and prepare a variety of inorganic nanofibers. Larsen et al. [53] successfully electrospun inorganic gels with a certain viscosity to prepare TiO_2/SiO_2 and Al_2O_3 nanofibers. Chang's research group prepared a series of inorganic oxide nanofibers by electrospinning, including ZrO_2, NiO, Co_3O_4, CuO, Mn_2O_3, and Mn_3O_4 nanofibers. In addition, SiO_2, Al_2O_3, V_2O_5, ZnO, Nb_2O_5, MoO_3,

Table 4.4 Preparation of various polymer nanofibers by electrospinning [45–52].

Polymer	Solvent	Spinning process (spinning voltage, flow rate, take-up distance)	Fiber diameter/nm	Application
Polypropylene (PAN) (molecular weight 75000)	Dimethylacetamide	40 kV, 25 cm, 4%	100–1000	Used for ion exchange fiber, oil recovery and water shutoff agent, functional fiber modification, etc.
Polyvinyl alcohol (PVA) (molecular weight 75000 ~ 80000)	Water	15 kV, 0.5 mL/h, 15 cm, 10%	100–1000	Used for medical materials, pervaporation membranes, optical fiber sensor membrane substrates, etc.
Polylactic acid (PLLA) (molecular weight 160,000)	Dichloromethane	8 kV, 0.8 mL/h, 15.5 cm, 5.7%	50	Preparation of nano-scale cytoskeleton
Polyethylene terephthalate (PET)	Trifluoroacetic acid + dichloromethane (volume ratio 4:1)	22 kV, 10 cm, 0.2 g/L	400–600	Used as a cardiovascular implant material
Polylactide (PLA)	Dimethyl formamide + acetone	10 kV, 0.1 mL/h, 10 cm, 0.2 g/L	100–200	Extracellular matrix tissue repair for skin and cartilage
Polyurethane (PU)	Dimethylacetamide	10 kV, 3 mL/h, 21 cm, 20%	125–300	Preparation of rigid foam plastics, artificial leather, elastomers, etc.
Polyimide (PI)	Dimethylacetamide	20 kV, 16 cm, 12%	10–200	Used in aerospace, nuclear power and microelectronics fields, etc.
Polystyrene (PS) (molecular weight 190000)	Tetrahydrofuran	19 kV, 12 cm, 7%	800–1800	Used to prepare various filters, protective clothing, etc.

$MgTiO_3$, and other inorganic substances can also be made into nanofibers by electrospinning [54]. Table 4.5 [55–62] lists the process conditions and application examples of electrospinning to prepare inorganic nanofibers.

Coaxial fiber and hollow fiber: In recent years, while using electrospinning technology to prepare organic polymer nanofibers, inorganic nanofibers, and organic–inorganic hybrid nanofibers, people have also conducted coaxial electrospinning. Coaxial electrospinning is a method that can prepare coaxial and hollow fibers with a shell–core structure. A schematic of the fabrication of coaxial fibers by electrospinning is shown in Fig. 4.6 [63]. The Sun research group was the first to report the coaxial electrospinning method, which uses PEO and PLA as the leather materials, poly-dodecyl thiophene, and palladium acetate [Pd $(OAc)_2$], polysulfone, PEO is the core material to prepare composite nanofibers [64].

Jiang et al. used DMF and chloroform solutions of the PCL as the shell and the PEG aqueous solution containing protein as the core for coaxial electrospinning. The Loscertales research group used heavy mineral oil as the core layer solution, the ethanol solution of polyvinylpyrrolidone (PVP) and Ti-$(OiPr)_4$ as the shell material, made coaxial fibers by electrospinning, and solvent extraction of the core layer minerals and burn in oil or 500°C to obtain hollow nanoceramic tubes [65].

Zussman uses PAN as the shell and polymethyl methacrylate (PMMA) as the core to obtain coaxial fibers. The PMMA is decomposed at a high temperature to obtain hollow CFs with an outer diameter of about 2 μm [66]. The Sun research group used PAN solution as the shell and methyl silicone oil as the core. They used coaxial electrospinning technology to prepare coaxial PAN composite fibers with an outer diameter of 3 μm. After preoxidation and carbonization, a hollow CF with a diameter of about 1 μm can be produced [67].

4.5 Application of nanofibers

Due to the unique properties of nanofibrous materials, researchers have investigated their applications in the following fields: composite reinforcement, filtration and barrier, biomedicine, protective clothing, optoelectronic materials, acoustic materials, and aerospace.

Table 4.5 Various inorganic nanofibers prepared by electrospinning [55–62].

Inorganic fiber	Polymer/ solvent	Spinning process (spinning voltage, flow rate, take-up distance)	Calcining temperature /°C	Fiber diameter/mm	Application or potential application
TiO_2	PVP/ethanol/ acetic acid	20 kV, 2 mL/h, 20 cm, 5%	500	78 ± 9	Used to prepare photocatalysts, etc.
ZnO	PVA/ H_2O	13 kV, 3 mL/h, 13 cm, 10%	500	100	Used in varistor, sensor, photocatalysis and other fields
SiC	PCS/xylene	25 kV, 3 mL/h, 22 cm	1000–1200	300–2000	Used in various high-temperature structural parts materials, etc.
SiC	PVP/ethane	22.5 kV, 12.5 cm, 8.6%	1600	50–100	Used in various high-temperature structural parts materials, etc.
Y_2O_3	PVA/H_2O	18 kV, 12 cm, 7%	600	100–200	Used for laser crystals, microwave absorbing materials, fluorescent materials, etc.
NiO	PVA/H_2O		700	200–400	Used in catalysts, battery electrodes, photoelectric conversion materials, etc.
Mn_2O_3	PVP/乙醇		700	300–700	Used to prepare semiconductors and catalyst materials, etc.
ZrO_2	PVA/H_2O		800	50–200	Used in catalysts, oxygen sensors, fuel cells, etc.

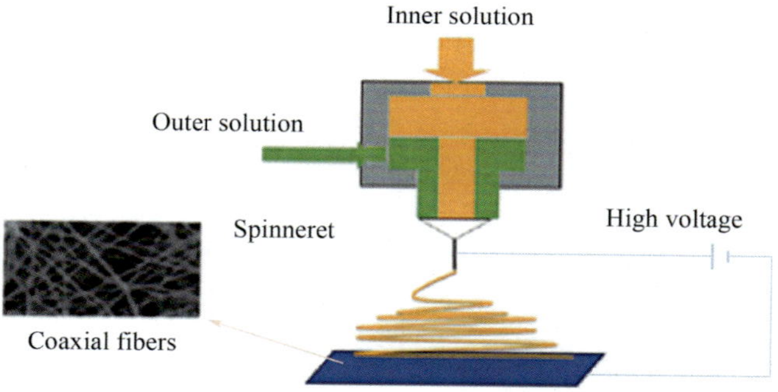

Figure 4.6 Schematic of the fabrication of coaxial fibers by electrospinning [63].

4.5.1 Composite reinforcement

Composite reinforcement here mainly refers to nanofiber-reinforced polymer composite materials prepared by electrospinning, which enables polymer materials to enhance their physical and chemical properties while maintaining mechanical properties. Bergshoef et al. added nylon-4, 6 nanofibers of 30–200 nm to epoxy resin. Because the fibers are smaller than the wavelength of visible light, the composite exhibits transparency, and the hardness and strength of the composite film are significantly increased compared with epoxy resin [68]. Kim et al. investigated the reinforcement of polybenzimidazole electrospun nanofibers in epoxy resins and rubbers. Then, they performed tensile, three-point flexure, double twist, and tear tests. They found that as the fiber increases in content, the flexural Young's modulus and breaking strength of epoxy nanocomposite increase slightly, but the fracture energy increases significantly; while Young's modulus of rubber nanocomposite is 10 times that of unfilled rubber materials, the breaking strength is twice the original [69]. The single-walled carbon nanotubes prepared by electrospinning are reinforced with polyimide composites in the form of nanofibers, which are expected to be used in spacecraft.

4.5.2 Filter and block

The ultrafine nanofiber film made by electrospinning has a large specific surface area and a small pore size, so it has strong adsorption, good filtration, barrier, adhesion, and thermal insulation. This filter material can easily capture tiny particles of 0.5 μm. In addition to satisfying conventional filtration, after this polymer nanofiber is composited with other selective coatings, it can also be applied to molecular filtration, chemical and biochemical drug isolation, and so on. Electrospinning military protective clothing of unique materials can diffuse moist steam and capture aerosol particles. It is lightweight, insoluble in organic solvents, and resistant to harmful chemical gases such as nerve gas. Research by Doshi found that the filter material made of nanofibers sandwiched between meltblown and spun-bond fabrics can filter ultrafine particles more effectively than traditional commercial filters. Even the filter material with nanofibers as the interlayer has a higher ratio. Only 1/15 of this composite filter material can achieve good performance with surface area characteristics and low quality [70].

4.5.3 Biomedicine

The use of nanofibers in biomedicine is also a current research focus. The main benefits are as follows:

1) Cell scaffold

Due to the particularity of the surface structure of electrospun nanofibers, new cells are easy to attach and multiply on the surface. At the same time, they will grow along the nanofibers. Therefore, the electrospun nanofiber structure is an ideal scaffold for tissue engineering. Li prepared a glycolide/lactide copolymer cell scaffold by electrospinning. It is believed that the structure of the cell scaffold is similar to that of natural tissue extracellular matrix, with porosity, wide pore size distribution, and good mechanical properties [71]. Matthews et al. studied the preparation method of collagen nanofiber scaffolds for tissue engineering. After the collagen is blended, it can be electrospun. Under appropriate temperature and pressure, the purified type I collagen is dissolved in a weak acid to neutralize polyoxyethylene. After mixing, it can be used to prepare nanofibers [72].

2) Bionic materials

The porosity and compliance properties of artificial blood vessels affect tissue response. Porous artificial blood vessels facilitate the growth of host tissue and make the inner wall of the artificial blood vessel better intima. Using electrospinning and raw materials such as PU and polytetrafluoroethylene, Dubson obtained a multilayer artificial blood vessel with good flexibility, high porosity, and excellent mechanical properties. From a biological point of view, almost all human organs (such as bone, periodontal tissue, collagen, skin, cartilage) exist in nanoscale fibers. Electrospinning can prepare nanoscale bionic fibers, so it has broad prospects for development [69].

3) Cell carrier

The degradable polymer ultrafine fibers made by electrospinning can be directly sprayed on the skin to heal the wound. It is conducive to normal skin growth and does not leave scars, so that ultrafine nanofibers can be used as a carrier for enzymes and catalysts. Benjamin et al. embedded cells in a double-layer fiber membrane prepared by electrospinning to form a fiber membrane/cell/fiber membrane structure. The bone cells were embedded in a double-layer fiber membrane made of polylactic acid, and it was found that this structure remained intact after being immersed in liquid nitrogen without becoming brittle. At the same time, due to the porous structure of the fiber membrane, it is conducive to transmitting oxygen, nutrients, and carbon dioxide. In addition, this structure is also conducive to cell release, but cell activity will not be affected [73].

4) Protective clothing

Due to the advantages of low density, high porosity, and large specific surface area, nanofibers can be used in multifunctional protective clothing. This kind of microfiber-laid net, with many microporous and membrane-like products, can allow vapor diffusion, the so-called breathability, which ensures the wearer's comfort. It can also block the wind and filter fine particles. The aerosols' barrier properties protect against biological or chemical weapons and biochemical toxicants [74–77].

5) Optoelectronic materials

Nanofiber membranes have a high specific surface area (about 1000 m^2/g), so using nanofiber membranes as sensor sensing membranes can improve sensitivity and be used in sensors, Schottky rectifier junctions, and regulators. It is well known that the electrochemical reaction rate is directly proportional to the surface area of the electrode. Therefore, conductive nanofiber membranes are suitable for developing high-performance batteries for porous electrodes [78,79]. Conductive film (according to its ability to conduct electricity, ions, and photoelectricity) can also be applied to antistatic, anticorrosion, electromagnetic shielding, and optoelectronic devices [80,81]. Waters et al. reported that liquid crystal devices using electrospun nanofibers as optical shutters are transparent under an electric field and opaque in the other state so that they can be switched. The main part of the liquid crystal device is a layer of nanofibers penetrated by the liquid crystal material, only tens of micrometers thick, placed between two electrodes to change the transmittance of the liquid crystal/nanofibers by an electric field. The size of the fiber determines the sensitivity of the refractive index difference between the liquid crystal and the fiber, which controls the light transmittance of the device. Nanoscale polymer fibers are necessary for such devices [82]. Electrospinning nanofibers have many other applications. For example, nanofibers made of polymers with a piezoelectric effect can be used as piezoelectric devices [83]. Electrospinning nanofibers with a high surface area can be used as chemical and biological sensors [84,85]. Susceptible optical sensors based on fluorescent electrospun nanofiber membranes have also been reported recently [86–88].

4.6 Preparation of high-performance shock-resistant vesicle

Here, we take the preparation of high-performance shock-resistant vesicles as an example of the application of aramid fiber. The research on the vesicle can be roughly divided into three stages: fiber selection, weaving verification, and performance testing.

4.7 Fiber selection

In the fiber selection stage, to meet the performance requirements of the impact vesicle and select the appropriate aramid fiber, we tested 12 kinds of aramid fibers for mechanical properties such as linear density, tensile strength, elastic modulus, and elongation. Some fiber samples and numbers are shown in Fig. 4.7. Among the 12 types of fibers, one is aramid 1313, one is aramid III, and the others are aramid 1414 [89–92].

The results show that the largest linear density is B (0.8487 m/g), the smallest is G (0.0895 m/g); the largest tensile strength is A (3051 MPa), the smallest is E (603 MPa); the largest elastic modulus is D (146 GPa), the smallest is E (13 GPa); the largest elongation is E (10.12%), and the smallest is D (1.85%). Therefore, starting from the performance requirements of the impact resistance, E can be excluded, D is the best, and F, A, C, and B can be selected according to the needs of knitting.

Upon request, we collected aramid fiber samples again and conducted stress and strain tests. According to the results, the tensile strength of G is closest to domestic aramid fiber (2.9 GPa). However, the D supplier provides three kinds of fibers, numbered D1, D2, and D3.

We again collected two aramid fiber samples named J and K. We tested the linear density, tensile strength, elastic modulus, elongation, and

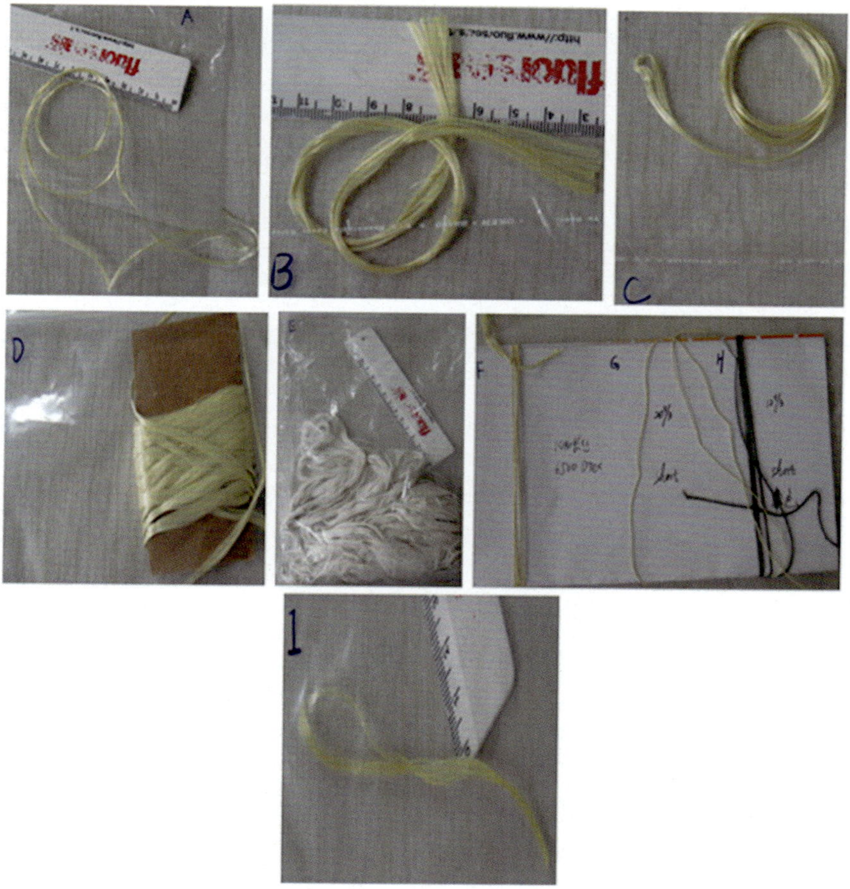

Figure 4.7 Some fiber samples and numbers.

other mechanical properties of these two new aramid fibers. The results compared with D1 are shown in Table 4.6:

Table 4.6 Comparison of mechanical properties of three fibers.

Mechanical properties	Maximum load (N)	Tensile strength (MPa)	Modulus of elasticity (GPa)	Elongation (%)
J	310	2719	121	2.51
K	312	2606	119	2.60
D1	196	2408	175	1.85

It can be seen from Table 4.6 that the tensile strengths of J and K are larger, but the elastic modulus is significantly reduced. According to the requirement of small deformation of the shock-resistant vesicle, the J and K fibers are not used. At that time, the price of Aramid III filament was about 30 times that of Aramid 1414. Its mechanical properties are shown in Table 4.7:

Table 4.7 Mechanical properties data of domestic aramid III fiber.

Tensile strength (MPa)	Elastic modulus (GPa)	Elongation (%)
3954	136	3.17

Compared with the aramid 1414 fiber, the tensile strength of aramid III is greatly improved, which is 1.64 times that of D1, and the elastic modulus is slightly decreased. Therefore, the load-bearing capacity of aramid III is much larger than aramid II's. However, the deformation is slightly larger. Therefore, aramid yarns D1 and G were selected as the raw material for the impact vesicle.

4.8 Weaving verification

The second stage of knitting verification can be roughly divided into four stages: test piece weaving, accessory production, shock-resistant vesicle sealing, and rubber soft vesicle preparation.

4.8.1 Sample piece weaving

Due to the high friction of G-aramid yarn, it is challenging to perform three-dimensional weaving. Therefore, the three-dimensional weaving of D1 fiber was first carried out, and the mechanical properties such as tensile strength, elongation, and Young's modulus were tested. The knitting form is three-dimensional and has six-directions. The knot length is 7.0 mm, the width is about 3.2 mm, and the fiber volume content is 60%. There are six test pieces, roughly divided into two groups according to thickness. Only the tearing force (KN) is machine-readable and accurate among the test results. The cross-sectional area is large because of the large and thick measurement factors (the test piece is soft, and the measurement force is different), so the reliability is not high.

Similarly, the elongation is also inaccurate because it is the estimated elongation after the measurement. The tensile strength and Young's modulus calculated based on the measurement results are unreliable. Because it was previously unknown how much tension the three-dimensional aramid braid could withstand, the first sample tried to load seven times before it broke. Due to the limitation of the equipment range, the above six samples, 1–3, 1–4, 2–3, and 2–4, were tested with one testing machine. The other two were tested with two test machines, respectively.

The initial test shows that the tensile force has a yield phenomenon due to the slippage between the epoxy resin at both ends of the aramid specimen and the jaws of the clamping. The mechanical data show that the strength, modulus, elongation, etc., are much worse than the data of monofilament. The main reasons are: (1) the fiber volume content in the test piece is 60%, meaning 40% is an aramid thread gap. (2) This weaving adopts a three-dimensional six-way structure. In the stretching direction, about 1/6 of the fiber is stressed. (3) The monofilament is tested with epoxy resin. When a bundle of monofilaments is tested, each filament is stressed. The force is even, while the braided sample is only treated with epoxy resin at both ends. No force transmission medium exists in the middle section because each filament is not connected to the tension. The extension direction is the same, so the mechanical test data are reduced. (4) The tensile test pieces are all broken from the root, inconsistent with the fracture from the middle described in the national standard.

Subsequently, four and three test pieces of different thicknesses were woven with D1 and G, respectively. The compounding process of Room-temperature vulcanized (RTV) and aramid braid were explored according to the needs. Finally, the silicone rubber with lower viscosity was selected. The composite experiment was carried out by the vacuum-assisted resin transfer molding method. The experimental results showed that the silicone rubber and the aramid braid were well composited, and the silicone rubber could penetrate the entire test piece.

The impact test piece is woven according to the requirements, as shown in Fig. 4.8, and the impact test data are shown in Fig. 4.9.

The maximum impact was carried out within the range of Instron's instrument, and the results showed that the aramid test piece was not damaged after completion, showing excellent impact resistance.

4.8.2 Attachment production

To inject fluid into the vesicle, we designed and produced the following parts:

1) Valve design

We designed a one-way valve to inject, withstand, and discharge high-pressure liquid according to the experimental requirements. Take the market valve (Fig. 4.10) as the sample, and the designed valve is shown in Fig. 4.11, based on the performance requirements.

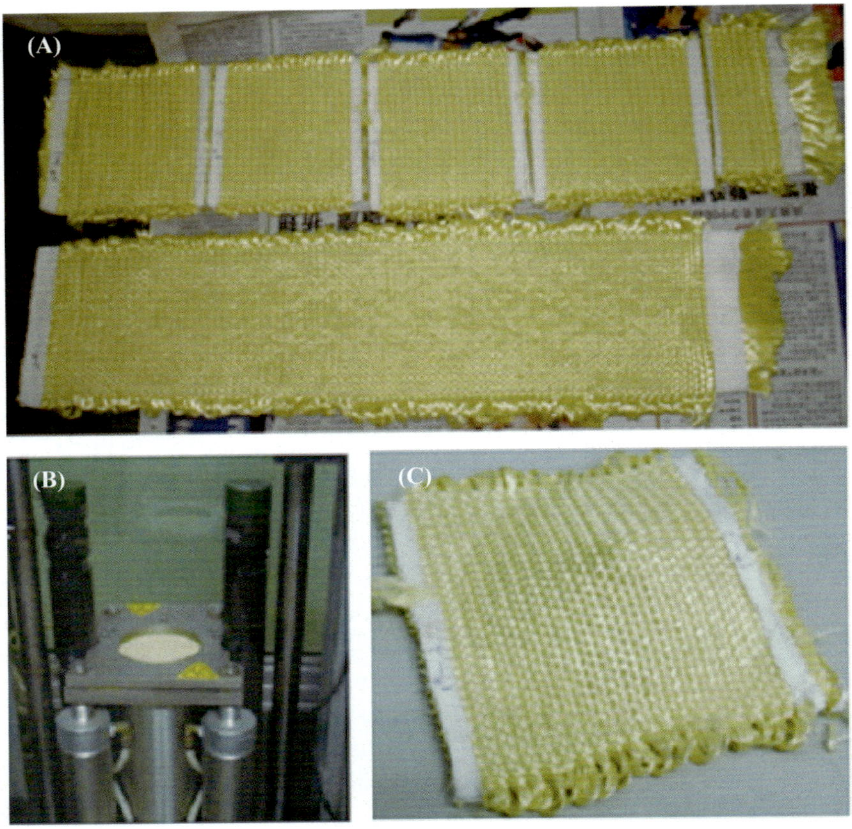

Figure 4.8 Braided impact test piece: (A) Weaved impact test piece; (B) aramid test piece on the impact bench; (C) test piece after impact.

The valve comprises six parts: the main body, connecting threaded body, shaft, spring, gasket, and pressure relief rod. The parts and assembly samples of the valve are shown in Figs 4.12 and 4.13, respectively. The pressure relief rod (not shown in the figure) is used after the experiment. Hold the shaft of the valve to drain out the internal high-pressure liquid.

According to the experiment's needs, 18 valves were produced in three batches.

2) Experimental protection box design

According to the requirements, we designed and customized the protection device for the pressure resistance test of the shock-resistant vesicle. The primary purpose is to protect the experimenter and the compression test machine if the liquid leaks during the pressure test on the

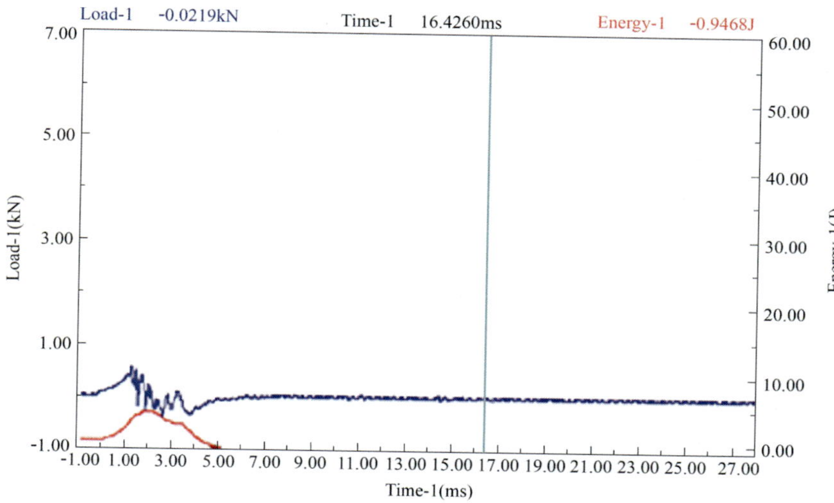

Figure 4.9 Impact test graph.

Figure 4.10 Small valves purchased on the market.

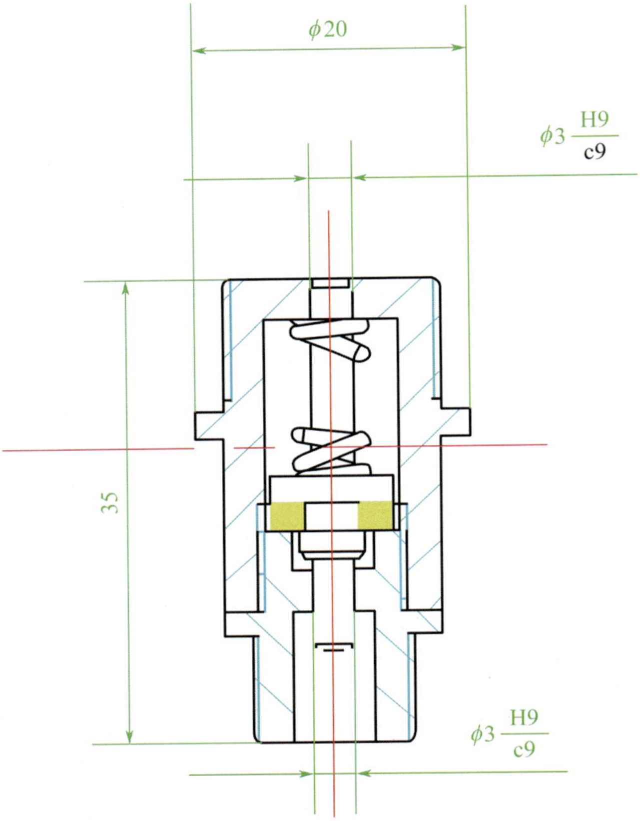

Figure 4.11 Assembly drawing of high-pressure check valve.

Figure 4.12 Valve parts.

Figure 4.13 The assembled valve.

compression test machine. We designed the upper and lower platforms using 45#steel with a smooth surface according to the experimental process requirements. The middle is composed of plexiglass with a wall thickness of 10 mm. The assembly drawing is shown in Fig. 4.14:

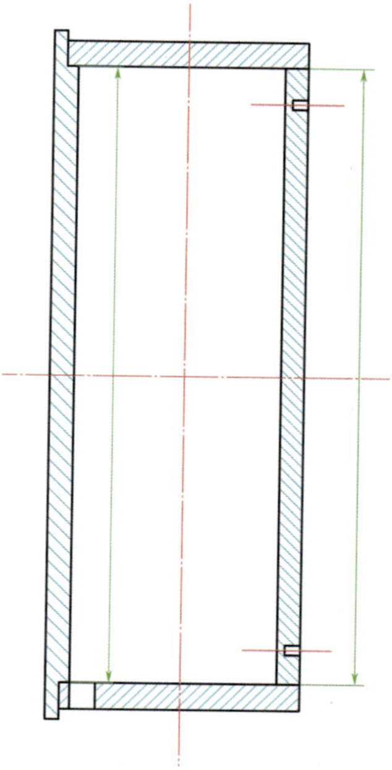

Figure 4.14 Assembly drawing of the protective device for the impact experiment.

Figure 4.15 Experimental protection device for antiflushing vesicle.

Two screw holes are machined on the upper cover, and the upper surface can be easily accessed after screwing on the screws. The physical map is shown in Fig. 4.15. The weight of the upper cover is 4.71 kg.

4.8.3 Research on the impact-resistant vesicle seal

To seal the three-dimensional woven aramid impact vesicle, we explored various ways.

1) Introduction to the glue-hanging process of fire-fighting water hose

The extrusion production site of the water delivery hose is shown in Fig. 4.16. When the fire-fighting water hose is glued, the fiber reinforcement layer is made of PU [93], and the polyester fiber is formed hundreds of meters long through a braiding machine. The round line enters the machine head after drying. There is a cone in the machine head, which is placed inside the tube. By adjusting the position of the cone, the thickness of polymer material (PU or PVC or other) in the inner and outer layers of the tube can be controlled.

Figure 4.16 Production site of water delivery hose extrusion.

As the glue-hanging process requires that both ends of the fiber-reinforced layer are open to place and take out the cone, it cannot be used directly for the glue hanging of the shock-resistant vesicle.

2) RTV silicone rubber

RTV silicone rubber is a new silicone elastomer type that emerged in the 1960s. The most notable feature of this rubber is that it can be cured in situ at room temperature without heat and pressure, making it extremely convenient to use. Due to its low molecular weight, RTV is liquid silicone rubber. Its physical form is usually a flowable fluid or a viscous paste, and its viscosity is between 100 and 1,000,000 centistokes. Now RTV has been widely used as adhesives, sealants, protective coatings, potting, and molding materials [94].

We first selected RTV for the glue-hanging experiment based on the weave density of the impact-resistant vesicle braid. We bought 1501 RTV and carried out the coating experiment on the aramid woven test pieces according to the different coating processes, curing time, and viscosity. After coating the silicone rubber on the test piece, the test piece was put in a vacuum. It can be seen that silicone rubber can penetrate aramid fibers. The curing is completed after 24 hours, and the surface is relatively smooth, but after stretching, it is found that the silicone rubber and aramid are not well combined, as shown in Fig. 4.17.

The following steps were carried out to improve the wetting of RTV on aramid fibers. Firstly, soak the aramid fiber sample in acetone for 72 hours to remove the coating material on the fiber surface. Then soak the aramid woven sample with 20% phosphoric acid for 2 hours, rinse with water until it is neutral, and coat it with silicone rubber. The tensile results of the sample after curing showed that the tensile strength and elongation of the test piece after phosphoric acid treatment decreased. A chemical reaction between phosphoric acid and certain components in the aramid fiber may cause it. It may also be improper handling. After the silicone rubber is cured, it is found that the strength is very low, which may be related to the low molecular weight of the selected silicone rubber. If the molecular weight is high, the viscosity and the process will increase the difficulty. The woven test piece made of aramid fiber G slightly differs from the aramid yarn in the gluing process. The test piece woven from aramid yarn has better wettability. Basically, under the action of gravity, it can penetrate from top to bottom. However, the test piece woven with aramid yarn may have too much aramid short fiber, forming a capillary effect. At the same time, the aramid yarn is woven relatively densely, which makes it difficult for the silicone rubber to penetrate. The infiltration process must be completed with the help of vacuum suction. There is no significant difference in the permeability of the silicone rubber between the phosphoric acid–treated and untreated test pieces.

Figure 4.17 The stretched photo of the aramid woven test piece after immersing in silicone rubber.

3) Liquid polyurethane

PU is a polymer that contains characteristic urethane units in the main chain. The mechanical properties of PU have excellent tunability. It can be controlled between the hard crystalline and noncrystalline soft segments. The ratio of PU can obtain different mechanical properties. Therefore, its products have excellent properties such as wear resistance, temperature resistance, sealing, sound insulation, good processing performance, and degradability. This polymer material is widely used in adhesives, coatings, low-speed tires, gaskets, car mats, and other industrial fields [95].

We immersed the aramid woven test piece in liquid PU, vacuumed it for 10 hours, took it out, cured it at high temperature for 72 hours, and folded it in half. It was found that the test piece of PU glue had no cracks, while the liquid silicone rubber was cured. The test piece is folded in half, with many fine cracks. In Fig. 4.18, the upper sample is impregnated with PU, the surface is relatively smooth, the strength is higher than that of silicone rubber, the elasticity is better, and it is not hard. The lower sample is not glued. Therefore, it was decided to use PU to treat the shock-resistant vesicle.

4) Making and hanging glue for resistant vesicle

To illustrate the shape of the shock-resistant vesicle, we made the aramid structure diagram shown in Fig. 4.19 and the schematic diagram in Fig. 4.20 for the weaver's reference.

Figure 4.18 Photographs of polyurethane dipped and nondipped test pieces.

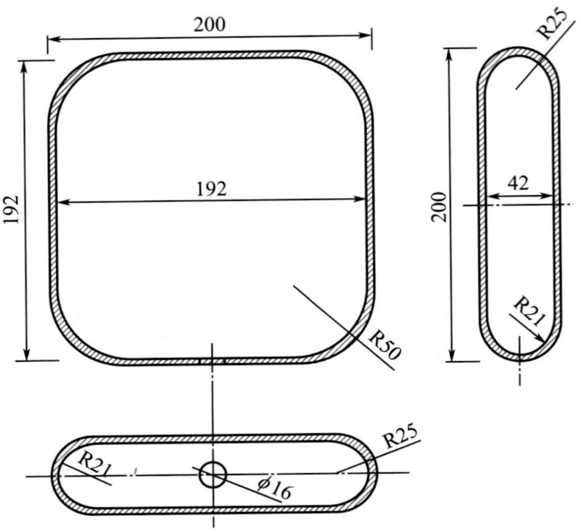

Figure 4.19 Structure diagram of the shock-resistant vesicle.

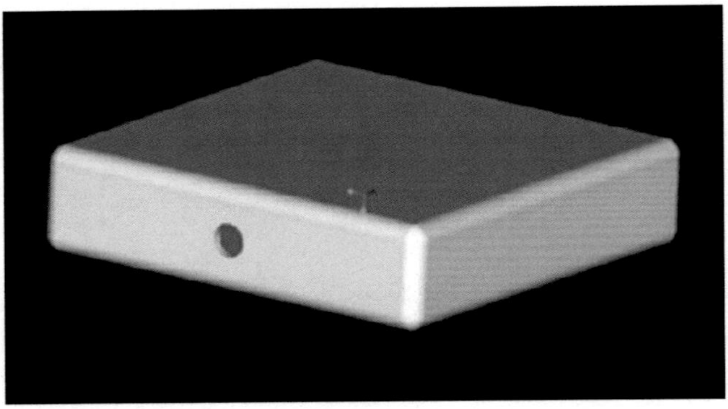

Figure 4.20 Schematic diagram of the shock-resistant vesicle.

As the shock-resistant vesicle manufacturing unit has not processed a completely enclosed product before, as the first piece, we need to explore the weaving method and process. Fig. 4.21 is the core mold we made and provided to the weaver.

The shock-resistant vesicle is trial-produced after many discussions on the weaving plan. The appearance of the first product is not good in quality. Uneven density can be seen on the surface, the valve connection is also relatively loose, and the corner transition is not smooth, as shown in Fig. 4.22. After reinforcing the valve with aramid bundles,

Figure 4.21 Impact-resistant vesicle core mold.

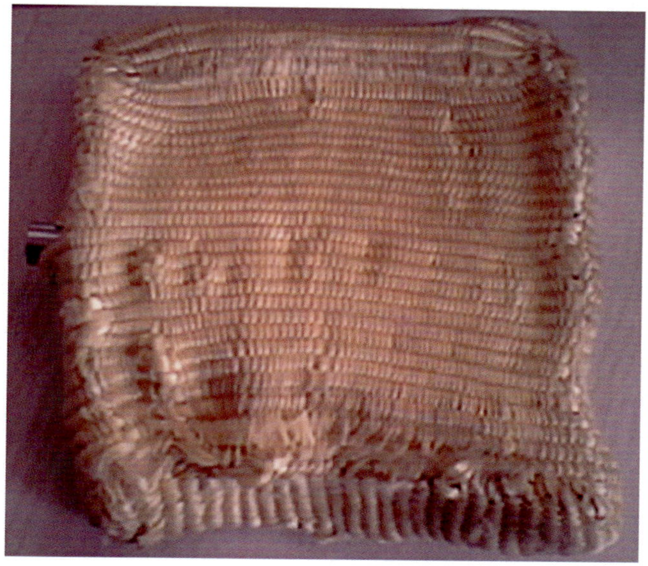

Figure 4.22 The first aramid weaved vesicle.

we used aramid bundles to pass through the upper and lower surfaces, respectively, so that the upper and lower surfaces were pulled apart after dipping.

Immerse the first vesicle in liquid PU, vacuum it for 12 hours, and take it out. After immersing in liquid PU, the aramid tow is relatively

Figure 4.23 Photo of the impact-resistant vesicle after curing by dipping PU.

slippery. There are eight points of aramid bundle used to pull apart the upper and lower surfaces. When it fails, the upper and lower surfaces are pulled apart by the remaining three aramid knots, hung in an oven, and cured at high temperature for 72 hours. Inspection after curing found that because the PU flows downward under the action of gravity, there is less glue on the upper part of the vesicle during the suspension. Later, the air leak was found in the inflation test, so we patched it up with super glue. As shown in Fig. 4.23, the leak was also found during the water-filled and pressure-resistant test. The reasons are as follows: First, during the curing process, most of the PU flowed to the lower side, resulting in a lack of glue on the top; second, for some parts of the shock-resistant vesicle, the knitting is too thin, resulting in incomplete gluing, leaving gaps or small holes.

However, the overall hardness of the vesicle after being glued exceeded expectations, although the PU surface was still very soft. The fibers of the vesicle cannot move among each other after being cured by PU. Hence, the overall hardness of the composite is relatively high. The water-filled and pressure-resistant experiment was done with the first piece of a rubber-coated vesicle. The water was tap water, and the pressure was not increased. No valve leakage was observed during the compression process. However, due to incomplete rubber coating, water leaked into the gap.

After discussion, it was decided to abandon the hanging glue solution and adopt the second solution, which is to make the outer shell of the aramid vesicle with a soft rubber vesicle with a valve inside. Although the program can ensure that the whole is soft and filled with high-pressure liquid, it may have the following disadvantages: Because the aramid fiber and the rubber inside are not compounded together, the force of each fiber may be different. The force that can withstand the vesicle's impact is much lower than the theoretical value.

We prepared the second shock-resistant vesicle. This time the quality is better than the first one, the density is more uniform, the transition is gentler, and the valve connection is more firm, as shown in Fig. 4.24, if a rubber vesicle is used, it is planned to disassemble the second vesicle, remove the valve, put the rubber vesicle with the valve, and sew it.

Figure 4.24 Photo of the second shock-resistant vesicle.

4.8.4 Production of soft rubber vesicle

The quality of the first rubber vesicle was not good, as shown in Fig. 4.25. There is apparent glue flow at the upper and lower interfaces, and the evident cracks near the valve are important.

After many tests, the fluidity and formula of the rubber compound were adjusted, and the valve was redesigned and processed to solve the problem of the gap (or groove) at the connection part between the valve and the rubber. And finally, it was successfully manufactured, as shown in Fig. 4.26. After the water filling experiment, no water leakage was found for the soft rubber vesicle. A good-looking rubber vesicle was selected from the vesicle for the three-dimensional weaving of the aramid shock-resistant vesicle.

We hope to disassemble the second woven aramid vesicle and put it in the soft rubber vesicle to speed up the experiment. However, the size of the second aramid vesicle is quite different from the soft rubber vesicle, so we weaved the third piece of the aramid vesicle. Fig. 4.27 shows the third aramid vesicle in the weaving.

4.9 Performance test

In this stage, the second aramid vesicle was used for different compression tests. Later, the rubber and aramid vesicle were remade, and the compression test was performed.

Figure 4.25 The first rubber vesicle.

Figure 4.26 The second processed rubber vesicle.

Figure 4.27 Aramid vesicle in the weaving process.

4.9.1 The second compression experiment

The compression experiment with the first PU-impregnated aramid fiber vesicle is the first, so this experiment is the second compression experiment. The third aramid and rubber vesicles (Fig. 4.28) were woven together in a specially designed and manufactured protection device for the second compression experiment. After being filled with water, the aramid vesicle expands and is relatively strong, indicating that the rubber in the soft vesicle fits well with the aramid vesicle.

Fig. 4.29 is a photo of the experiment during the compression process. The experiment was carried out until the aramid vesicle leaked. The experimental process curve is shown in Fig. 4.30. The maximum pressure is 8147.69 N. During the experiment, the valve and the aramid vesicle were intact. It may be that the quality of the rubber vesicle inside is not uniform. There are still invisible internal weld marks or weak adhesion at the junction of the valve and the rubber. The pressure is high, and the defect of the rubber vesicle leaks first.

4.9.2 The third compression experiment

The second compression test leaked earlier as the soft rubber vesicle may have quality defects and poor tensile or pressure resistance. Therefore, we bought some balloons and performed the third compression experiment using the second aramid vesicle that had not been tested. The procedure is as follows: First, put the balloon into the aramid vesicle, then fill the balloon with water until the aramid vesicle is inflated and can no longer

Figure 4.28 Vesicle resistant to flushing after being filled with water.

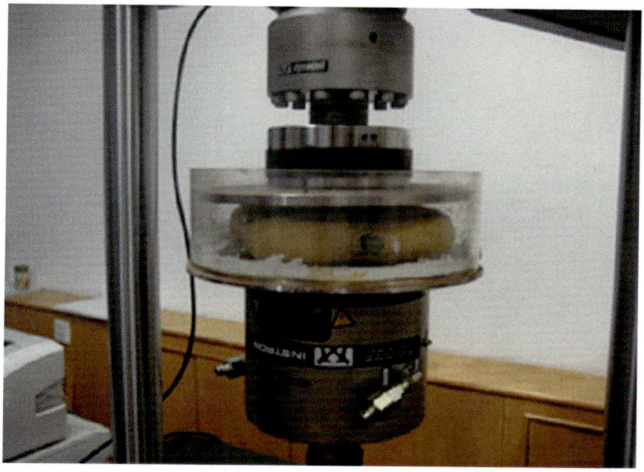

Figure 4.29 Photo of the compression test of the shock-resistant vesicle.

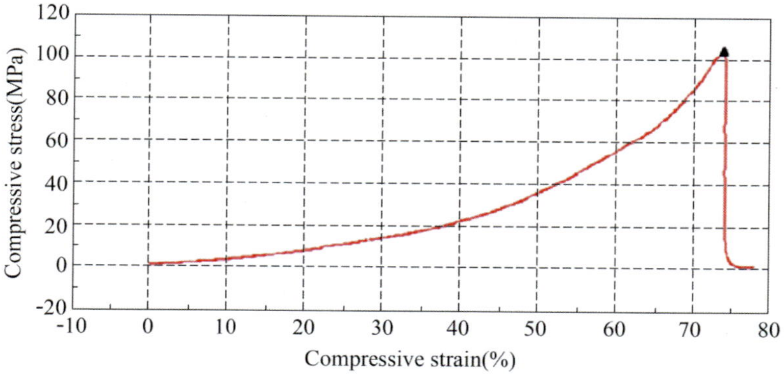

Figure 4.30 Compressive stress–strain curve.

be filled with water; thirdly, tie the balloon mouth tightly with aramid silk thread, and stitch the mouth of the aramid vesicle tightly with aramid silk thread. After the compression experiment, the balloon was taken out, and it was found that there was much wooden sawdust in the aramid fiber vesicle (as shown in Fig. 4.31). It was thought that the wooden sawdust had punctured the balloon. Another reason is the small mouth of the vesicle. Because the mouth is too small, the sawdust is closely attached to the inner wall of the aramid vesicle. It is impossible to remove all sawdust from the vesicle and leave some small pieces of sawdust in it.

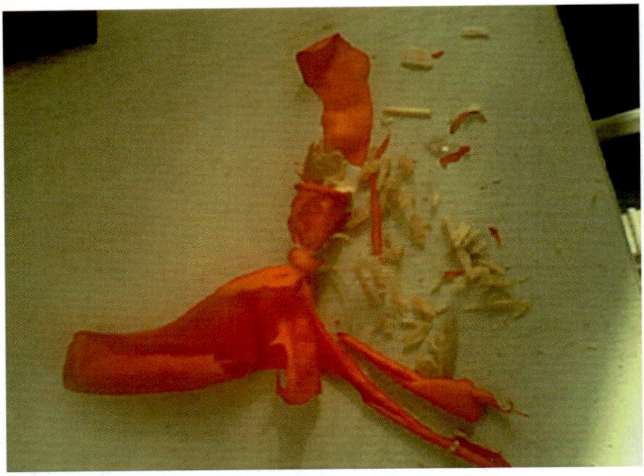

Figure 4.31 Much wooden sawdust in the aramid vesicle.

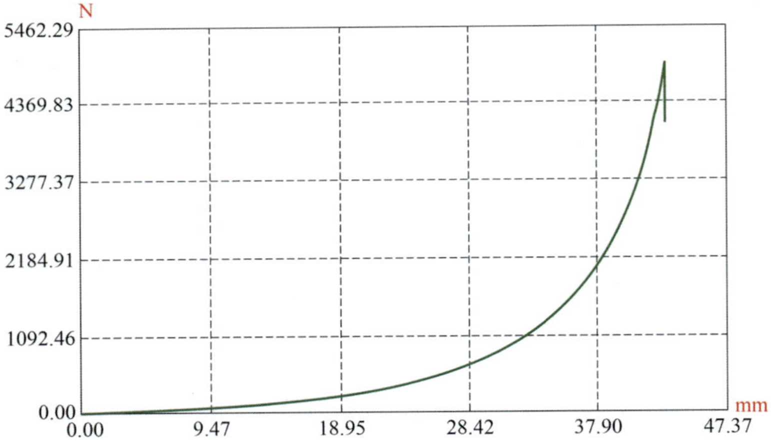

Figure 4.32 Compressive stress–strain curve.

4.9.3 The fourth compression experiment

According to the previous experience, remove the sawdust in the aramid vesicle as much as possible, stuff some cotton inside the aramid vesicle, and then put in two balloons overlapping inside and outside. The vesicle is sutured, and the experimental curve is shown in Fig. 4.32. The maximum pressure is 4938.57 N. After the experiment, the balloon and cotton were taken out, and it was found that there was still sawdust in the cotton.

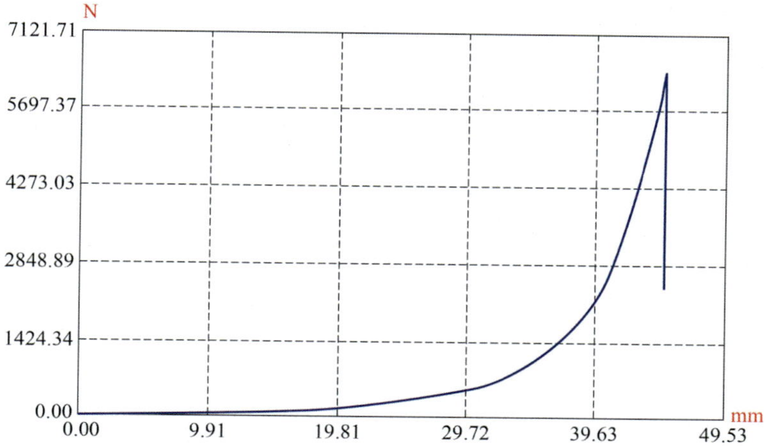

Figure 4.33 Compressive stress–strain curve.

4.9.4 The fifth compression experiment

Clean the sawdust in the aramid vesicle again, refill the cotton, use four balloons, and repeat the previous experiment. The curve obtained is shown in Fig. 4.33. The maximum pressure is 6451.43 N.

The maximum pressure increased significantly, possibly related to the sawdust's decrease or the balloons' increase.

4.9.5 The sixth compression experiment

The balloon used in this experiment is larger than the aramid vesicle. Fig. 4.34 shows two balloons placed in the aramid vesicle, filled with water, and compressed. The experimental data are shown in Fig. 4.35. The maximum pressure is 8414.29 N, which exceeds the maximum pressure of the secondary aramid vesicle plus the soft rubber vesicle. The rupture is relatively small, possibly related to the small wooden sawdust in the aramid vesicle.

4.9.6 The seventh compression experiment

We continue to use small balloons (4). This time, the balloons are about the same size as the aramid vesicles, as shown in Fig. 4.36. However, the quality is not so good. The experimental data are shown in Fig. 4.37. The maximum pressure is 6037.14 N. After the balloon is taken out, it is observed that the gap is large. It may not be easy to reach the corner of the aramid vesicle after the balloon is filled with water.

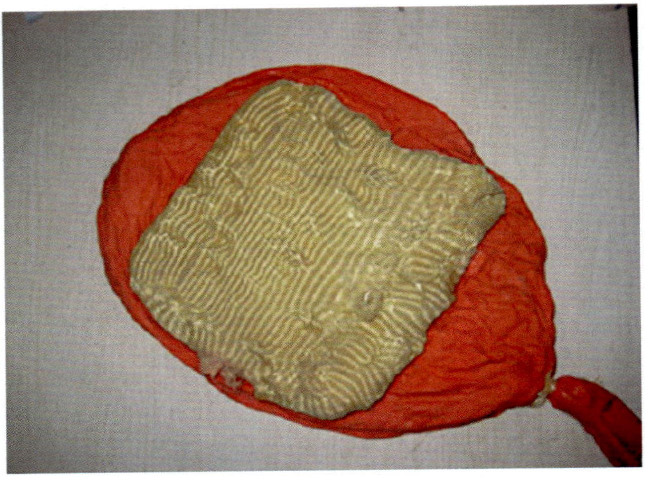

Figure 4.34 Aramid vesicle and balloon size comparison.

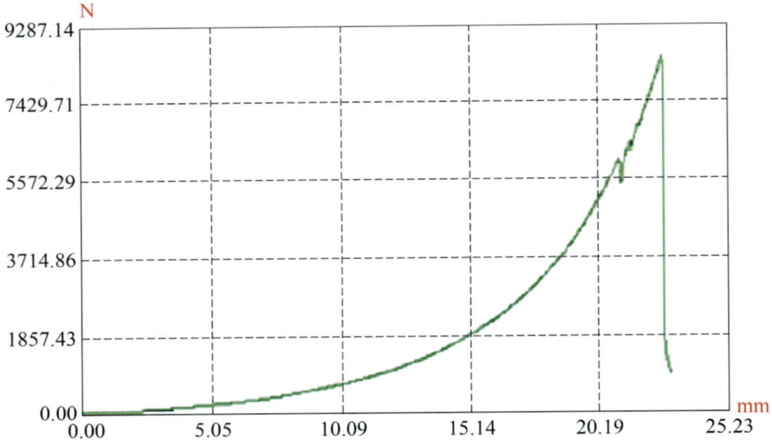

Figure 4.35 Compressive stress–strain curve.

The same aramid vesicle (the second piece) is used from the third to the seventh, and the appearance of the aramid vesicle is not as good as the third aramid vesicle. After many use, the aramid vesicle has no damage, indicating that: (1) the three-dimensional woven aramid vesicle has good pressure resistance. (2) The pressure resistance is mainly related to the liquid sealing layer in the aramid vesicle. (3) The aramid vesicle and the sealing layer should have the same shape. (4) The aramid vesicle should be smooth and free of rigid and foreign objects.

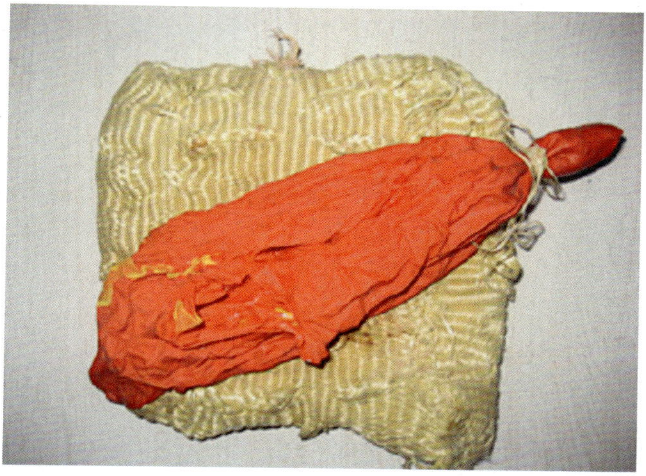

Figure 4.36 Balloon size.

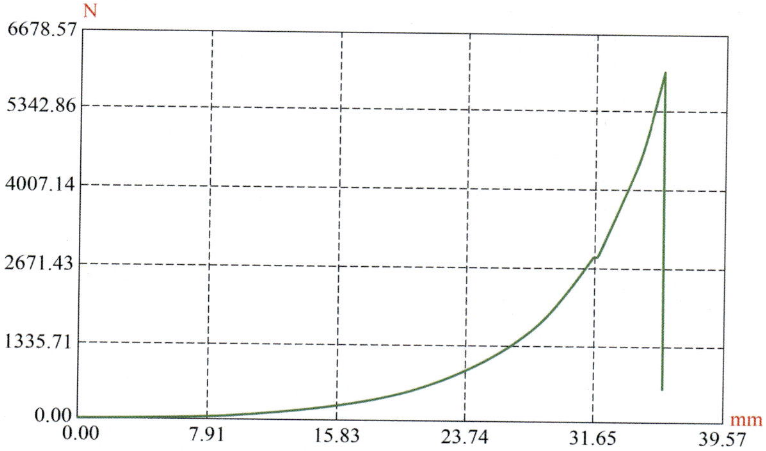

Figure 4.37 Compressive stress–strain curve.

4.9.7 The eighth compression test

After many discussions, it was decided to make another rubber vesicle with a small hole in the middle. The valve was not needed. After filling with water, the small hole was blocked. As shown in Figs. 4.38 and 4.39, three-dimensional aramid weaving was carried out. The fourth woven vesicle is shown in Fig. 4.40. On the one hand, the rubber and aramid vesicles were in complete contact. On the other hand, the vesicle is stronger than the balloon, and there will be no foreign matter in the aramid

Figure 4.38 The shock-resistant vesicle is frozen and sealed after being filled with water.

Figure 4.39 The shock-resistant vesicle is sealed twice.

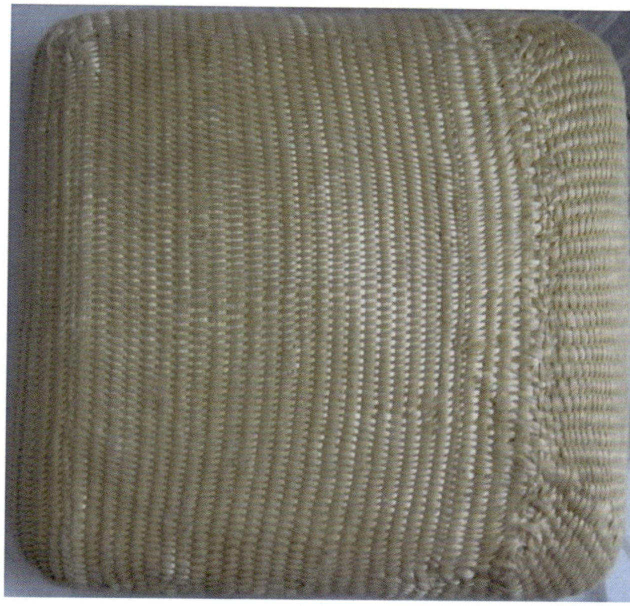

Figure 4.40 The woven fourth shock-resistant vesicle.

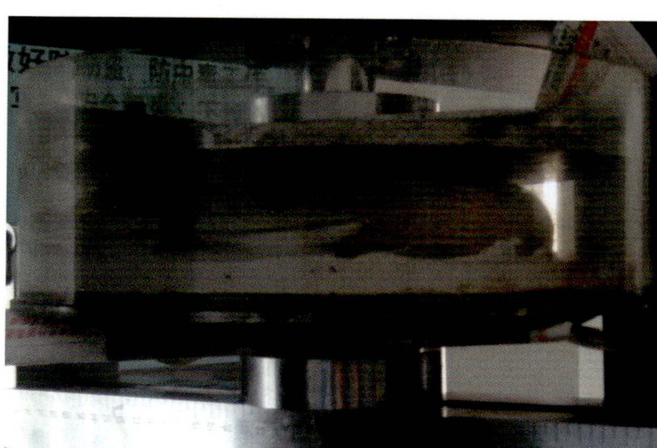

Figure 4.41 Compression test of 10 t press.

vesicle—stress concentration. The rubber vesicle is frozen and sealed after being filled with water, as shown in Fig. 4.38.

Subsequently, the compression test was carried out on the fourth piece of the shock-resistant vesicle, as shown in Fig. 4.41. As the maximum load of the previous several times was 8 tons, the purpose of this test is to withstand the pressure of about 8 tons or to exceed that of the last time. The results show that the shock-resistant vesicle can withstand the compression of greater load. Even if it exceeds the maximum limit of the experimental equipment, the shock-resistant vesicle is still intact. Its pressure deformation curve is shown in Fig. 4.42.

We used a 50 t pressure testing machine. Due to the large compression platform, a pressure-resistant aluminum alloy block was added to the upper cover plate during the experiment, as shown in Fig. 4.43. We first

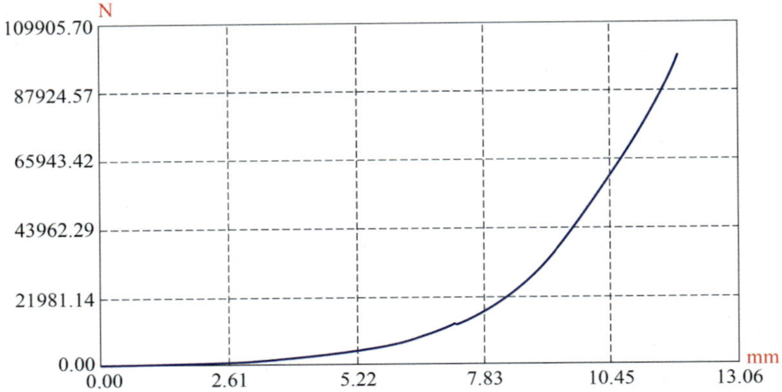

Figure 4.42 Compression load–displacement curve of 10 t compression test.

Figure 4.43 Compression test of 50 t press.

carried out compression according to the calculation's maximum load of 22 tons, and the compression curve is shown in Fig. 4.44. After reaching the set load, there is no water leakage, which indicates that it can withstand a greater load.

Then we conducted a compression experiment on the vesicle again, hoping to crush it to see how much compression load it could bear. The curve of the compression process is shown in Fig. 4.45. The maximum load at the beginning of the water leakage of the shock-resistant vesicle is 245,400 N, and the compression displacement is 8.31 mm. The outer aramid capsule is not damaged, as shown in Fig. 4.46, which indicates that the shock-resistant vesicle inside is broken. As it is not opened, it is

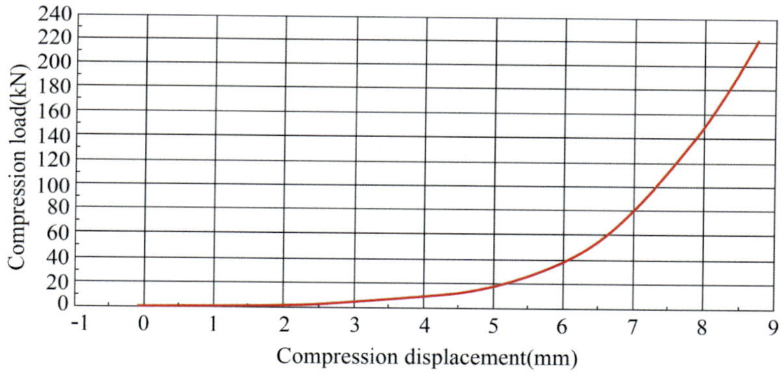

Figure 4.44 Compression load–displacement curve of 50 t press under constant load compression.

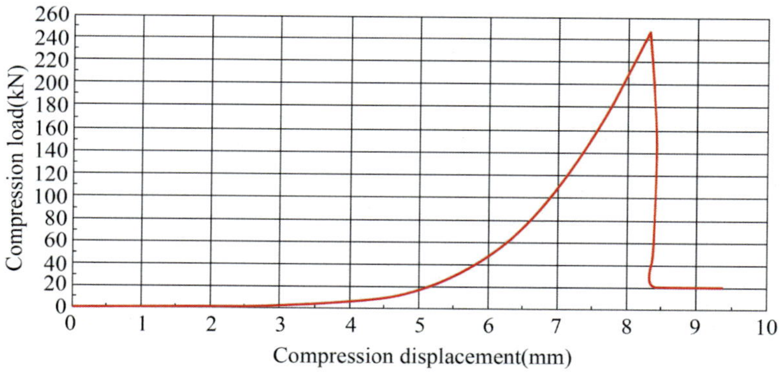

Figure 4.45 Compression load–displacement curve of 50 t press during destructive compression.

Figure 4.46 The appearance of the shock-resistant vesicle after water leakage.

unclear what part of the shock-resistant vesicle is broken. The crack is estimated to be very small because it has little water leakage and cannot be squeezed out by manpower.

So far, the project has been completed. As this project is an unprecedented exploratory subject, it involves many contents and problems such as high pressure, sealing, materials, composite, surface treatment, manufacturing process, and supporting devices. Through unremitting efforts, the task has been successfully completed.

References

[1] Xiong J, Huang Y, Wang QJ. Development and application of high performance fiber. FRP/Composite Materials 2004;005:49–52.

[2] China Petroleum and Petrochemical Engineering Research Association. Synthetic Fiber. 3rd ed. Beijing: China Petrochemical Press; 2012.

[3] Zou HT, Meng JG. Properties and applications of high performance fibers. Textile Science Research 2001;04:23–31.

[4] Zhao JX. Production, market and application of high-performance carbon fibers in the world. New Materials Industry 2003;000(003):18–22.

[5] Zhao ZG. Development and application of carbon fiber and its composite materials. Petrochemical Technology and Application 2002;55(04):58–76.
[6] Zhao JX. Status and progress of carbon fiber in the world. Fiberglass/Composite Materials 2003;2:40–3.
[7] Zhao JX, Wang MX. Development and application of high-performance carbon fibers for composite materials. New carbon materials 2000;015(001):68–75.
[8] Hong H, Bae KJ, Jung H, et al. Preparation and characterization of carbon fiber reinforced plastics (CFRPs) incorporating through-plane-stitched carbon fibers. Composite Structures 2022;284:115198.
[9] Meng J, Xiao B, Wu F, et al. Co-axial fibrous scaffolds integrating with carbon fiber promote cardiac tissue regeneration post myocardial infarction. Materials Today Bio 2022;16:100415.
[10] Xu XZ, Zhou J, Jiang L, et al. Lignin-based carbon fibers: carbon nanotube decoration and superior thermal stability. Carbon 2014;80(1):91–102.
[11] Shin HK, Park M, Kim HY, et al. An overview of new oxidation methods for polyacrylonitrile-based carbon fibers. Carbon Letters 2015;16(1):11–18.
[12] Wang SC, Li Y, Xiang HX, et al. Low cost carbon fibers from bio-renewable lignin/poly blends. Composites Science and Technology 2015;119(Nov. 23):20–5.
[13] Jiang XF, Ouyang Q, Liu DP, et al. Preparation of low-cost carbon fiber precursors from blends of wheat straw lignin and commercial textile-grade polyacrylonitrile. Holzforschung 2018;72(9):727–34.
[14] Jiang F, Sui KY, Yuan H, Zhang XY, Tan YQ. Exploration of virtual simulation teaching and course ideological and political model of wet spinning experiment. Education and Teaching Forum 2021;39:144–7.
[15] Ouyang Q, Xia KQ, Liu DP, et al. Fabrication of partially biobased carbon fibers from novel lignosulfonate–acrylonitrile copolymers. Journal of Materials Science 2017;52(12):7439–51.
[16] Xia KQ, Ouyang Q, Chen YS, et al. Preparation and characterization of lignosulfonate–acrylonitrile copolymer as a novel carbon fiber precursor. ACS Sustainable Chemistry & Engineering 2016;4(1):159–68.
[17] Sun FJ, Chai ZH. High-performance new chemical materials and their applications (2) Aramid fibers and their applications. New Chemical Materials 1998;05:41–3.
[18] Ding L, Xiao D, Zhao Z, et al. Ultrathin and ultrastrong Kevlar Aramid nanofiber membranes for highly stable osmotic energy conversion. Advanced Science 2022;9 (25):2202869.
[19] Tran TT, Le VA, Nguyen CH, et al. Thermoconductive ultralong copper nanowire-based aramid nanofiber nanopapers for electromagnetic interference shielding. ACS Applied Nano Materials; 2022.
[20] Ma XG, Liu Y. Development of high-performance fibers and their application in advanced composite materials. Fiber Composite Materials 2000;04:14–18.
[21] Liu L, Zhang X, Huang YD, et al. Research status and development trend of aramid surface and interface modification technology. High-tech Fibers and Applications 2002;027(004):12–17.
[22] Kong HJ, Qin ML, Ding XM, et al. Effect of *n*-hexane solvent on the polymerization of poly. Synthetic Fiber Industry 2017;40(003):27–30.
[23] Zhang K. Synthesis and Characterization of Aromatic Polyamide Copolymers Containing Benzimidazole. Synthetic Fibers 2017;46(002):22–6.
[24] Liang MS. Synthesis of poly (p-phenylene terephthalamide) polymers. Changchun University of Technology; 2016.
[25] Liu GW, Xu J, Luo XF, et al. Synthesis of poly-m-phenylene isophthalamide resin. Fine Chemical Intermediates 2017;47(004):46–50.

[26] Ao YH, Geng J, Chen G. Preparation of poly-m-phenylene isophthalamide by interfacial polycondensation. Polymer Materials Science and Engineering 2012;028(012):21–3 28.
[27] Fu Q, You XL, Liu ZF, et al. Preparation of pulp-like aramid fibers by direct polycondensation of low temperature solution. Synthesis Technology and Application 2002;017(004):8–10.
[28] Shin H. Vapor-phase preparation of aromatic polyamides. U.S. Patent 1977;4009153 2-22.
[29] Luan XN, Yu JR, Liu ZF. Ultra-high molecular weight polyethylene fibers and their applications. High-tech Fibers and Applications 2003;03:23–7.
[30] Lv SH, Liang GZ, He Y, et al. Research progress of ultra-high molecular weight polyethylene fibers. New Chemical Materials 2002;(8)).
[31] Li SJ, Li HS. Effects of high-energy radiation on the structure and mechanical properties of ultra-high molecular weight polyethylene fibers. Journal of Radiation Research and Radiation Technology 1998;016(003):135–40.
[32] Chen ZL, Yu JR, Liu ZF, et al. Effect of UV irradiation crosslinking on the structure and properties of ultra-high molecular weight polyethylene fibers. Polymer Materials Science and Engineering 2001;3:62–5.
[33] Zhao GL. Advances in the preparation and application of ultra-high molecular weight polyethylene fibers. Journal of Beijing Institute of Fashion Technology 2019;(2)).
[34] Zheng YC, Yang ZK, Wang XC, et al. Study on Melt Spinning UHMWPE/Polyolefin Blend System and Its Fiber Structure Properties. Journal of Beijing Institute of Fashion Technology (Natural Science Edition) 2017;01:12–21.
[35] The principle of melt spinning process, <https://bbuy.dye-ol.com/Solutions/53476.aspx>. Mar 31, 2021 [accessed 19.06.23].
[36] Lou LH, Osemwegie O, Ramkumar SS. Functional nanofibers and their applications. Industrial & Engineering Chemistry Research 2020;59(13):5439–55.
[37] Drawing a single nanofibre over hundreds of microns. Europhysics Letters 1998;42(2):215.
[38] Feng L, Li SH, Li HJ, et al. Super-hydrophobic surface of aligned polyacrylonitrile nanofibers. Angewandte Chemie International Edition 2002;41(7):1221–3.
[39] Martin CR. Membrane-based synthesis of nanomaterials. Chemistry of Materials 1996;8(8):1739–46.
[40] Ma PX, Zhang RY. Synthetic nano-scale fibrous extracellular matrix. Journal of Biomedical Materials Research 1999;46(1):60–72.
[41] Liu G, Ding J, Qiao L, et al. Polystyrene-block-poly (2-cinnamoylethyl methacrylate) nanofibers—preparation, characterization, and liquid crystalline properties. Chemistry—A European Journal 1999;5(9):2740–9.
[42] Munir MM, Iskandar F, Khairurrijal, et al. High performance electrospinning system for fabricating highly uniform polymer nanofibers. Review of Scientific Instruments 2009;80(2):1151.
[43] Li D, Mccann JT, Xia YN, et al. Electrospinning: a simple and versatile technique for producing ceramic nanofibers and nanotubes. Journal of the American Ceramic Society 2006;89(6):1861–9.
[44] Yang DX, Li EZ, Guo WL, et al. Research progress of electrospinning nanofibers and their industrialization. Materials Review 2011;15:64–8.
[45] Qin XH, Wang XW, Hu ZM, et al. Study the relationship between process parameters and fiber diameter of electrospinning polyacrylonitrile nanofibers. Journal of Donghua University: Natural Science Edition 2005;031(006):16–22.
[46] Li J. Preparation of polyvinyl alcohol nanofibers by electrospinning. Ship Chemical Defense 2009;04:6–11.
[47] Zhao ML, Sui G, Deng XL, et al. Polylactic acid nanofiber nonwoven felt by electrospinning. Synthetic Fiber Industry 2006;29(001):5–7.

[48] Wang PJ, Li CJ, Chang M, et al. Effect of solvent on spinnability of electrospun polyester nanofibers. Synthetic Fiber Industry 2007;30(002):11–13.
[49] Dong CH, Duan B, Yuan XY, et al. Preparation of polylactide microfibers by electrospinning. Journal of Biomedical Engineering 2005;22(006):1245–8.
[50] Gupta P, Wilkes GL. Some investigations on the fiber formation by utilizing a side-by-side bicomponent electrospinning approach. Polymer 2003;44(20):6353–9.
[51] Wang ZL, Zhang MY, Zhang YJ. Polyimide nanofibers prepared by electrospinning. Insulation Materials 2006;39(6):7–8.
[52] Zhu DD, Li CJ, Wang PJ, Chang M, Li XN, Gao XS. The effect of solvent on the spinnability of electrospun polystyrene and the formation of "bead" morphology. Journal of Beijing Institute of Fashion Technology 2006;01:1–7.
[53] Larsen G, Velarde-Ortiz R, Minchow K, et al. A method for making inorganic and hybrid (organic/inorganic) fibers and vesicles with diameters in the submicrometer and micrometer range via sol-gel chemistry and electrically forced liquid jets. Journal of the American Chemical Society 2003;125(5):1154–5.
[54] Shao CL, Guan HY, Liu YC, et al. A novel method for making ZrO_2 nanofibres via an electrospinning technique. Journal of Crystal Growth 2004;267(1/2):380–4.
[55] Chen C, Yu Y, Xue G, et al. TiO_2/PAN nanofiber felt prepared by electrospinning and its properties. Environmental Science and Management 2009;34(002):105–9.
[56] Liu Y, Xia N, Chen RY, et al. Study on the preparation of ZnO nanofibers by electrospinning and their photocatalytic properties. Journal of Fujian Normal University: Natural Science Edition 2008;01:66–9.
[57] Yang DX, Xiao P, Zhao XF, et al. Preparation of continuous SiC submicron/nanofibers by electrospinning. China Surface Engineering 2010;23(001):39–44.
[58] Li JY, Zhang YF, Zhong XH, et al. Single-crystalline nanowires of SiC synthesized by carbothermal reduction of electrospun PVP/TEOS composite fibers. Nanotechnology 2007;18(24):3999–4002.
[59] Liu L, Dong XT, Wang JX, et al. Preparation of Y2O3 nanofibers by electrospinning technology. Journal of Changchun University of Technology 2007;03:98–101.
[60] Shao CL, Guan HY, Wen SB, et al. Preparation and characterization of NiO nanofibers by electrospinning method. Journal of Chemistry of Universities 2004;25(6):1013 1013.
[61] Zhai GJ, Wang PJ, Li CJ, et al. Preparation of microporous Mn_2O_3 micro/nanofibers by electrospinning and characterization of fiber structure. Journal of Beijing Institute of Fashion Technology 2006;026(001):8–11.
[62] Guan HY, Shao CL, Liu YC, et al. Preparation of ZrO_2 nanofibers by electrospinning. Journal of Chemistry of Universities 2004;25(8):1413–15.
[63] Han ZY, Li YJ, Chen FT, Tang SP, Wang P. Preparation of ZnO/Ag_2O nanofibers by coaxial electrospinning and their photoelectric catalytic properties. Chemical Journal of Chinese Universities 2020;41(02):308–16.
[64] Reneker D, Yarin A, Zussman E, et al. Electrospinning of nanofibers from polymer solutions and melts. Advances in applied mechanics 2007;41:43–346.
[65] Mit-uppatham C, Nithitanakul M, Supaphol P. Effects of solution concentration, emitting electrode polarity, solvent type, and salt addition on electrospun Polyamide-6 fibers: a preliminary report//macromolecular symposia. Weinheim: WILEY-VCH Verlag 2004;216(1):293–300.
[66] Theron SA, Zussman E, Yarin AL. Experimental investigation of the governing parameters in the electrospinning of polymer solutions. Polymer 2004;45(6):2017–30.
[67] Sun LK, Cheng HF, Chu ZY, et al. Preparation of PAN-based hollow carbon fibers by coaxial electrospinning and two-step post-treatment. Acta Polymerica Sinica 2009;1(001):61–5.

[68] Bergshoef MM, Vancso GJ. Transparent nanocomposites with ultrathin, electrospun nylon-4, 6 fiber reinforcement. Advanced Materials 1999;11(16):1362−5.
[69] Dubson A, Bar E. Improved vascular prosthesis and method for production thereof. CA Patent 2001;CA2432164.
[70] Doshi J. Nanofiber-based nonwoven composites: properties and applications. Nonwovens World 2001;10(4):64−8.
[71] Laurencin LIW-J, Caterson E J CT, et al. Electrospun nanofibrous structure: a novel scaffold for tissue engineering. Journal of Biomedical Materials Research 2002;60 (4):613−21.
[72] Matthews JA, Boland ED, Wnek GE, et al. Electrospinning of collagen type II: a feasibility study. Acoustics Speech & Signal Processing Newsletter IEEE 2003;18 (2):125−34.
[73] Chu B, Hsiao BS, Hadjiargyrou M, et al. Cell storage and delivery system. US Patent, US20050014252, 2005.
[74] Gibson P, Schreuder-Gibson H, Rivin D. Transport properties of porous membranes based on electrospun nanofibers. Colloids and Surfaces A: Physicochemical and Engineering Aspects 2001;187−188:469−81.
[75] Gibson P, Schreuder-Gibson H, Pentheny C. Electrospinning technology: direct application of tailorable ultrathin membranes. Journal of Coated Fabrics 1998;28 (1):63−72.
[76] Schreuder-Gibson H, Gibson P, Senecal K, et al. Protective textile materials based on electrospun nanofibers. Journal of Advanced Materials 2002;34(3):44−55.
[77] Thomas HT, Auerbach M, Li HY. Non-woven fabrics in military combat suit system. Technical textiles 2001;09:36−42.
[78] Norris ID, Shaker MM, Ko FK, et al. Electrostatic fabrication of ultrafine conducting fibers: polyaniline/polyethylene oxide blends. Synthetic Metals 2000;114(2):109−14.
[79] Lee WS, Jo SM, Chun SW, et al. Method for preparing thin fiber-structured polymer web. US Patent, US10014550, 2002.
[80] Senecal K, Samuelson L, Sennett M, et al. Conductive (electrical, ionic, and photoelectric) polymer membrane articles, and method for producing same. US Patent, US10722213, 2006.
[81] Senecal KJ, Ziegler DP, He J, et al. Photoelectric response from nanofibrous membranes. Materials Research Society Symposium Proceedings, Organic Optoelectronic Materials, Processing and Devices 2002;708:285−9.
[82] Waters CM, Noakes TJ, Pavey I, et al. Liquid crystal devices. US Patent, US5088807, 1992.
[83] Scopelianos AG. Piezoelectric biomedical device. US Patent, US5522879, 1996.
[84] Kwoun SJ, Lec RM, Han B, et al. A novel polymer nanofiber interface for chemical sensor applications. In: Proceedings of the 2000 IEEE/EIA International Frequency Control Symposium and Exhibition, IEEE, 2000, 52−57.
[85] Kwoun SJ, Lec RM, Han B, et al. Polymer nanofiber thin films for biosensor applications. In: Proceedings of the IEEE 27th Annual Northeast Bioengineering Conference, IEEE, 2001, 9−10.
[86] Lee SH, Ku BC, Wang X, et al. Design, synthesis and electrospinning of a novel fluorescent polymer for optical sensor applications. MRS Online Proceedings Library 2001;708.
[87] Wang X, Lee SH, Drew C, et al. Highly sensitive optical sensors using electrospun polymeric nanofibrous membranes. MRS Online Proceedings Library 2001;708.
[88] Wang X, Drew C, Lee SH, et al. Electrospun nanofibrous membranes for highly sensitive optical sensors. Nano Letters 2002;2(11):1273−5.
[89] Xiong J, Shi MW, Zhou GT, et al. Study on mechanical properties of aramid fibers under high strain rate. Chinese Journal of Textiles 2000;06:7−10.

[90] Gao XL, Zhu JZ, Gu FC. Testing and analysis of aramid 1313/aramid 1414 fiber properties. Journal of Henan Institute of Technology 2017;29(001):1–5.
[91] Yang DL. Comprehensive research on the properties of domestic para-aramid fibers. Journal of Zhongyuan University of Technology 2009;20(05):47–51.
[92] Liu Y, An Y, Yan H. Comparison of mechanical performance of aramid fiber and its sort. Materials Engineering 2010; 4–2.
[93] Niu JF, Tang YF, Zhang YC. Development of polyurethane adhesive for fire hoses. Chemical Propellants and Polymer Materials 2010;03:55–8.
[94] Anonymous. Room temperature vulcanized silicone rubber. Organosilicon Fluorine Information 2005;7:21–3.
[95] Wu BY, Wang YP. Polyurethane elastic sealing material for building joints (1) Synthesis of liquid polyurethane rubber. Journal of Wuhan University of Technology 1987;001:1–5.

CHAPTER 5

New lens materials and processing methods

Contents

According to the Vision Council of America, approximately 75% of adults use some vision correction. About 64% of them wear eyeglasses, and about 11% wear contact lenses, either exclusively or with glasses. Therefore glasses or lens materials are an interesting topic to most people. According to the materials used, lenses are divided into three categories: natural crystals, optical glass, and optical resin.

5.1 Natural crystals

The natural crystal is crystal stone, a kind of transparent quartz composed of silicon dioxide, and its refractive index is higher than glass. It is divided into natural crystal and artificial crystal. The colors are white crystal, tea crystal, colored citrine, etc. Natural crystal is the ancient lens material. In the past, due to undeveloped science and technology, most lenses were made with natural crystals. This material contains impurities, which easily cause refraction disorder, poor ultraviolet (UV) absorption function, high

New Polymeric Products
DOI: https://doi.org/10.1016/B978-0-443-19407-8.00006-3

hardness, brittle quality, and difficulty in the process. In nature, the stock is small, and the price is high. In terms of optical performance, it is not as good as glass materials [1].

5.2 Optical glass

Optical glass is made of pure silicon dioxide as the primary raw material and contains calcium, aluminum, sodium, potassium, and other refined substances, also called inorganic glass. The advantage of optical glass is that it has stable chemical and physical properties, can resist the influence of weather for a long time and is not colored or faded. The disadvantage is that the lenses are heavy and easy to break, and modern resin lenses have replaced them [1].

5.3 Optical resin

Optical resin is made of polymer organic compounds through compression or injection molding. The hardness is close to the inorganic glass, lightweight, and has strong impact resistance, but it is easy to wear. There are three common raw materials for resin lenses: AC is PMMA (plexiglass), CR-39 is Columbia resin, and PC is polycarbonate [1].

Compared with glass lenses, resin lenses have their unique advantages. The following three common resin lenses are introduced.

1. CR-39 material

   CR-39 is a new type of spectacle lens material developed by the Columbia Chemical Branch of PPG Company in the United States. CR is the abbreviation of Columbia Resin. Its chemical name is allyl diethylene glycol carbonate, an optical thermosetting plastic. CR-39 optical resin lens is made of CR-39 monomer as the main material, mixed with several other chemicals, and is directly cast in a special glass mold by heating and temperature control. It does not require grinding and polishing. It is light and not easy to break [2].

   Table 5.1 compares the CR-39 resin and the glass lens. The main advantage of a resin lens is lightweight and is not fragile. This safety is especially suitable for teenagers and people who love sports. It is also very suitable for drivers because it can resist UV rays. Therefore it can effectively protect the cornea, lens, and retina of the eye. In addition,

Table 5.1 Comparison of CR-39 resin lens and glass lens [3].

Project	CR-39 resin lenses	Optical glass lens
Light transmittance	92%	91%
Proportion	1.32 g/cm^3	2.54 g/cm^3
Impact resistance test	Not broken	Broken
Fragmentation	Larger blocks are broken and have no sharp corners	Relatively finely divided
Ultraviolet blocking	350 mm	290–300 mm
Refraction	1.502	1.523
Wear resistance	Poor	Good
Atomization trend	Good	Bad
Thickness	Thicker	Thinner
Dyeing	Easy	Difficulty

the resin lens is also easy to dye into the color required by the wearer and processed into a variety of complex optical curved surfaces to make a bifocal lens, which is more suitable for the needs of patients. The disadvantage is that the lens is thick and has poor wear resistance, but its wear resistance can be significantly improved with the hardening technology [3].

2. AC material

The AC component is polymethylmethacrylate (PMMA), a thermoplastic material. The trade name is also known as PMMA. The refractive index is 1.49, the Abbe number (known as the V-number or constringence of a transparent material, is a measure of the material's dispersion) is 58, and the specific gravity is 1.19 g/cm^3.

Advantages: (1) Lightweight, 1/2 of the crown glass; (2) good impact resistance and safety.

Disadvantages:

It is a thermoplastic material with a softening point of 75°C. Therefore it is easily deformed by heat, resulting in a change in diopter;

1. Insufficient surface hardness and poor wear resistance;
2. Compared with CR-39, it is more brittle.

As AC sheets are relatively brittle in quality, the biggest problem lies in surface scratches. The surface hardness and scratch resistance can be improved by organo-siliconizing the inner and outer surfaces [1]. PMMA has been making spectacle lenses since the 1950s. Still, it is easily deformed by heat and has extremely poor wear resistance, so it has not been widely promoted in the spectacles industry. Nowadays, it is primarily used in low-price sunglasses.

5.4 PC material

PC (polycarbonate) is a thermoplastic polymer material that in injection molding processing methods can apply to yield PC products. PC has excellent creep resistance, low shrinkage (0.5%—0.7%), good dimensional stability, and is suitable for processing high-precision products.

Advantages:

1. Lightweight. The specific gravity of PC is 1.209 g/cm^3, which is less than that of CR-39 and glass. PC lenses have the lowest specific gravity for the same specifications and geometric dimensions, while the lens's weight is the lightest.
2. The lens is thin. The refractive index of CR-39 is 1.499, and the refractive index of PC is 1.586. The higher the refractive index, the thinner the lens. Therefore the PC lens is relatively thinner than the CR-39 lens.
3. The safety is excellent. PC has excellent impact resistance. It can be widely used in manufacturing aviation windows, bulletproof "glasses," riot masks, and shields. The impact strength of PC is as high as 87 kg/cm^2, which exceeds that of cast zinc and cast aluminum and is 12 times that of CR-39. Spectacle lenses made of PC are the only lenses that are "not broken" when falling on the concrete floor.

In addition, PC has many advantages, such as high heat resistance, strong practicability, and environmental protection, which is very suitable for glasses. Of course, PC also has disadvantages such as high processing difficulty, low Abbe number, and low surface hardness. Still, it can be overcome by modification and unique processing methods, and satisfactory results have been achieved [4].

5.4.1 Resin lens processing and molding process

5.4.1.1 Methods to realize resin lens processing on conventional injection molding machines

The mechanical structure of existing injection molding machines often determines specific molding process parameters, such as injection speed and pressure. The speed and pressure of this injection molding machine are determined when the screw diameter, the diameter of the injection cylinder, and the hydraulic system's working pressure and flow rate are also determined. Therefore a good and reasonable mechanical structure is necessary to produce qualified resin lenses successfully. Taking the 125 g injection molding machine produced by Wenzhou Guangming Plastic Machinery Factory as an example, according to the processing and molding characteristics of Acrylic, the machine has been modified in the following aspects, and excellent technical effects and economic benefits have been obtained [5].

1. The clamping part is where the mold is installed and the lens is formed. To ensure that the lens can be formed smoothly, the rigidity and clamping force of the clamping mechanism must be increased. The maximum clamping force is increased from 1000 to 1250 kN. The following improvements have been made to achieve the expected results: the diameter of the tie rod has been increased from the original φ60–φ70 mm. Currently, most of the clamping parts of horizontal injection molding machines are inverted five-hole obliquely arranged double-toggle mechanisms. The reaction force of this mechanism acting on the mold clamping force acts on the four tie rods [5]. According to the (Eq. 5.1):

$$Pc = \frac{L_1(\cos\alpha - \cos\alpha_0) + L_2(\cos\beta - \cos\beta_0)}{\frac{1}{ZC_p} + \frac{1}{C_k} + f_1 + f_2} \tag{5.1}$$

$$C = EF/L(2) \tag{5.2}$$

In the formula:

a. Pc is the deformation force of the clamping mechanism;
b. L_1 is the length of the rear link;
c. L_2 is the length of the front link;
d. α is the angle between the rear connecting rod and the horizontal axis when the mechanism is deformed;
e. α_0 is the angle between the rear connecting rod and the horizontal axis when the mechanism is just deformed;
f. β is the angle between the front connecting rod and the horizontal axis when the mechanism is deformed;
g. β_0 is the angle between the front connecting rod and the horizontal axis when the mechanism is just deformed;
h. $1/ZC_p$ is the stiffness of the tie rod;
i. $1/C_k$ is the total stiffness of the compressed parts;

j. f_1 and f_2 are the bending deflections of the front and rear templates;
k. E is the elastic modulus of the part;
l. F is the cross-sectional area of the force surface of the part;
m. L is the length of the part.

It can be seen from the above formula that the length of the tie rod is much longer than the length of other parts, and its cross-sectional area is much smaller than that of other stressed parts. The stiffness of the tie rod is dozens of worse than the rigidity of other parts. It can be considered that the pull rod has the most significant influence on the deformation force of the entire toggle mechanism. Therefore increasing the rigidity of the tie rod is the most effective way to increase the clamping force and is also the most economical way [5]. Enlarge the inner diameter of the clamping cylinder, and increase the inner diameter of the clamping cylinder from $\phi 70$ to $\phi 85$ mm. Taking a single toggle lever as an example, the formula for calculating the thrust of the clamping cylinder (because the principle of a double-toggle lever is similar to that of a single-toggle lever) is:

$$P_0 = L_1(1+\lambda)^2 C\alpha_0^3 \times 10^{-16} \tag{5.3}$$

In the formula:

a. P_0 is the thrust of the clamping cylinder;
b. L_1 is the length of the rear link;
c. λ is the piece length ratio;
d. α_0 is the critical angle;
e. C is the total stiffness of the clamping mechanism.

It can be seen from the formula [Eq. (5.3)] that L_1, λ, and α_0 are fixed values, and the total stiffness C of the mechanism is proportional to the thrust of the clamping cylinder. The tie rod's diameter increases, the mechanism's rigidity increases, and the thrust of the clamping cylinder increases correspondingly [5].

2. The injection part is key to whether the lens is molded. Two improvements have been made to the injection parts [5]:

The screw has been redesigned. According to the characteristics of high melt viscosity and poor fluidity of PMMA, while strictly controlling the screw speed, the length of the feed section is shortened, and the compression section is lengthened. Avoid the increase of resistance due to sudden changes in the flow channel. According to the fluid theory and fluid thermal power formula, the number of barrier ribs is increased, and the length of the mixing head runner is lengthened. In this way, the shear dispersibility of the plastic is increased, and the temperature and pressure fluctuations of the melt at the end of the barrel are reduced to facilitate the molding of the product.

Increase the maximum injection pressure. Injection pressure is an important parameter to ensure the smooth molding of PMMA lenses. Typically, PMMA lenses require higher injection pressure during molding. As mentioned earlier, the holding pressure for PMMA lens molding is 140–185 MPa. Increasing the diameter of the injection cylinder increases the maximum injection pressure from 174 to 215 MPa to ensure that the machine has sufficient pressure reserves. In addition, when the injection molding machine is working, the screw speed, nozzle temperature, and the conveying section should be strictly controlled. If necessary, a screw speed control device should be installed. Based on the above analysis and the existing models, the high-quality resin lenses have been mass-produced by adopting six molds. The machine's technical and molding process parameters after the specific transformation are shown in Tables 5.2 and 5.3 [5].

Finally, there are two points to be explained. First, because the moisture absorption rate of PMMA is 0.3%, it must be dried before molding. Second, heat treatment to eliminate internal stress is required after the lens is molded.

Table 5.2 Machine manufacturing parameters after transformation [5].

Project		**Numerical value**
Injection parts	Screw diameter /mm	35
	Screw length-to-diameter ratio	20
	Theoretical volume/cm^3	144
	Injection weight/g	125
	Injection rate/g/s	77
	Plasticizing ability/g/s	17
	Injection pressure/MPa	215
	Screw speed/r/min	200
Clamping parts	Clamping force/kN	1 250
	Move mold stroke/mm	267
	Tie rod inner distance/mm	343
	Maximum mold thickness/mm	400
	Minimum mold thickness/mm	80
Remaining parts	Maximum oil pump pressure/MPa	14
	Motor power/kW	13
	Electric heating power/kW	7
	Oil pump flow/L/min	76

Table 5.3 Molding process parameters after transformation [5].

Name	Numerical value
Barrel temperature/°C	First stage second stage third stage nozzle 220 240 230 215
Injection pressure/MPa	Level 1 Level 2 Level 3 6.5 8.0 7.5
Injection time/s	2.7
Holding pressure/MPa	9.5
Compress time/s	25
Cooling time/s	15
Clamping pressure/MPa	10
Screw speed/r/min	70
Mold temperature/°C	55

Conclusion: Through some mechanical structural transformations of conventional injection molding machines, resin lenses can be processed with high quality and quantity. This will significantly alleviate the pressure of introducing specially imported machines and promote the effective economic growth of the lens manufacturing industry [5].

5.4.1.2 Design of local spherical parallel machine tool for resin lens processing

Aiming at the complicated processing technology of the traditional injection and casting methods of optical resin lenses, a method of realizing lens processing using a local spherical parallel machine tool is proposed. The basic structure of the local spherical parallel machine tool is designed according to the theory of the degree of freedom of the space mechanism. By analyzing the working space of the local spherical parallel machine tool, a theoretical model of the rod's length change and the hinge angle is established, and the guarantee lens is researched on this basis. The design method of a local spherical parallel machine tool for resin lens processing is provided [6].

The parallel machine tool is developed based on the principle of the Stewart platform of the parallel spatial mechanism. It is a new concept machine tool that has only appeared recently. It is the product of the parallel robot mechanism and the machine tool. It combines space mechanism, mechanical manufacturing, numerical control technology, computer software and hardware technology. High-tech products are highly integrated with computer-aided design/manufacturing and CAD/CAM technology. It overcomes the inherent shortcomings of traditional machine tool tandem mechanisms: the tools can only be fed along a fixed guide rail, the degree of freedom of tool operation is low, and the equipment processing flexibility and maneuverability are insufficient. It can realize multicoordinate computerized numerical control machining, assembly and measurement functions. It can better meet the processing of complex and particular parts [6].

This design is to comply with the development trend of machine tools, starting from the advantages of parallel machine tools, innovatively proposed the development and design of local spherical parallel machine tools. The research and development of the local spherical parallel machine tool will effectively improve the processing efficiency of the lens, reduce labor intensity, minimize the space occupied by the machine tool, and improve the processing accuracy of the lens. It has a considerable development prospect [6].

1. Structural analysis and working principle

   Structural analysis: The local spherical parallel machine tool is composed of four parts: a tool, a lens holder, a worktable, and a telescopic rod (as shown in Fig. 5.1):

   Telescopic rod: one end of telescopic rod 1 is connected with a ball hinge, a rotating pair connects the other end, and a ball hinge connects both ends of telescopic rod 2.

   Workbench: In addition to the connection with the telescopic rod, the center position of the lower surface is also connected to the frame with a ball pin pair. It restricts the worktable from rotating in two directions so that the contour of the worktable is spherical.

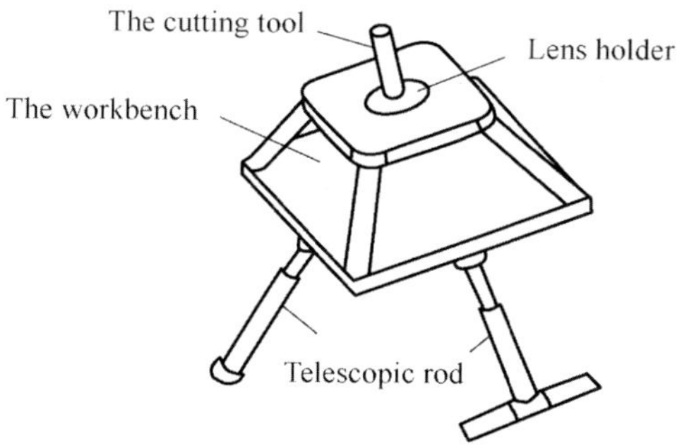

Figure 5.1 A simplified diagram of the spatial structure of a partial spherical parallel machine tool [6].

Movement and drive of the telescopic rod: It is driven by a stepping motor. Driven by the motor, the length of the telescopic rod is changed through the transmission of gears. The stepping motor is fixedly connected to the telescopic rod and moves with the movement of the telescopic rod.

Lens installation: directly fix the lens by air adsorption. The lens holder adopts a material with good flexibility and certain elasticity. The local spherical parallel machine tool can be used to process various contours of resin lenses, and the design and processing range of the lenses is 40 × 40 mm [6].

2. Brief description of the working principle

When processing the lens, the tool is located directly above the center of the upper surface of the working platform. It is fixed in the horizontal direction, and in the vertical direction, it moves up and down with the angle of the worktable and the requirements of the workpiece. This movement guarantees the complete processing of the lens periphery and ensures that the worktable and the cutter do not interfere. The workbench is in different postures under the coordinated action of the two telescopic rods, and the workbench can only rotate in two directions under the restraint of each hinge. It is a workbench with two degrees of freedom of movement. This workbench realizes the peripheral contour processing of the glasses. Different lens contours can be processed by controlling the telescopic speed of the length of the two telescopic rods [6].

5.4.2 New contact lenses and preparation process

The contact lens (CL) market continues to grow, and a quick search of patent documents shows that more than 100 patents have been applied since 2000, which shows that CL is becoming more and more popular. CL's applications range from vision correction and treatment to cosmetic appearance. In these applications, the needs of the end user of the lens include wearing length, comfort, durability, operational practicability, and visual stability. This also means requirements from manufacturers in the application of contact lenses, such as the material cost of CLs, ease of production, and reliability. Finally, the needs of manufacturers determine the parameters of the material, and scientists must focus on the development of CL materials. This premise guided material scientists from the glass scleral lens of the 1930s to the rigid, nonbreathable PMMA of the 1940s. Hydrogel (polymer and silicone) lenses ushered in the 1960s and 70s, and silicone hydrogel proved to be the essential CL material today [7].

Usually, "hard" or "soft" labels are used as a general definition of CL. Hard CLs are rigid (durable), breathable lenses, while soft contact lenses (SCLs) are made of flexible, high-water-content materials. Hard lenses are often interchangeably called rigid gas-permeable lenses (RGP). However, this is not entirely true. The original PMMA lenses can be classified as hard lenses. In contrast, modern RGP lenses are more flexible due to the addition of low-modulus components—therefore they are more rigid. Another defining feature is that PMMA hard lenses have no oxygen permeability, while RGP lenses have oxygen permeability [7].

On the other hand, a SCLs is a highly flexible, oxygen-permeable material, usually with high water content. This flexibility means that SCL adapts to the shape of the user's eye faster than rigid lenses. SCL can be processed daily, weekly, or monthly. These comprehensive definitions of CL can imply its material properties but cannot be determined. The materials used between hard and soft lenses usually overlap, such as silicone hydrogel and RGP. Although both use silicone materials, the gel network and water content differ. The derivatives can further diversify the range of possible CL and its properties. The requirements for CL are extensive, and many existing CLs are on the market to reflect this: daily disposable lenses, weekly/monthly lenses, special wear lenses, and even lenses that can be worn overnight. The user's needs can be described by several general parameters, such as comfort, wearing time, handling (cleanliness, ease of use), cost, and visual specifications [7].

PMMA is a durable, optically transparent polymer with limited hydrophilic properties. However, the oxygen permeability of PMMA is negligible, which may cause some eye health problems, such as hypoxia [8,9]. Researchers soon discovered polymer hydrogels, usually based on hydroxyethyl methacrylate (HEMA) [10,11]. These polymer hydrogels are composed of hydrophilic monomers, which means they contain electrochemical polarity that allows interaction with water. This provides much more excellent biocompatibility than PMMA. These hydrogels are also oxygen-permeable and flexible materials that can hold large amounts of water in the polymer network. These factors have improved CL's comfort, oxygen permeability, and wearing time, resulting in many HEMA-based CL derivatives [12,13].

The subsequent development of CL is silicon-based rigid lenses and hydrogels. These materials have very high air permeability, so long-wear lenses can be made. However, silicone materials are inherently hydrophobic, which makes them uncomfortable due to poor wettability and abrasion. Polymer scientists overcome this by combining silicone monomers (for example, siloxymethacrylates and fluoro) with hydrophilic comonomers to add the required hydrophilicity [14–17]. This is particularly successful for silicone hydrogel lenses, the essential CL material on the market today.

Newer CL materials appeared in the 1990s, such as polyvinyl alcohol (PVA), a low-cost and very hydrophilic polymer [18,19]. PVA hydrogel was the subject of research as a potential CL material in the early 1990s [20]. New surface coatings, such as polyethylene glycol (PEG), have recently improved silicone-based CL's hydrophilicity [21].

5.4.3 Manufacturing of contact lens materials

To make CL, there must be suitable polymer materials. This opens up countless possibilities, not only from the range of polymers but also from the formulation of ingredients in a given formulation. In addition, different types of polymerization mechanisms can be considered to form the same polymer, such as free radical polymerization (FRP) and catalytic polymerization and derivatives. The polymerization conditions (temperature, initiator type, the container used, etc.) can be changed to produce the same polymer but with different characteristics. Finally, the material must be suitable for the manufacturing stage, including the synthesis, inspection, and packaging processes. Fig. 5.2 contains some of the most important factors when designing a CL from a materials science point of view. Therefore Table 5.4 assesses the advantages and disadvantages of current CL materials [7].

From a material science point of view, CL lenses depend on many parameters. According to the specific needs of CL, more emphasis is needed on specific characteristics. The final CL material takes into account wear time and comfort. These characteristics usually depend on the material but include manufacturing processes such as plasma treatment [7].

For clarity, polymers are macromolecules composed of hundreds or thousands of repeating molecules called monomers. Each monomer has a covalent bond between each linking unit. Some typical monomers and polymers used to produce CL are shown in Fig. 5.3. The process of forming polymers is called polymerization. Another reactive molecule, the

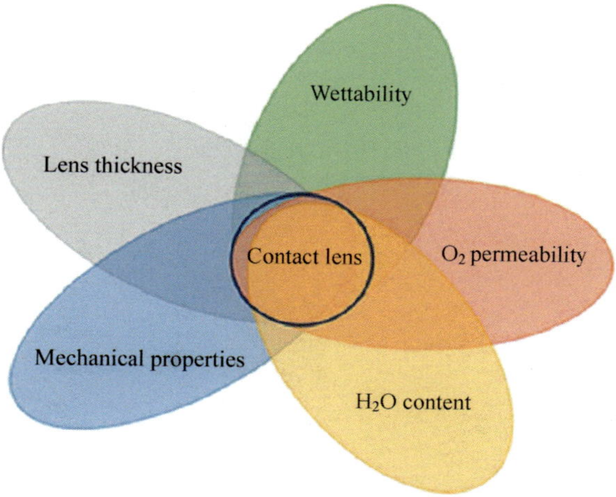

Figure 5.2 Requirements of contact lens [7].

Table 5.4 General pros and cons of current CL material classes.

CL material	Pro	Cons
PMMA	Inexpensive, well-understood polymer	No oxygen permeability, inflexible on the eye
RGP	High oxygen permeability, durable	Expensive regents require hydrophilic comonomer and can be abrasive
HEMA hydrogel	Inexpensive, biocompatible, abundant copolymer possibilities	Low oxygen permeability, protein deposition issues
Silicone hydrogel	High oxygen permeability, durable, comfortable	Expensive regents require hydrophilic comonomer and can be abrasive
PVA	Inexpensive, straight-forward manufacturing biocompatible	Low oxygen permeability, fixed water content

initiator, is used to start the polymerization. The selection of the polymerization initiator is usually based on the interaction with the reactive functional groups in the monomer. When the initiator decomposes, free radicals are formed, and then the polymerization reaction is initiated by stripping the free radicals from the monomer functional groups. Now there is a free radical inside the monomer, which reacts with the adjacent

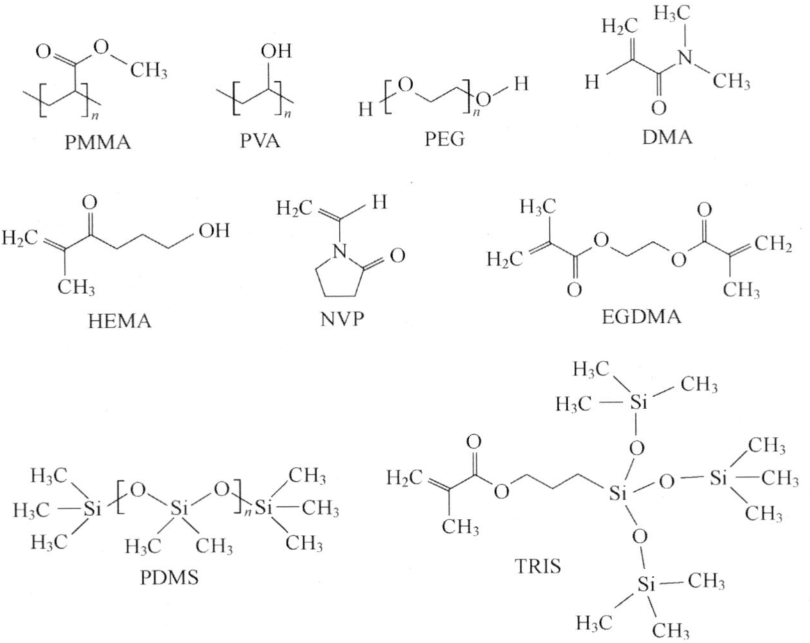

Figure 5.3 The chemical structures of common monomers and polymers used to produce CLs. This includes some macromonomers and cross-linking agents. *DMA*, dimethyl methacrylate; *EGDMA*, ethylene glycol dimethacrylate; *HEMA*, hydroxy ethyl methacrylate; *NVP*, N-vinyl pyrrolidone; *PDMS*, poly dimethyl siloxane; *PEG*, polyethylene glycol; *PMMA*, poly methyl methacrylate; *PVA*, polyvinyl alcohol; *TRIS*, 3-[tris(trimethylsiloxy)silyl]propyl methacrylate [7].

monomer functional group to strip a free radical from the group. The original monomer will form a new bond with the new monomer unit and now has a new free radical to continue polymerization. This growth step is repeated continuously, forming a polymer over time, and the polymer will disappear when no more monomer is consumed. When the free radicals are quenched in another way, for example, by a free radical scavenger, the polymerization is terminated. The initiator, such as UV or thermal initiator, can be selected according to its practicality, which is an essential consideration in the manufacturing method of producing CL [7].

5.4.4 Polymerization mechanism

FRP is a chain growth polymerization mechanism. FRP is usually not easy to control, resulting in high dispersibility (significant molecular weight changes) of the resulting polymer. This leads to a distribution of polymer properties. The polymer chain length usually determines high-strain and tensile-modulus polymers [22].

One advantage of FRP is that it easily forms a gel network because polymerization has many initiation points in the container. This allows many chains to grow simultaneously, physically entangled or cross-linked to form a gel network. Cross-linking depends on one of the monomers containing two functional groups, which allows chemical bonding to two different polymer chains. Cross-linking polymer chains generally improve gelation and increase the modulus of the material. Most CLs are produced using FRP [16,23–27]. It is easy and does not require expensive reagents such as catalysts. In addition, unwanted/unreacted chemicals can be removed from the material during the cleaning process after manufacturing. Catalysts are usually heavy metals, which should be avoided for human health.

Today, most CLs are made by polymerizing two or more monomers [25,28]. Copolymers combine the characteristics of a single polymer; therefore copolymerization is usually the first way to solve the problem of a single polymer. For many years, this principle has dominated the development of CL. For example, silicone polymers are very hydrophobic despite their high oxygen permeability. Therefore they are not ideal as homopolymer CL materials. However, silicone copolymers and hydrophilic (highly polar) monomers can solve this problem. Copolymerization can also improve physical properties by adding molecular weight to the chain by cross-linking polymer chains. Soft polymers can sometimes reduce the modulus of rigid materials (usually silicone).

Table 5.5 contains the properties of common CL materials and related copolymers, including oxygen permeability, water content, modulus, and wear time. The wearing time is based on the maximum wearing time of CL before the appearance of eye health problems. The copolymerization effect of rigid materials derived from PMMA is particularly significant, and its modulus is significantly reduced. In contrast, the stable range of modulus of hydrogel lenses is much wider, but the water content and oxygen permeability vary greatly.

Full-density polymers use molecular weight and intermolecular forces to strengthen the material, partly because PMMA has a high modulus. Other factors, such as the thickness of the final lens, are also important. The characteristics of very low-density polymers (10% of total density, 90% air), such as porous materials, are sensitive to small changes in a

Table 5.5 General characteristics of some common CL materials.

Material	Oxygen permeability (Dk/t)	Water content (wt. %)	Modulus (MPa)	Wear time (days)[a]
PMMA	0	0	1000	<1
PMMA-silicone	15	0		
Silicone-HEMA (rigid)	10–00	0	10	
HEMA hydrogel	10–50	30–80	0.2–2	1–7
HEMA-NVP	60–200	20–55	0.2–2	About 7–28
HEMA-MMA				
Silicone (PDMS)				
Hydrogel				
TRIS-DMA				
PDMS-HEMA				
PVA	10–30	60–70		<1

[a]Maximum wear time without extensive complications to the eye before lens disposal.

crossover [29,30]. Another study showed that Young's modulus of 90% of porous materials was reduced by 40% due to low cross-linking efficiency [31]. From the final material properties and manufacturing considerations, this may be related to the hydrogel with very high water content [32].

Maldonado-Codina and Efron emphasized improving the process between polymerization batches and manufacturing methods. Other important factors include the uniformity of the wall's apex, uniform pore size, etc. These cannot be guaranteed to be consistent with FRP. In addition, these factors must be considered in future designs if we consider SCL a porous solid. Although FRP has always been the leading force in manufacturing high-quality CL materials, it may not be able to produce a fully effective network for functional materials [33]. In addition to the many reasons stated by Dixon et al., this may be another reason why no drug delivery CL is on the market today [34].

Alternative polymerization mechanisms may be of interest to improve the physical properties of CL materials. Other kinds of methods include catalysts or controlled FRP. A controlled method of FRP is chain transfer polymerization. This has been a hot spot for the growth of polymer science, including research in other biological applications [35–37]. The chain transfer agent accurately mediates the growth of the polymer chain so that the molecular weight can be predesigned [22,38,39]. This results in a low-dispersion polymer, which means that the resulting performance and structure are more reliable than FRP. Many examples of hydrogels use chain transfer agents [40–42]. A specific reverse addition-fragmentation chain transfer (RAFT) polymerization example includes silicone-based polymers [43]. Other researchers have also used RAFT to modify polyacrylic acid pH-responsive hydrogels for drug delivery [42]. This opens up the potential of RAFT-synthesized silicones and traditional hydrogels such as CL. Recently, Zhang et al. synthesized a promising RAFT polymerized SCLs based on polyallyl methacrylate and PEG [44]. The lens has a low contact angle (<80 degrees), a high Dk value. The Dk value measures oxygen permeability, and the Dk/t value measures the oxygen transmissibility of a contact lens. Dk/t is calculated using the oxygen permeability (Dk) of the material and the thickness (t) of the contact lens (> 100 barriers), and a modulus of elasticity ranging from 0.5 to 1.5 MPa, which are all within the CL parameter range given in Table 5.5.

5.4.5 Preparation process

The manufacturing method of contact lenses is related to the material properties. The use of polymethyl methacrylate (PMMA) is used to manufacture rigid contact lenses for car polishing. The casting method is often used to manufacture SCLs with poly HEMA and silicone rubber [45].

5.5 Car throwing method

The turning polishing method is mainly used for complicated contact lens processing. It is first turned on a dedicated bench lathe with a diamond knife and then polished on a bench polishing machine. The edge grinding is mainly shaped and polished manually on the rotating tip. However, the PMMA sheet can also be cut into small disks, hot-pressed in a well-polished mold, and then ground and polished to make a hard contact lens. The turning method can also process SCLs. However, when processing HEMA by turning method, it should be noted that the polishing accessories cannot be mixed with water. The size during processing and after swelling should be accurately controlled, and the processing environment should be kept dry [45].

5.6 Casting method

There are two casting methods: mold casting and centrifugal casting. The molding and casting method can be divided into two steps: molding and casting. The former uses heating and pressure to inject solid thermoplastic into a molding mold to solidify it into a molding plastic mold. The latter refers to injecting the monomer material into a plastic mold of a certain shape and transforming the liquid hydrophilic organic monomer into a solid plastic through a FRP process. The production process of the general hydrophilic SCL is first to make a plastic mold through molding, then use the plastic mold to pour the liquid monomer into the mold, place the cross-linked HEMA monomer in the concave mold (front in the curved mold), and close the aluminum mold (back curved mold). Flanges on the aluminum mold's periphery limit the mold's closing and control the lens's thickness. Finally, the mold is placed in a furnace to heat and cure [45].

The centrifugal casting method injects the HEMA monomer into a rotating mold, making it flow and solidify in the mold by centrifugal force. This process accurately calculates and mixes the two monomers in a particular container. Under the action of gravity, they are injected into a rotating mold. The centrifugal force causes the monomer to flow outward, and the surface tension of the monomer limits its possibility of being thrown out. When the droplet begins to extend, the polymerization reaction begins until it is completed in the mold. In addition, the rotation speed of the mold has a significant influence on the thickness of the lens. The coordinated control of related process parameters in this process, such as speed, material density, specific gravity, dosage, surface tension, centrifugal force, and gravity, can be controlled by a computer [45].

References

[1] Chen D. Analysis of lens materials. China Journal of Optical Science and Technology 2013;11:115–16.
[2] Chen JD. An ideal new type of spectacle lens—CR39 optical resin lens. Optical Technology 1986;04:28–9.
[3] Chen H, Luo LQ, Li MR. Analysis of the advantages and disadvantages of resin lenses and glass lenses. Journal of Baotou Medical College 2003;19(2):125–6.
[4] Su YM. PC lens—the dominant spectacle lens in the 21st century. China Optical Science and Technology Journal 2005;5:83–5.
[5] Pan XS. The method of realizing resin lens processing on conventional injection molding machine. Journal of Wuhan University of Technology 2001;23(10):56–8.
[6] Lin L. Design of partial spherical parallel machine tool for resin lens processing. Journal of Harbin University of Science and Technology 2012;17(3):115–19.
[7] Musgrave CSA, Fang F. Contact lens materials: a materials science perspective. Materials 2019;12(2):261.
[8] Bruce A, Brennan N, Lindsay R. Diagnosis and management of ocular changes during contact lens wear, part 1. Ciln Signs Ophthalmol 1995;16(4):2–11.
[9] Mcmahon TT, Zadnik K. Twenty-five years of contact lenses: the impact on the cornea and ophthalmic practice. Cornea 2000;19(5):730–40.

[10] Wichterle O, Lim D. Hydrophilic gels for biological use. Nature 1960;185 (4706):117–18.
[11] Wichterle O. Method of manufacturing soft and flexible contact lenses. US Patent, US3496254; 1970.
[12] Lim D., Kopecek J., Nee S.B., et al. Hydrophilic N, N-diethyl acrylamide copolymers, US Patent, US05/657602; 1978.
[13] Yokoyama Y., Masuhara E., Kadoma Y., et al. Soft contact lens. US Patent, EP82304053.0; 1987.
[14] Mitchell D.D. Wettable silicone resin optical devices and curable compositions therefor. US Patent, US06/475270; 1984.
[15] Chromecek R.C., Deichert W.G., Falcetta J.J., et al. Polysiloxane/acrylic acid/polcyclic esters of methacrylic acid polymer contact lens. US Patent, US06/075365; 1981.
[16] Gaylord N.G. Oxygen permeable contact lens composition, methods and article of manufacture. US Patent, US06/215486; 1983.
[17] Keogh P.L., Kunzler J.F., Niu G.C.C. Hydrophilic contact lens made from polysiloxanes which are thermally bonded to polymerizable groups and which contain hydrophilic sidechains. US Patent, US06/102009; 1981.
[18] Buehler N, Haerri HP, Hofmann M, et al. Nelfilcon A, a new material for contact lenses. CHIMIA International Journal for Chemistry 1999;53(6):269–74.
[19] Goldenberg M. Polyoxirane crosslinked polyvinyl alcohol hydrogel contact lens. US Patent, US06/693484; 1986.
[20] Kita M, Ogura Y, Honda Y, et al. Evaluation of polyvinyl alcohol hydrogel as a soft contact lens material. Graefe's Archive for Clinical and Experimental Ophthalmology 1990;228(6):533–7.
[21] Imafuku S. Silicone hydrogel soft contact lens having wettable surface. US Patent 2019;10(241):234.
[22] William G, Perkins, et al. The effect of molecular weight on the physical and mechanical properties of ultra-drawn high density polyethylene. Polymer Engineering & Science 1976;16(3):200–3.
[23] Santos J, Alvarez-Lorenzo C, Silva M, et al. Soft contact lenses functionalized with pendant cyclodextrins for controlled drug delivery. Biomaterials 2009;30(7):1348–55.
[24] Seidner L., Spinelli H.J., Ali M., et al. Silicone-containing contact lens polymers, oxygen permeable contact lenses and methods for making these lenses and treating patients with visual impairment. US, US5331067; 1994.
[25] Hahn D., Johansson G.A., Ruscio D.V., et al. Method for manufacturing hydrophilic contact lenses., US Patent, 4,983,332; 1991.
[26] Tanaka K., Takahashi K., Kanada M., et al. Copolymer for soft contact lens, its preparation and soft contact lens made thereof. US Patent, 4139513; 1979.
[27] Keeley E.M. Method of manufacturing soft contact lens buttons. US Patent, 4931228; 1990.
[28] Lin CH, Yeh YH, Lin WC, et al. Novel silicone hydrogel based on PDMS and PEGMA for contact lens application. Colloids and Surfaces B: Biointerfaces 2014;123:986–94.
[29] Zhao J, Mayumi K, Creton C, et al. Rheological properties of tough hydrogels based on an associating polymer with permanent and transient crosslinks: effects of crosslinking density. Journal of Rheology 2017;61(6):1371–83.
[30] Narita T, Mayumi K, Ducouret G, et al. Viscoelastic properties of poly (vinyl alcohol) hydrogels having permanent and transient cross-links studied by microrheology, classical rheometry, and dynamic light scattering. Macromolecules 2013;46(10):4174–83.
[31] Musgrave CSA, Nazarov W, Bazin N. The effect of para-divinyl benzene on styrenic emulsion-templated porous polymers: a chemical Trojan horse. Journal of Materials Science 2017;52(6):3179–87.

[32] Maldonado-Codina C, Efron N. Impact of manufacturing technology and material composition on the mechanical properties of hydrogel contact lenses. Ophthalmic and Physiological Optics 2004;24(6):551−61.
[33] Luo Y, Wang AN, Gao X. Pushing the mechanical strength of PolyHIPEs up to the theoretical limit through living radical polymerization. Soft Matter 2012;8(6):1824−30.
[34] Dixon P, Shafor C, Gause S, et al. Therapeutic contact lenses: a patent review. Expert Opinion On Therapeutic Patents 2015;25(10):1117−29.
[35] Fairbanks BD, Gunatillake PA, Meagher L. Biomedical applications of polymers derived by reversible addition-fragmentation chain-transfer (RAFT). Advanced Drug Delivery Reviews 2015;91:141−52.
[36] Pai TSC, Barner-Kowollik C, Davis TP, et al. Synthesis of amphiphilic block copolymers based on poly (dimethylsiloxane) via fragmentation chain transfer (RAFT) polymerization. Polymer 2004;45(13):4383−9.
[37] Abdollahi E, Khalafi-Nezhad A, Mohammadi A, et al. Synthesis of new molecularly imprinted polymer via reversible addition fragmentation transfer polymerization as a drug delivery system. Polymer 2018;143:245−57.
[38] Junkers T, Lovestead TM, Barner-Kowollik C. The RAFT process as a kinetic tool: accessing fundamental parameters of free radical polymerization. Handbook of RAFT Polymerization 2008;105−49.
[39] Krstina J, Moad G, Rizzardo E, et al. Narrow polydispersity block copolymers by free-radical polymerization in the presence of macromonomers. Macromolecules 1995;28(15):5381−5.
[40] Bivigou-Koumba AM, Görnitz E, Laschewsky A, et al. Thermoresponsive amphiphilic symmetrical triblock copolymers with a hydrophilic middle block made of poly (N-isopropylacrylamide): synthesis, self-organization, and hydrogel formation. Colloid and Polymer Science 2010;288(5):499−517.
[41] Hemp ST, Smith AE, Bunyard WC, et al. RAFT polymerization of temperature- and salt-responsive block copolymers as reversible hydrogels. Polymer 2014;55 (10):2325−31.
[42] Liu J, Cui L, Kong N, et al. RAFT controlled synthesis of graphene/polymer hydrogel with enhanced mechanical property for pH-controlled drug release. European Polymer Journal 2014;50:9−17.
[43] Guan CM, Luo ZH, Qiu JJ, et al. Novel fluorosilicone triblock copolymers prepared by two-step RAFT polymerization: synthesis, characterization, and surface properties. European Polymer Journal 2010;46(7):1582−93.
[44] Zhang C, Liu Z, Wang H, et al. Novel anti-biofouling soft contact lens: l-cysteine conjugated amphiphilic conetworks via RAFT and Thiol−Ene click chemistry. Macromolecular Bioscience 2017;17(7):1600444.
[45] Zhao JM. Review of contact lens manufacturing. Optical Technology 1987;04:44−6.

CHAPTER 6

Biomedical applications of polymer materials

Contents

6.1 Introduction

"How to realize the localization of products related to people's livelihood as soon as possible is the direction that people who study materials, especially biomedical materials, need to work hard. If basic research is not

New Polymeric Products
DOI: https://doi.org/10.1016/B978-0-443-19407-8.00007-5

done well, applied research is not done well. But if we only do basic research, we will not be able to do well. When doing applied research, we just publish articles, which may not solve practicable problems [1]." Chen Xuesi, the academician of the Chinese Academy of Sciences, the researchers of Changchun Institute of Applied Chemistry of the Chinese Academy of Sciences, made a statement at the "seminar on the current and development trend of advanced materials" at the 22nd annual meeting of the China Association for Science and Technology.

Most human tissues and organs are composed of polymer compounds, so biomedical polymer materials have unique characteristics when applied in medicine. The development stages of medical polymer materials are shown in Table 6.1.

Classification of medical polymer materials:

1. According to the source, medical polymer materials can be divided into synthetic and natural.
2. According to their properties, medical polymer materials can be divided into nondegradable and biodegradable.

   Nondegradable polymer materials must be stable for a long time in the physiological environment without degradation, cross-linking, and physical wear and have good mechanical properties. Although there is no tough polymer, it is required that itself and its degradation products

Table 6.1 Development stage of medical polymer materials.

Stage	Time/sign	Characteristic
Phase I	Since 1937	The polymer materials used are ready-made materials
Phase II	Since 1953, it has been marked by the emergence of medical silicone rubber	Optimize synthetic polymers' composition, formula, and process at the molecular level and purposefully develop the required polymer materials
New stage	From looking for synthetic materials to replace biological tissues, we have turned to a new kind of materials that can actively induce and stimulate the regeneration and repair of human tissues	The organic combination of living tissue, cells, and artificial materials generally forms this material. In vivo, it promotes the growth of surrounding tissues and cells

will not produce apparent toxic side effects on the body. At the same time, the material will not be catastrophically damaged. Such materials include polyethylene, polypropylene, polyacrylate, aromatic polyester, polysiloxane, and polyoxymethylene, mainly used for human hard and soft tissue repair and artificial organs, artificial blood vessels, contact lenses, and adhesives.

Biodegradable polymer materials can cause structural damage in the physiological environment, and the body can excrete the degradation products through regular metabolism. Some materials such as collagen, aliphatic polyester, chitin, cellulose, polysaccharides, polyvinyl alcohol, and polycaprolactone are mainly used as drug release carriers and non-permanent implants.

3. Medical polymer materials can be divided into cardiovascular, hard tissue, and soft tissue repair materials according to the purpose or use.

Medical polymer materials are the earliest and most widely used biomedical materials. This chapter introduces several applications of polymer materials in biomedicine, including artificial blood vessel walls, anti-*Mycobacterium tuberculosis* composite drug-carrying fiber, dragon blood wound dressing, fibrous plaster, preparation and properties of the controlled-release fiber.

6.2 Artificial blood vessel

In modern society, cardiovascular disease threatens human health. According to statistics, by 2030, the number of people dying of cardiovascular diseases worldwide will increase to 23.3 million [2] every year. Such conditions can be treated by improving living and eating habits, drugs, or surgery. Severe cases need vascular surgery, such as vascular transplantation. Because of the limited source of autologous blood vessels and the problem of immune rejection of allogeneic blood vessels, researchers are committed to studying an artificial blood vessel that can replace autologous blood vessels.

6.2.1 Development of artificial blood vessels

The development of vascular grafts has roughly experienced five stages (as shown in Table 6.2): biological tissue type, synthetic type, artificial physical hybrid type, tissue engineering type, and stent artificial blood vessels [5].

Table 6.2 Development stages of vascular grafts.

Stage	Characteristic	Description of representative events or details
Exploration period	Researchers have initially explored vascular grafts but are limited to conditions; most graft materials are rigid plastic tubes—severe coagulation after transplantation	a. In the early 20th century, scholars from various countries tried to carry out animal vascular transplantation experiments with glass, metal and rubber b. In the 1940s, Hufnaged [3] conducted a blood vessel transplantation experiment with a hard plastic tube, which caused a severe coagulation reaction In an experiment in 1952, Voorhees accidentally found that the silk suture implanted in the organism was covered with a layer of endothelial cells. He imagined that assuming that the same phenomenon occurred in the fabric embedded in the organism, the direct contact between blood and the implant could be avoided to prevent the occurrence of coagulation, which provided ideas for the development of artificial blood vessels
Biological tissue type	Blood vessel transplantation extracted from patients, others or animals is also called autologous, allogeneic, and xenotransplantation. The latter two immune rejection is difficult to rule out. Few blood vessels are available for autologous transplantation	a. Gluck (1898) and Carrel (1906) successfully used autologous veins instead of artery transplantation [4] b. Autologous internal thoracic artery and radial artery are widely used in coronary artery bypass grafting, and the great saphenous vein is used to establish peripheral artery bypass [5]

Synthetic type	Artificial blood vessels need mesh	a. The turning point from self to artificial. In 1952, Voorhees was transplanted into the abdominal aorta of dogs with polyester as material b. Critical theory: The mesh principle is that artificial blood vessels must have appropriate mesh c. In 1957, polyester artificial blood vessels were first used in surgery d. In 1959, Edwards developed polytetrafluoroethylene (PTFE) artificial blood vessels [6]
Artificial biological hybrid	A layer of biomaterial is coated on available synthetic materials to improve biocompatibility. Four kinds of coating materials are reported in the literature: albumin, fibronectin, collagen and gelatin (new silk fibroin)	a. Albumin is not easy to adhere to platelets and does not participate in blood coagulation. It can improve anticoagulant performance b. Fibronectin is a double-chain macromolecule that promotes intimal formation, transplantation, and coagulation c. Collagen: promote fibroblast migration and intimal formation d. Gelatin: promote cell adhesion e. Silk Fibroin [7]: Antithrombotic and good mechanical strength f. The inner and outer walls of bovine collagen-impregnated polyester artificial blood vessels were fixed by thermal crosslinking to reduce the water permeability of blood vessel walls [8]

(*Continued*)

Table 6.2 (Continued)

Stage	Characteristic	Description of representative events or details
Tissue engineering	The artificial blood vessel prepared by tissue engineering technology has no rejection, thrombosis, growth or plasticity [9]	In 1978, Herring et al. obtained the endothelialization lumen by implanting endothelial cells into the surface of artificial blood vessels for tissue culture
Stent artificial blood vessel	Stent vessels are more suitable for clinical application	In 1991, Argentine vascular surgeon Parodi et al. [10] carried out the first successful treatment of abdominal aortic aneurysm with a stent artificial vascular complex and began a new era of AAA endovascular treatment. Wang et al. [11] in China first cured subrenal AAA with a self-made endovascular stent graft in 1998

6.2.2 Main functions and structures of natural blood vessels

Blood vessels transport blood from the heart to various tissues and organs to provide them with nutrition. In the circulatory, blood vessels are divided into arterial and venous vessels. Natural blood vessels are mainly composed of three layers: intima, media, and adventitia. The innermost layer is the endothelial cell layer, which is attached to the connective tissue bed of the basement membrane and can prevent the activation of coagulation and complement factors and inhibit the adhesion of leukocytes and platelets. This layer participates in vasoconstriction, expansion, growth, and vascular remodeling. The vascular mesh membrane is composed of a large number of circumferentially oriented smooth muscle cells (SMCs), which can contract or relax under the stimulation of some signals to complete the physiological function of blood vessels. The adventitia is composed of loose connective tissue, collagen, and other extracellular matrix (ECM) dominated by fibroblasts [12].

As the main component of extracellular, collagen can provide mechanical support for blood vessels. In contrast, SMC, elastin fibers, and collagen in the middle membrane ensure the mechanical strength and elasticity of blood vessels.

6.2.3 Common materials of artificial blood vessels

Physical properties, chemical properties, and biocompatibility must be considered when designing and manufacturing artificial blood vessels. The ideal artificial blood vessel should have a three-layer structure similar to the natural blood vessel. It needs good durability, fine disinfection and preservation, and great biocompatibility. Biocompatibility includes blood compatibility and tissue compatibility. Blood compatibility means that it does not react with platelets and red blood cells in the blood to cause thrombosis after an artificial blood vessel is implanted into the human body. Histocompatibility is that when injected into the human body as a vascular scaffold outside the human tissue, it does not produce physical and chemical stimulation with tissue cells. It does not compel an explosive reaction, causing cell adhesion or proliferation. The commonly used artificial blood vessel materials mainly include polyester, polytetrafluoroethylene (PTFE), polyurethane (PU), and natural mulberry silk [13].

Polyester (polyethylene terephthalate) has excellent mechanical properties, stable chemical properties, poor blood compatibility, and good tissue compatibility. After being implanted into the body as an artificial blood vessel, the blood flows into the micropores of the blood vessel wall to form a coagulation layer, where SMCs and endothelial cells cover and grow. However, coagulation cannot be absorbed and decomposed by the body. Polyester artificial blood vessels are only suitable for transplantation and replacing large-diameter blood vessels. Application to small-caliber blood vessels causes blood vessel blockage and thrombosis.

PTFE is one of the most durable polymer fibers known. It has good biocompatibility and antithrombotic capacity. It was often used to prepare medium- and small-caliber artificial blood vessels. However, it is not easy to suture because of its hard texture. After implantation, the patency of blood vessels is reduced. The research focuses on the surface modification of expanded PTFE artificial blood vessels to improve their anticoagulant performance.

PU has excellent compliance [6] (the contraction or relaxation of artificial blood vessels with the pulsation of human blood flow) and elasticity. Both PTFE and PU artificial blood vessels have good biocompatibility and antithrombotic resistance. However, PU is more suitable for small-diameter artificial blood vessels because when implanted into the body, the endothelialization of artificial blood vessels can be quickly observed, and the inner membrane is thicker.

The diameter of natural silk artificial blood vessels can be as low as 1 mm, making it challenging to form thrombosis. Because of its crude protein structure and better biocompatibility, the prepared artificial blood vessel network structure is poor, difficult to maintain, and has low strength, limiting its clinical application [14].

The common materials described above have significant advantages and disadvantages. We can modify the fabric to make the most of it and improve its application performance. Considering the above commonly used materials, we summarized the characteristics that should be possessed as vascular grafts and sorted them into the following Table 6.3.

6.2.4 Preparation method of artificial blood vessel

Human vascular transplantation has a history of 100 years. Large- and medium-sized artificial blood vessels have been successfully used in clinical applications. People are devoted to the study of small-caliber vessels with an inner diameter of ≤ 6 mm [15]. The common method for preparing artificial blood vessels is the fiber connection method (which can be subdivided into knitting, three-dimensional weaving, and nonwoven). Representative

Table 6.3 Performance requirements of vascular grafts [6,13,14].

Performance required	Performance requirement
Biocompatibility	Histocompatibility: when the blood vessel is transplanted into the body, it will not react adversely to biological tissues and blood. It has stable performance, is nontoxic and harmless Blood compatibility: no coagulation, hemolysis, and other phenomena in contact with blood
Biomechanical strength	Grafted blood vessels should have the same elasticity, strength, and bending properties as healthy natural blood vessels
Morphological characteristics	The transplanted blood vessel should have a three-layer structure of natural blood vessels with the same inner diameter
Easy to sterilize and process	For better clinical application, transplanted blood vessels should be easy to disinfect and sterilize, the preparation method and operation process should be simple and can be mass-produced
Other properties	With the progress of science and technology, materials and methods for preparing blood vessels for transplantation emerge endlessly. While having the above basic properties, promoting vascular regeneration will become the focus of future research

processing technologies include electrospinning, wet spinning, tissue engineering, and 3D printing. The knitting or woven technology has matured, so the current research direction is mainly to prepare artificial blood vessels through electrospinning, tissue engineering, and 3D printing. The three are new artificial blood vessel technologies. In addition, there are biological coating technologies, which are briefly introduced in this chapter.

6.2.4.1 Electrospinning technology

Electrospinning is a simple and economical method for the large-scale preparation of one-dimensional nanomaterials. The one-dimensional fiber material prepared by electrospinning has the advantages of a large specific surface area, adjustable chemical composition, adjustable morphology, large fiber diameter, and high porosity. Electrospun fibers have been widely applied in many fields, including nanocatalysts, tissue engineering scaffolds, protective clothing, filtration, biomedicine, pharmacy, drug loading, optoelectronics, and environmental engineering.

An electrospinning device usually consists of three parts: high-voltage power supply, spinneret, and receiving device. As shown in Fig. 6.1, it is a classic electrospinning device. In the electrospinning process, the high-voltage power supply generates a charged polymer solution jet directed by the electrostatic force to the collector to produce interconnected fiber membranes. Its working principle can be described as the spinning liquid from the spinneret receiving the action of two forces: the droplet's surface tension and electrostatic force. When the electrostatic force is large enough to overcome the droplet's surface tension, the droplet becomes an eruptive trickle. During the eruptive, the excess solvent of the trickle evaporates or solidifies and finally falls on the collector. Therefore the characteristics of electrospun films depend on the properties of precursor solutions (such as conductivity, surface tension, viscosity, and solvent selection), processing variables (such as flow rate, voltage, and distance between capillary and collector), and

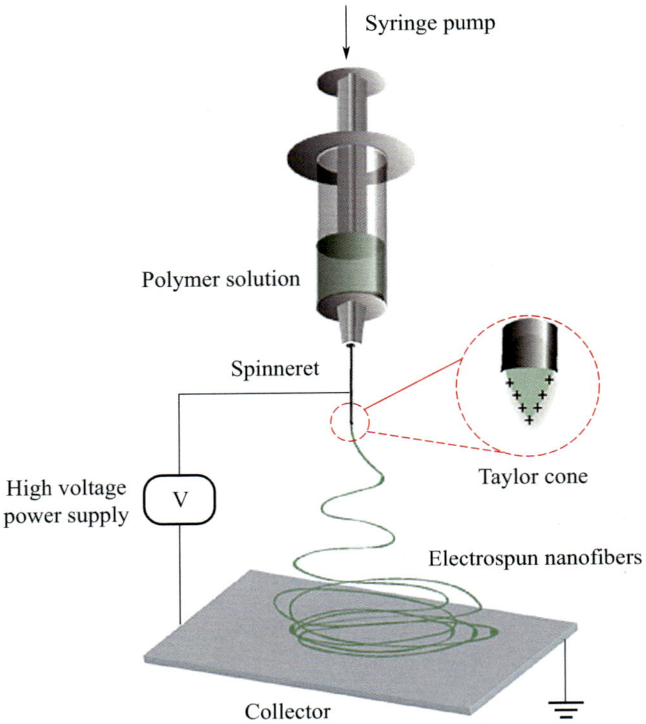

Figure 6.1 Classic electrospinning technology and equipment [16].

environmental conditions (such as temperature and humidity). The control of these parameters directly affects nanofibers' average diameter and arrangement [17].

6.2.4.1.1 Advantages and application of electrospinning in preparation of artificial blood vessels

The fibers prepared by electrospinning have a good application prospect in vascular tissue engineering. The material's mechanical properties can be adjusted by controlling the technical parameters of electrospinning. The size and arrangement of the fibers can be precisely controlled by changing the processing conditions. Therefore the production problem of small-diameter artificial blood vessels can be solved. According to specific application scenarios, artificial blood vessels with specific structures, mechanical properties, and biocompatibility can be designed.

The advantages of electrospinning technology mainly include the following points: It can produce a structure similar to a natural ECM. The preparation process is simple. Fibers can be prepared by electrospinning technology, whether natural or synthetic polymers or single or composite polymers. Highly interconnected porous structure [18] with appropriate porosity provides a structure easy for cell adhesion and value-added. The structure realizes the exchange of gas, nutrients, and liquid. According to the research, the material used in tissue engineering should have a porosity higher than 90%. The tubular structure of natural blood vessels can be well simulated by adjusting spinning solution parameters and conditions. The tubular frame of an artificial blood vessel prepared by electrospinning technology is shown in Fig. 6.2.

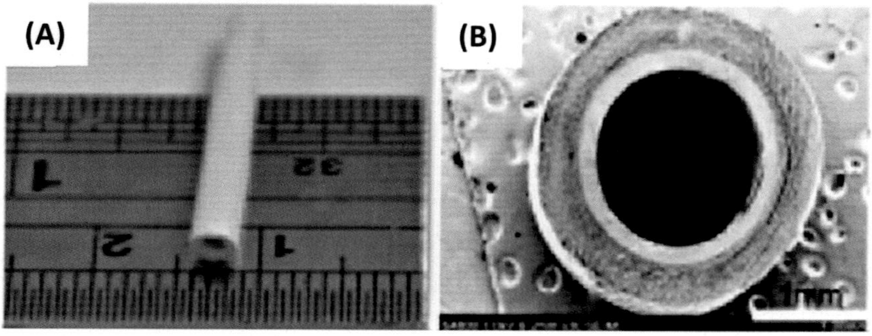

Figure 6.2 Digital photo (A) and scanning electron microscope image (B) of the vascular stent [19].

6.2.4.1.2 Examples of research results

A scaffold material simulating the morphology and mechanical properties of natural blood vessels was fabricated by electrospinning. The scaffold was prepared by continuous double-layer electrospinning on a rotating mandrel collector. The tubular scaffold (inner diameter 4 mm, length 3 cm) consists of an outer layer of PU fiber and an inner layer of gelatin heparin fiber. The composition, structure, and mechanical properties of the support can be controlled. The scaffolds' microstructure, fiber morphology, and mechanical properties were studied using a scanning electron microscope (SEM) and a tensile tester. PU/Gelatin heparin tubular scaffold has a porous structure. The scaffold has reached the breaking strength [(3.7 ± 0.13) MPa], and elongation at break [(110 ± 8)%] is suitable for artificial blood vessels. After soaking in water for 1 hour, the breaking strength decreased slightly to [(2.2 ± 0.3) MPa], but the elongation at break increased to [(145 ± 21)%]. The gelatin heparin fiber scaffold significantly inhibited platelet adhesion in the test. Heparin was released from the scaffold quite uniformly from day 2 to day 9. This scaffold is expected to mimic the complex matrix structure of the natural artery. The scaffold has good biocompatibility due to the release of heparin [20].

A small-caliber tissue engineering vascular scaffold with bioactivity, artificial synthesis, degradability, high histocompatibility, and ECM characteristics was prepared using poly (L-lactic acid)/poly (caprolactone) (PLLA/PCL). The new vascular scaffold was transplanted into animals, and its function for in situ vascular reconstructions was studied through in vitro experiments. It was found that this new vascular scaffold attracted two kinds of stem cells simultaneously during reconstruction [21].

Studies have shown that the formation of a complete endothelial cell monolayer on the lumen surface of artificial blood vessels can effectively curb thrombosis and lessen intimal hyperplasia. Zhang [22] constructed zein films with different micro and nanostructure surfaces based on static spinning technology to increase endothelial cell adhesion ability on the material surface to reduce the risk of thrombosis of artificial blood vessels.

Han [23] prepared a small-caliber vascular graft loaded with Pueraria. Pueraria is a practical component extracted from Pueraria lobata. It has antiinflammatory and antioxidant effects. Pueraria can inhibit the recruitment and activation of monocyte chemotactic protein-1 (MCP-1) in mice with myocardial infarction, reduce transforming growth factor (TGF) in myocardial tissue-β (TGF-β), TGF-β_1 expression [24], which can significantly reduce myocardial fibrosis in mice after myocardial infarction. Here, myocardial fibrosis, also known as myocardial calcification, results from persistent and repeatedly aggravated myocardial ischemia and hypoxia caused by moderate-to-severe coronary atherosclerotic stenosis, resulting in IHD (chronic ischemic heart disease) gradually developing into heart failure. MCP-1 and TGF-β_1 are chemokines related to the inflammatory response. Pueraria can also improve vascular function and hematological abnormalities [25]. Kudzu vascular grafts are prepared by coaxial electrospinning method using lactic acid as the carrier material and multiwalled carbon nanotubes as the reinforcing material. The results showed that the obtained fibers had enhanced hydrophilicity, hydrophobicity, mechanical properties, and degradable core—shell structure, which was very consistent with the properties of natural blood vessels. In addition, the biocompatibility test of kudzu-containing vascular grafts showed that the fibers containing kudzu root have no side effects on endothelial cells, do not cause hemolysis, and comply with the national medical material safety regulations.

6.2.4.2 Wet spinning

Unlike electrospinning technology, wet spinning injects the prepolymer solution into the coagulation bath (a place where the polymer continues to polymerize. For the polymer, the coagulation bath usually be a harmful solvent or nonsolvent) to form a long fiber. It produces a wide range of yarns.

The wet spinning process is as follows:

Step 1: Dissolve the polymer in an appropriate organic solvent and mix evenly to prepare the spinning solution.

Step 2: The spinning solution is extruded from the nozzle and enters the coagulation bath.

Step 3: After the spinning solution enters the coagulation bath, the solvent part slowly diffuses into the coagulation bath, and the coagulant penetrates with the drip to the critical concentration. The fibers precipitate out.

Taking the Polycaprolactone-PU (PCL-PU) blood vessel preparation process as an example, let's understand the equipment and operation process of wet spinning.

As shown in Fig. 6.3a, the syringe absorbs an appropriate amount of PCL solution and is connected to the microflow syringe pump. The needle is combined with a special syringe nozzle and extends to the coagulation bath. The cylindrical receiving rod is fixed on the rotating motor,

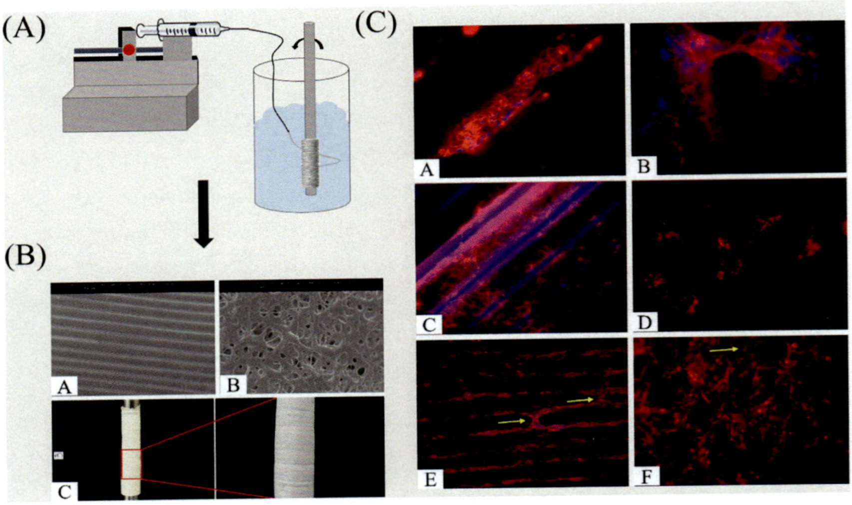

Figure 6.3 (a) Preparation of artificial blood vessel PCL fiber by wet spinning. (b) Scanning electron microscopy (SEM) of the 2-ply scaffolds luminal (A) and antiluminal (B) surface topography. The luminal surface (50 ×) shows originated PCL fibers with a fiber-to-fiber gap of about ~ μm. The antiluminal surface (400 ×) offers a highly porous topography with a pore size ranging from 10 to 30 μm. (C) Digital picture of the tubular 2-ply scaffold. (c) HUVECs were seeded on a 2-ply scaffold and seeded at 1×10^5. (A) after 24 h at 20 × magnification and stained for PECAM-1 (*red*) and nuclei (*blue*). Cells showing alignment along the fibers of the scaffold luminal surface. (B) After 24 h at 40 × magnification and stained for PECAM-1 (*red*) and nuclei (*blue*). Cells cross the gaps between fibers. (C) After 7 days at 20 × magnification and stained for PECAM-1 (*red*) and nuclei (*blue*). HUVECs form a monolayer on the surface. (D) After 7 days, 40 × magnification was stained for vWF (*red*) and nuclei (*blue*). HUVECs produce stored granules of vWF. (E) Confocal image at 10 × magnification and stain for PECAM-1 (*red*) and nuclei (*blue*) after 24 h. (F) Aortic SMCs (1×10^5) after 7 days seed on the antilumen surface, staining for F-actin (*red*) [26].

and the bottom end goes deep into the coagulation bath. After the spinneret is stable, start the engine, the receiving rod starts to rotate, and adjust the distance between the needle and the receiving rod to wind the fiber in the receiving rod. So far, the inner PCL fiber [27] with circumferential orientation has been prepared. The electrospun PU fiber was coated with the PCL monofilament for about 2 hours to form a film with a 250 μm PCL-PU composite scaffold thickness. PCL-PU composite scaffolds have different surface morphologies. The luminal surface (Fig. 6.3b A) showed oriented PCL fibers, and the antiluminal surface (Fig. 6.3c B) showed a pore size of 10–30 μm.

Human umbilical vein cells (HUVECs) on the scaffolds were observed. It was found that the cells were first arranged along the fibers on the surface of the scaffold (Fig. 6.3c A), then formed bridging grooves (Fig. 6.3c B), and then formed a monolayer cell layer. HUVECs also produced antihuman von Willebrand factor (vWF), indicating that they did not alter the phenotype (Fig. 6.3c D). (Fig. 6.3c E) shows that HUVECs attached to PCL-PU composite scaffolds successfully expressed endothelial cell adhesion molecule (PECAM-1). (Fig. 6.3c F) can clearly see the elongated F-actin stress fibers (indicated by arrows), indicating that the antiluminal surface favors SMC [26,27].

6.2.4.3 Tissue engineering

The artificial blood vessel composed of living tissue is an ideal vascular graft. Vascular tissue engineering can achieve this goal. Artificial blood vessels constructed by this method need three essential elements: structural scaffolds made of collagen or biodegradable polymers, vascular cells, and a nourishing environment [28]. Synthetic blood vessel scaffolds prepared by tissue engineering artificial blood vessel technology have good biocompatibility, antiinfection ability, and certain biodegradability through tissue cell culture. Since then, the concept has been put forward for more than 30 years and has made significant progress.

Aper [29] prepared tissue-engineered blood vessels using a highly compressed fibrin matrix. The blood vessels were implanted into sheep for 6 months. The results showed that the structure of the tissue-engineered blood vessels was highly similar to that of natural arteries.

Fibrin is an almost ideal scaffold material because it can make cells adhere, proliferate, and arrange in three dimensions [29]. It has high biocompatibility and can be easily separated from the blood. One major disadvantage is its poor biomechanical stability. Aper [29] compacted fibrin with the centrifugal force generated during high-speed rotation, added XIII factors, inoculated endothelial and SMCs, and cultured blood vessels in vitro. Let's briefly understand the preparation of tissue-engineered blood vessels. As shown in Fig. 6.4, 100 mL of plasma was separated to obtain fibrin preparation and then moved into a high-speed rotating mold to rotate and compact fibrin at high speed. At the same time, a 100 mm long fibrin tube with an outer diameter of 7 mm was produced in a particular mold. The SEM observation results of the prepared fibrin tube (Fig. 6.5) show that the compaction of fibrin leads to

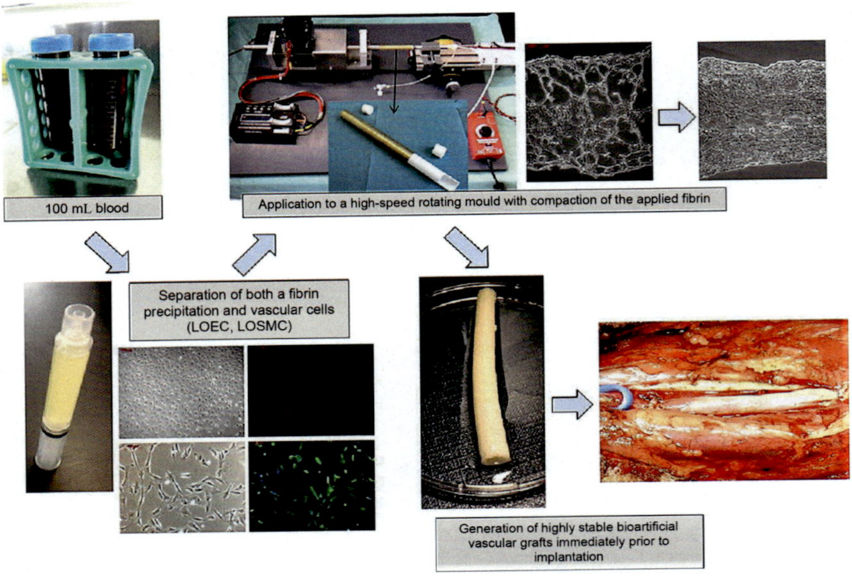

Figure 6.4 Tissue-engineered vascular grafts produced by highly dense fibrin matrix [29].

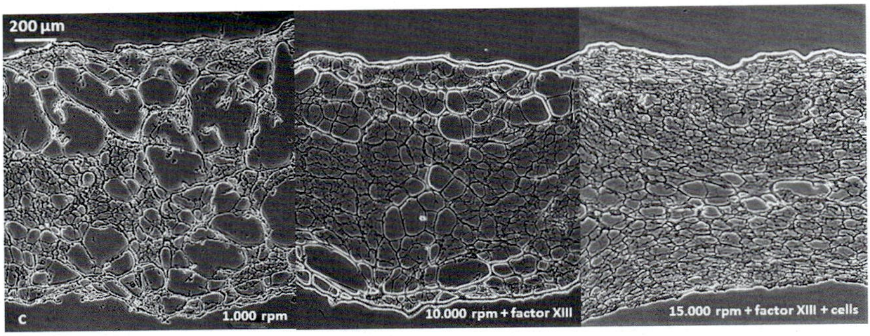

Figure 6.5 Scanning electron microscope observation of fibrin tube [29].

increased cross-linking between fibrin fibrils, and the fibrin fibrils become more acceptable branched. This effect reached the highest level when adding factor XIII and cell (c).

6.2.4.4 3D printing

Generally speaking, 3D printing technology is a forming method of printing and stacking materials into three-dimensional objects layer by layer in a set way. In recent years, 3D printing technology has been widely used in the biomedical field. Experts in related fields have used it to obtain better artificial blood vessels [30].

3D printing technology is widely used in the medical field, such as 3D printing medical models for teaching demonstrations, case discussion, 3D printing orthopedic medical devices, 3D printing living tissues or organs: artificial liver, artificial kidney, artificial blood vessels, 3D printing skin, artificial ear [30]. In addition, 3D printing technology can also tailor the structure of tablets to control drug release. The application in cardiovascular surgery [31–33] mainly involves the following three aspects: 3D printing vascular models, bioprinting vascular stents, and printing artificial blood vessels.

6.2.4.4.1 3D printing vascular model

CT angiography, magnetic resonance imaging, and other imaging results are often used for preoperative vascular surgery evaluation. However, because of the differences in machine performance, doctors' personal medical quality and expression ability, they constantly cannot comprehensively evaluate the situation of patients. 3D printing technology can transform the image into actual vascular conditions, so that doctors can more accurately and intuitively observe vascular lesions and then diagnose and treat them.

6.2.4.4.2 Bioprinting vascular stent

Because of the defects of biological activity in the printing model, the flawless operation of 3D printing has not been realized. The price of the bioprinted vascular stent is higher than that of the traditional stent, so it has less clinical application. The advantage is that personalized stents for different patients can be designed by adjusting the 3D printing parameters.

6.2.4.4.3 Print artificial blood vessel

The key to 3D printing artificial blood vessels is to find suitable materials that have the biocompatibility and other conditions required for blood vessel transplantation and meet the ideal processability for 3D printing. There are two trends in the 3D printing of artificial blood vessels: one is to use cells as raw materials to print tubular structures directly. The second is to print the three-dimensional cell network as a scaffold to grow into blood vessels. Let's take absorbable vascular stents as an example to understand 3D printing technology.

Given the shortcomings of the existing equipment and technology for preparing absorbable vascular stents, a new method for preparing composite bioabsorbable vascular stents (CBVSs) by combining 3D bioprinting and electrospinning was proposed. The inner layer of CBVS uses poly-p-dioxanone (PPDO) for 3D bioprinting, and the outer layer uses chitosan (CS/polyvinyl alcohol (PVA) composite electrospun fiber membrane [34]). As shown in Figs. 6.6 and 6.7, there are two kinds of workstations.

The preparation of absorbable vascular stents can be divided into three processes: inner layer preparation, outer layer preparation, and cell seeding.

Inner layer preparation As shown in Fig. 6.6, the test platform of the 3D bioprinting complex-forming CBVS is used to prepare the inner layer of the scaffold. In short, PPDO was inserted into a stainless-steel syringe and heated with wires. When the polymer reaches the molten state, PPDO is extruded through the nozzle and deposited on a continuous mobile platform controlled by the computer. The inner layer of the stent is obtained by depositing PPDO fibers along a predetermined path (see Fig. 6.8). Finally, the inner layer of the scaffold is manufactured by 3D bioprinting, as shown in Fig. 6.9.

Outer layer preparation Fix the 3D bioprinting support on the rotary collection device shown in Fig. 6.7. A mixed solution of CS PVA was

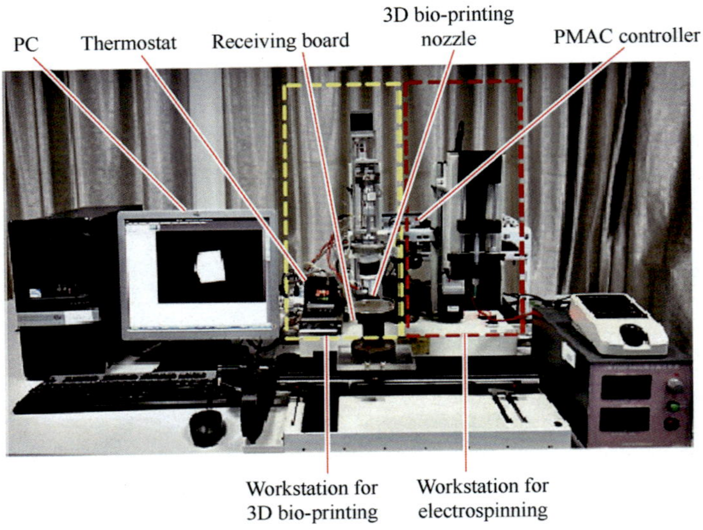

Figure 6.6 Workstation for 3D bioprinting [34].

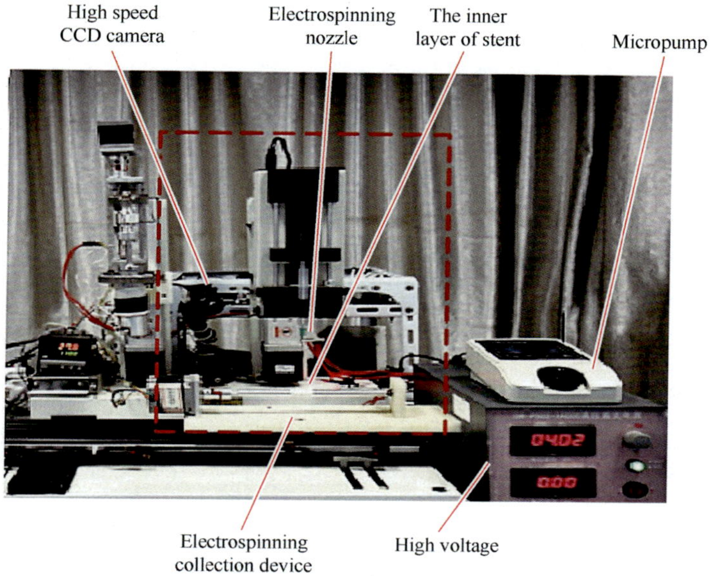

Figure 6.7 Workstation for electrospinning.

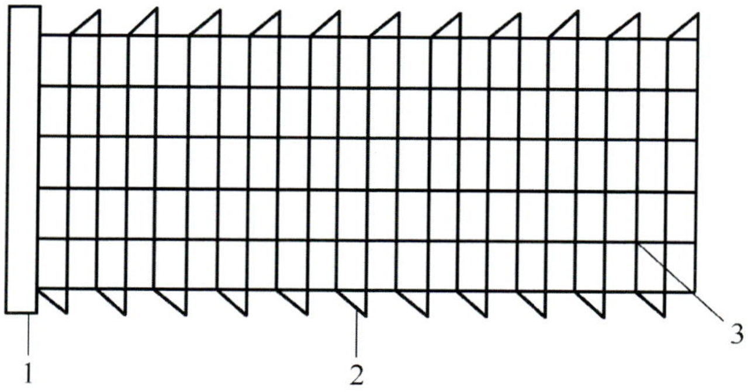

Figure 6.8 Schematic illustration of a PPDO CBVS with the sliding-lock structure. 1—Framework, 2—Barbs, 3—Lamellar mesh structure [34].

placed in a syringe to make the outer layer of the scaffold. The needle was connected to a micropump and dispensed the solution at a rate of 20 μL/min. Apply a voltage of 15 kV to the pin and ground collector (8 mm diameter spindle at 300 r/min). After 3 hours of rotation, the outer layer of the support can be obtained. The composite backing of the two-layer structure is shown in Fig. 6.10.

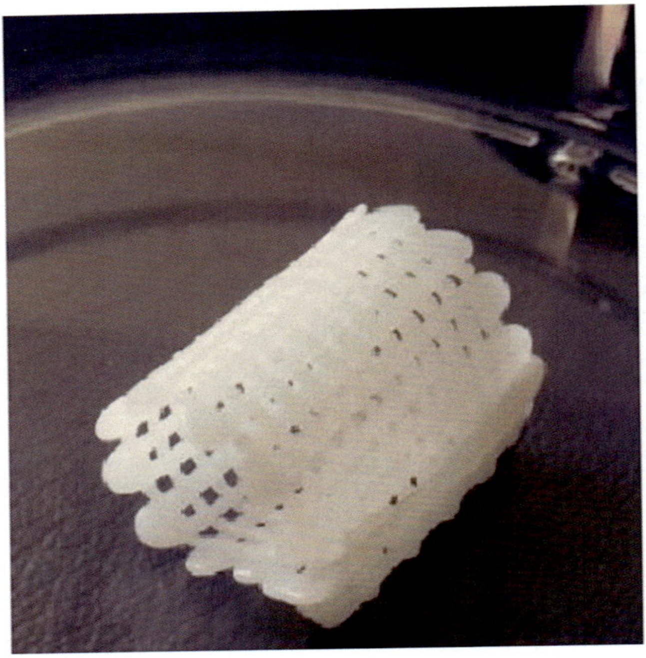

Figure 6.9 The inner layer of the stent [34].

Figure 6.10 Composite support of the two-layer structure [34].

Cell inoculation CBVS was soaked in alcohol for 1 hour and then washed three times with phosphate buffer with pH 7.4 containing 0.05% Tween-20, abbreviated as PBS, to remove the alcohol in CBVS. The excess solution remaining in the scaffold is removed by massive suction. Rat fibroblasts were harvested and suspended at a 5.0×10^7 cells/mL medium density. Then, 0.05 mL cell suspension was inoculated into CBVS to form a cell scaffold structure. 8 mL of growth medium was added to the cell scaffold construct and placed in the Petri dish to make the cells adhere. The cell scaffold constructs in the culture medium were cultured in vitro, and the culture medium was changed twice a week.

From this session, we can say that although 3D printing of blood vessels still has many problems, such as the time-consuming, expensive selection of raw materials, and unsatisfactory biocompatibility of printed blood vessels, the combination of 3D printing and vascular surgery is a breakthrough. Science and technology are bound to mature, bring convenience to doctors, and solve patients' pain. As a constantly improving cutting-edge technology, it will make more significant contributions to human medicine and the development of medicine.

6.2.4.5 Biological coating technology

This technology is not so much new for the preparation of artificial blood vessels as a technology to improve the performance of artificial blood vessels. To enhance the performance of artificial blood vessels and better apply them to clinical medicine, researchers have done much work to improve the performance of artificial blood vessels. Its modification mainly starts from material, structure, and biocompatibility. Biological coating technology belongs to material modification.

When the polyester mentioned above is used as artificial blood vessel material, it has poor blood compatibility and coagulation, limiting its use by coating a layer of biological coating on the surface of the polyester artificial blood vessels and combining specific active substances. Commonly used natural coating materials [14] include albumin and gelatin. Albumin can effectively improve the anticoagulant performance of artificial blood vessels, inhibit coagulation, and promote the formation of fibronectin in vascular intima; Gelatin can promote cell adhesion and growth, induce intimal formation after vascular implantation, and prevent clotting.

6.3 Anti-*Mycobacterium tuberculosis* composite drug–loaded fiber

Tuberculosis is a chronic infectious disease caused by *M.* or *M. tuberculosis* infection. The death rate is second only to Acquired Immune Deficiency Syndrome. The traditional treatment method is chemotherapeutic drugs, which can achieve an excellent therapeutic effect. However, in recent years, the drug resistance of *M. tuberculosis* has gradually increased. The drug cannot be fully absorbed, making the treatment more difficult. It is crucial to develop new medicines.

6.3.1 Introduction to *Mycobacterium tuberculosis*

M. tuberculosis does not produce internal and external toxins. Its pathogenicity comes from the inflammation caused by mass reproduction in tissue cells, the toxicity of bacterial components and metabolites, and the immune damage caused by immune cells related to the capsule, lipid, and protein. It can invade the susceptible body from the respiratory and digestive tract and cause skin damage. It causes injury to various tissues and organs, among which pulmonary tuberculosis is most caused through the respiratory tract. Because there are too many bacteria in the intestinal tract, *M. tuberculosis* must compete to survive and adhere to susceptible cells. At the same time, there are sterile groups in the lung, so pulmonary tuberculosis is the most standard [35].

6.3.2 Antituberculosis drugs

Antituberculosis drugs are divided into antituberculosis first-line medicines, second-line antituberculosis drugs, and new drugs.

6.3.2.1 First-line antituberculosis drugs

Sound therapeutic effects characterize the first-line antituberculous drugs. It can eliminate mycobacteria with fewer side effects in rapid and slow propagation periods. Representative drugs include rifampin, isoniazid (INH), pyrazinamide, ethambutol, and fixed-dose compound preparations. The information on common first-line drugs is shown in Table 6.4 [36].

6.3.2.2 Second-line antituberculosis drugs

Second-line antituberculosis (TB) drugs mainly treat multidrug-resistant TB patients. Multidrug-resistant tuberculosis refers to tuberculosis that is resistant to both INH and rifampicin (RIF). The second-line

Table 6.4 First-line antituberculosis drugs [36].

Drug	Chemical description	*M* (daily adult dose)/mg	Efficacy	Toxicity	Antibacterial activity
Rifampin	Rifamycin derivative	600	Bactericidal	Low	++++
Isoniazid	Nicotinic acid hydrazide	300	Bactericidal	Low	++++
Ethambutol	Ethylene diamino di-1-butanol	1200	Bacteriostatic	Low	++
Pyrazinamide	Nicotinamide derivative	2000	Bacteriostatic	Low	+++

The more " + ," the more robust the antibacterial activity of second-line antituberculosis drugs.

antituberculosis drugs are selected because they are immune to the two main drugs in the first line. Second-line antituberculosis drugs include quinolones (ofloxacin, levofloxacin, moxifloxacin, gatifloxacin), macrolides (azithromycin, clarithromycin, roxithromycin), aminoglycosides (amikacin, kanamycin), isonicotinic acid derivatives (ethylthioisonicotinamide, propylthioisonicotinamide), phenothiazines (clofazimine), and cyclic polypeptides (capreomycin).

6.3.2.3 New drugs

Natural products are important sources of new drugs, and their structures are diverse. The natural products with antitubercular bacilli obtained from medicinal plants differ from existing therapeutic drugs' designs, which are of great significance for the research and treatment of tuberculosis. Minimum inhibitory concentration (MIC) of anti-*M. tuberculosis* (MIC: the minimum drug concentration that can inhibit the significant growth of a microorganism after incubation in a specific environment for 24 hours, that is, the MIC, which is used to determine the antibacterial in vitro quantitatively) is less than 1 μ. See Appendix A [37] for the name and structural formula of a small molecule of G/mL.

6.3.3 Composite drug-loaded fiber

As mentioned earlier, *M. tuberculosis* can cause systemic tissue lesions; for example, common pulmonary tuberculosis, bone tuberculosis, and spinal tuberculosis. There are different treatment methods for other parts. In addition to different age stages, the body's immune capacity is infected with varying degrees, such as disseminated tuberculosis in children and tuberculosis. Traditional treatment methods, such as oral and surgical treatment, have toxic side effects of antituberculous drugs. With various medicines and long-term treatment, the systemic medication's harmful side effects increase, medication is taken every day, the patient's compliance is poor, and the bone is destroyed by surgical treatment. Therefore, local administration, drug sustained release, and bone tissue repair have become research hot spots. Noninvasive approaches such as lung, nose, brain, colon, buccal, and percutaneous drug delivery are being explored.

Composite fiber means two or more immiscible polymer components on the cross section. The advantage is to combine the characteristics of various materials. In treating tuberculosis, the antituberculosis drugs are loaded into composite fibers and used in multiple forms of tuberculosis, such as fibrous scaffolds, fiber particles, hollow fiber, and hydrogel fibers. It has a prospect and research value. Let's learn about the preparation of gel-loaded, drug-loaded fiber, and drug-loaded scaffolds.

6.3.3.1 Hydrogel fiber

Hydrogel is a highly absorbent polymer material with a slow-release 3D network structure. Hydrogel fibers have fiber or fibrous morphology. This fiber morphology increased its specific surface area and aspect ratio, which improved the swelling property, slow-release capability, and immobilization ability of hydrogel fibers and gave many new uses [38]. For example, hydrogel fibers have broad application prospects in tissue engineering, biomedicine, textile materials, and other fields. It can be administered percutaneously or sent into the lung through the fiberoptic bronchoscope.

The preparation of hydrogel fibers can be summarized in physical and chemical cross-linking, forming and preparing hydrogel fibers.

There are many methods for preparing hydrogel fibers (Table 6.5): extrusion filament [39], hydrodynamic spinning, electrospinning [40], gel spinning [41−43], wet spinning [41,44], biological scaffold 3D printing method [45], electrostatic and mechanical traction method, and comparison of preparation methods of various hydrogel fibers.

Preparation methods of hydrogel fibers: extrusion into filaments, electrospinning, and microfluidic technology.

6.3.3.1.1 Extrusion spinning preparation of hydrogel fibers

The forming principle of extrusion filament is to extrude, cure, and cross-link the material. The different properties of raw materials can be divided into "extrusion before cross-linking" and "cross-linking before extrusion." The commonly used equipment includes the injection machine, motor providing extrusion power, and collection: beaker or Petri dish.

Sodium alginate is a polysaccharide carbohydrate extracted from algae. It has good water solubility and biocompatibility. It can be quickly cross-linked under the action of divalent cations. Zhang [39] selected calcium chloride as a cross-linking solution and continuously formed silk in the way of "extrusion before crosslinking." The forming device and results of hydrogel fiber are shown in Fig. 6.11. The winding and assembly of yarns were completed in the rotating flow field to facilitate the extraction and collection of fibers from the cross-linking solution. The parameters such as material concentration, extrusion speed, the inner diameter of the

Table 6.5 Preparation methods of various hydrogel fibers [38].

Preparation method	Preparation of matrix	Fiber diameter	Swelling and mechanical properties	Application
Hydrodynamic spinning	Gelatin, p-hydroxybenzoic acid	The outer diameter is 278 μm. The inner diameter is 69 μm (hollow fiber)	–	Tissue engineering
Electrospinning	Isopropyl acrylamide	Three hundred μm	–	Tissue engineering
Extrusion spinning	PVA	210 ± 10 nm	90 times water absorption	Drug delivery system
Wet spinning	PVA/TiO_2	155 ± 25 nm	It can absorb 1/4 times of dye solution	Photocatalysis
Gel spinning	Sodium alginate	200–250 μm	–	Tissue engineering
3D printing of biological scaffolds	PVA, PAA	1.0 mm	40 times water absorption	Material science
Electrostatic and mechanical traction method	PVA, PIPA	1.2 mm	Up to 13 times, tensile strength 250 cn/dtex, elongation at break 25%	Material Science
Preparation method	UHMW-PAN	–	Tensile strength 70.8 cn/dtex, elongation at break 10.4%	Material science
Hydrodynamic spinning	PVA, sericin	1 mm	Tensile strength 200 MPa, elongation at break 50%	Artificial skin
Electrospinning	UHMW-PAN	–	Tensile strength 108.4 cn/dtex, elongation at break 16.4%	
Extrusion spinning	Sodium alginate	Five hundred μm	–	Artificial skin
Wet spinning	Sodium alginate	17–116 μm	The maximum dry and wet tensile modulus is 104 and 105 MPa	Tissue engineering

needle, crosslinking attention, and stirring process are controlled in the molding process. The molding quality of fiber is comprehensively analyzed to determine the best experimental scheme.

In the molding material, the greater the sodium alginate concentration, the more sodium alginate particles contained in the needle tube with the same inner diameter, the greater the rigidity of the material, and the more elastic deformation energy stored. When the cross-linking conditions are the same, the high concentration slows the cross-linking, the noticeable extrusion swelling effect, and the larger the wire diameter. The larger the extrusion speed, the smaller the inner diameter of the needle, and the smaller the fiber diameter. Cross-linking concentration and stirring process have little effect.

Similarly, we can mix the antituberculosis drugs and polymer materials into silk to produce hydrogel drug–loaded fibers, which can be absorbed through the skin or sent to the lungs by fiberoptic bronchoscopy.

6.3.3.1.2 Electrospinning preparation of hydrogel fibers

Electrospinning is the most commonly used and suitable for preparing ultra-nanofibers. The principle and equipment of electrospinning technology are described in section I of the artificial blood vessel wall. Zhou [46] made a new hydrogel material. PEO (polyethylene oxide) and GelMA blend solutions are prepared using electrospinning technology to improve spinning performance because GelMA (methacrylic anhydride gelatin) has good biocompatibility and stability. Then, the three-dimensional electrospinning platform was designed and built. The preparation process of micro nanofibers in the far-field and near-field conditions was systematically studied. The effects of concentration, voltage, and liquid velocity on the diameter and morphology of the hydrogel fiber were mainly investigated.

6.3.3.1.3 Microfluidic preparation of hydrogel fibers

Microfluidic technology [47] refers to using microchannels to control microfluids. The microfluidic technology's basic principle for preparing microfibers is in the microfluidic channel. When the interfacial tension between two-phase liquids is less than the shear effect of external liquid on internal liquid, fibers are formed. On the contrary, droplets are created. The spinning liquid has laminar flow property when flowing. With this property, the rapid cross-linking and stable flow of multiphase fluid can be realized, and the fiber has good morphology and structure.

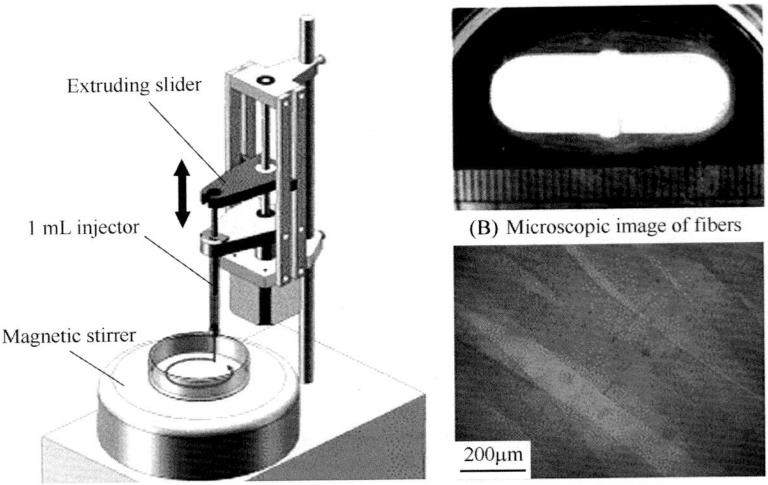

(A) Forming device of hydrogel fibers (B) Microscopic image of fibers (C) Microscopic image of fibers

Figure 6.11 Forming device and forming results of hydrogel fibers [39]. (A) Hydrogel fiber filature device. (B) The macromorphology of the hydrogel fiber. The white material is the magnetic stirrer, and the surrounding filamentous material is the hydrogel fiber. (C) The micro morphology of the hydrogel fiber.

The microfluidic device mainly includes the spinning solution injection, spinning, and fiber collection.

Microfluidic devices and curing methods can be designed according to different products. In the past research, the microfiber structures that have been designed and fabricated include solid/porous micro fibers—the most common and easy to prepare, and centrally controlled microfibers, which can be used in vascular bionics and tissue engineering, multilayer microfibers, microgroove fibers, micro spindle fibers, spiral microfibers. The hydrogel microfiber prepared by this technology has long, thin, and flexible characteristics and can be used to construct 3D structures by folding. Fig. 6.12 shows a common microfluidic spinning device. The device includes a polydimethylsiloxane (PDMS) chip formed by photolithography, a coaxial tube assembled by metal needles, and a glass chip assembled by glass capillaries. Because of designability of microfluidic devices, they have broad application prospects in biomedicine [35].

6.3.3.2 Drug-loaded antituberculosis bone scaffold

M. tuberculosis [48] fibrous stents are widely used to treat bone tuberculosis. Pathological changes characterize bone tuberculosis, mainly

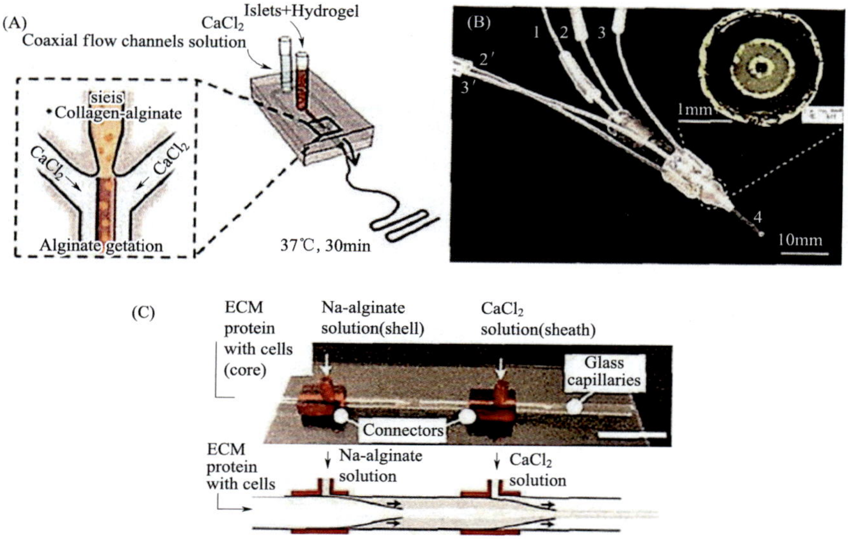

Figure 6.12 Standard microfluidic spinning devices: (A) PDMS chip, (B) coaxial needle, (C) glass tube chip [35].

manifested as bone damage. In the later stage, bone defects are formed because of the destruction of typical bone structure, leading to limb deformities and paraplegia [49]. The treatment of bone tuberculosis is complicated. One is that its pathology is changeable, and the drug concentration is difficult to reach an adequate treatment level. The other is that the focus is difficult to remove. In the treatment, if oral or injection is used, the drug concentration may be too low and unsustainable because of the poor blood circulation caused by the disease. The drug dose increases the load of the liver and other organs. Therefore, the local administration has become the focus of treating bone tuberculosis. Bone transplantation plus drug treatment is often used to repair bone defect.

RIF and INH are commonly used, with an excellent bactericidal effect. They have the same bactericidal effect on *M. tuberculosis* whether in the reproductive period, dormant period, or in vitro. Sterilization has a wide range of types and functions on various bacteria, stable at room temperature. The characteristics of slow-release graft carrier materials are good biocompatibility and no cytotoxicity. The carrier material must have good biodegradability to achieve sustained release of the drug. The compound method is simple and effective and does not affect the drug; it is a comprehensive source, low price, and easy to disinfect and store. As a slow-release antituberculosis bone repair material, it should also have good guiding osteogenesis, bone induction, specific porosity, and mechanical strength [50].

INH is a water-soluble drug [51]. If the water-soluble drug is only prepared in simple blending, there are many problems, such as initial burst release. Therefore it is necessary to change the drug loading, and the methods that can improve drug blending include the chemical reaction method and microsphere encapsulation method. The chemical reaction method involves loading the drug and polymer scaffold material into the carrier through a chemical combination. The microsphere encapsulation method wraps the drug in the microsphere and loads the microsphere on the scaffold. Different drug loading methods lead to other drug release mechanisms. There are three main drug release mechanisms of drug-loaded bone scaffolds: the first is that drugs are released with the degradation of carrier materials, including the chemical reaction method. For this mechanism, the nature of carrier materials is the decisive factor. The second release mechanism is through the pores of the scaffold. The third is released through drug-loaded microspheres. Because it is an oil-soluble drug, RIF does not suddenly release many medications in the initial stage. Still, the late release speed is slow because of the carrier material. Therefore for oil-soluble drugs, the priority of carrier material is more important.

We will list the preparation and performance evaluation of two kinds of antituberculosis bone scaffolds according to the introduction of drug-loading methods. They are bone scaffolds prepared by a chemical combination of drugs and carriers and coaxial tissue-engineered bone scaffolds loaded with microspheres made of drugs [52]. 3D printing technology has the characteristics of personalized design. The bone scaffold prepared by this technology has good biocompatibility and no immune rejection. It is the method mainly introduced in this section.

1. Coaxial tissue engineering bone scaffold

   Materials and procedures:

	Main materials	Step
Preparation of rifampicin silk fibroin microspheres.	Rifampicin, silk fibroin (SF).	Emulsification preparation: the SF aqueous solution and rifampicin were fully compatible, then slowly and evenly poured into the emulsion, then heated and stirred, then the organic solvent was removed after curing, then the SF drug-loaded microspheres were prepared by freeze-drying
Preparation of SF drug-loaded engineering bone.	Silk fibroin, polyvinyl alcohol, hydroxyapatite (the main component of HA natural bone).	The inner core material of the support is HA, which is mixed with a PVA hydrogel with a concentration of 10% in a particular proportion. The outer core material is SF composite powder with good biocompatibility and PVA composite hydrogel prepared by mass ratio. It is mixed with HA in proportion and made into a coaxial tissue engineering bone scaffold by printing through the solid high system and biological 3D printer. Drug-loaded microspheres and drugs are embedded in the inner core

Performance: The micrograph shows the drug is well dispersed and compounded in the inner core material for such scaffolds, so the sustained-release effect is good. The pore size and porosity of the coaxial skeleton's outer core structure affect its slow-release energy. Similarly, the study on the degradation performance and drug release of engineered bone is also essential to evaluate the comprehensive implementation of the prepared bone scaffold.

2. *Drug-loaded composite bone scaffold* [53]

   The materials required for the preparation of bone scaffolds are mesoporous bioactive glass (MBG), poly 3-hydroxybutyric acid co-3-hydroxycaproic acid (PHBHHx), RIF, and INH. MBG is an inorganic material that can deposit hydroxyapatite, but the pure MBG scaffold is too fragile. Compounding with polymer material PHBHHx acid can improve this disadvantage and the bioactivity and slow-release performance of the composite scaffold.

First, MBG powder was prepared. Then, hydrophilic carboxyl groups and hydrophobic methyl groups were grafted on the surface of MBG mesoporous channels for chemical modification, which were used to carry antituberculosis drugs INH and RIF. The composite bone scaffold was printed by configuring PHBHHX and treated MBG into a slurry suitable for 3D printing.

Performance:

Physical and chemical properties: the ideal composite bone scaffold should have appropriate porosity and compressive strength to ensure the stable and slow release of drugs and a durable structure.

Bioactivity: It has an excellent ability to deposit hydroxyapatite. Bioactive glass MBG not only has the power of bone conduction but also can induce bone regeneration, [54] unique osteoblast adhesion ability, that is, good biological activity (the adhesion and growth of cells on the implant surface are the key to the interaction between the implant and surrounding tissue, affecting the combination of bone and tissue [55]). The activity of alkaline phosphatase ALP on drug-loaded and nondrug-loaded composite bone scaffolds was detected. Alkaline phosphatase is a membrane-bound enzyme. Its action can reflect the activity and state of osteoblasts, primarily newly formed osteoblasts. Alkaline phosphatase is expressed in the early stage of osteoblast differentiation, marking the beginning of osteoblast differentiation and ECM maturation. The activity of alkaline phosphatase can represent the early differentiation of osteoblasts. When the cells enter the calcification stage, the movement of alkaline phosphatase in cells begins to decline [56,57].

Drug release performance: The amount and rate must reach the MIC.

6.3.3.3 Drug-loaded antituberculous fibrous membrane

Liu [58] used RIF and INH to prepare RIF/INH polyvinylpyrrolidone (PVP) to blend fiber with a rapid release by electrospinning.

INH has good water solubility, and RIF has poor water solubility and low bioavailability. Studies have shown that combining the two can reduce drug resistance, produce superimposed antibacterial ability, and significantly improve the therapeutic effect [59]. However, direct contact between the two drugs accelerates the hydrolysis of RIF. Electrospinning combined with hydrophilic carriers and hydrophobic drugs is an effective method to improve drug bioavailability [60]. The RIF/PVP, INH/PVP fiber membrane, and RIF/INH/PVP fiber membrane were prepared using PVP as the carrier. Fig. 6.13 shows the preparation process of the RIF/INH/PVP fiber membrane.

As shown in Fig. 6.13, a fiber membrane containing a drug can be obtained if only one side spinning is performed.

Comprehensive analysis showed that INH PVP nanofibers and RIF PVP microfibers were mixed in RIF/INH/PVP with two drugs coexisting in one fiber. Both INH and RIF changed from a crystal state to an amorphous state, which was conducive to the release of drugs from the fiber. The addition of hydrophilic carrier PVP increased the hydrophilicity of RIF fiber. The increase of hydrophilicity and the transformation of crystal morphology make the release of RIF-loaded fiber faster than that of RIF powder. The release of RIF is not affected by INH. In addition, RIF-loaded fibers and drug powders showed the same antibacterial properties, and their medicinal properties were not involved in the preparation

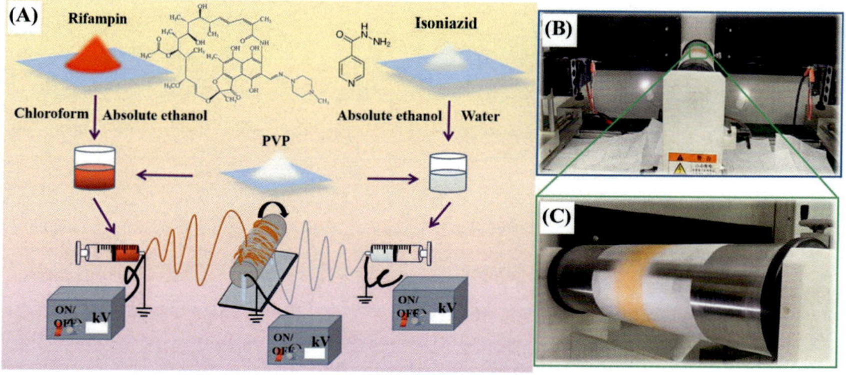

Figure 6.13 The preparation process of RIF/INH/PVP [58]. (A) Schematic diagram of solution preparation and spinning process. (B) Schematic diagram of electrospinning receiving device. (C) Enlarged view of electrospinning receiving device and obtained fiber diagram.

process. Therefore, the results show that RIF fiber and RIF sustained-release membrane enhance the release rate of RIF, effectively improving the bioavailability and reducing drug waste. RIF/INH/PVP can effectively separate RIF and INH and avoid the rapid degradation of RIF caused by direct contact between the two drugs.

6.4 Wound dressing containing dragon blood

Dragon's blood is a traditional Chinese medicine in Asia. It has many functions, such as antiinflammatory, analgesic, hemostatic and blood activating, antibacterial and fungal, deprivation and muscle regeneration, and promoting blood circulation [61]. However, it is insoluble in water, is difficult to inject, and has a low utilization rate. Loading dragon blood on nanofiber wound dressing can significantly improve its utilization.

6.4.1 Introduction to Dracaena Draconis

Dragon's blood, also known as "Kirin dragon," is made from the resin exuded from the fruit of the Kirin dragon, a palm plant. It promotes blood circulation, eliminates pain, removes blood stasis, removes corruption, and regenerates muscle [61]. The physical image and chemical structure of dragon blood are shown in Fig. 6.14.

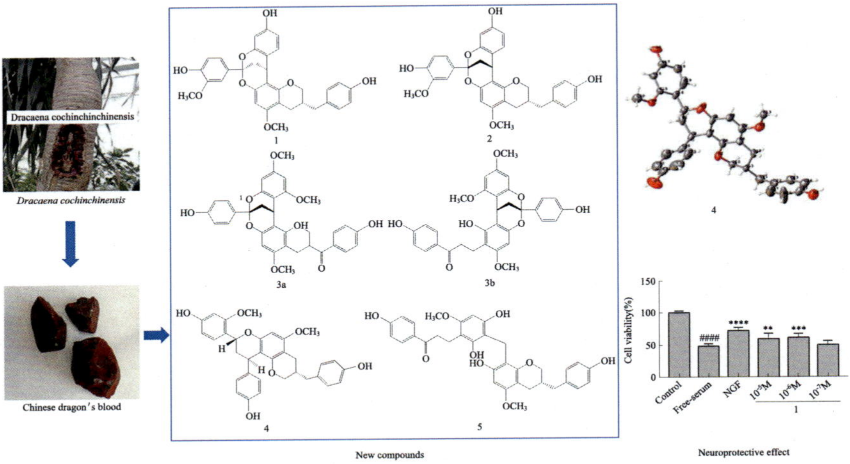

Figure 6.14 Dragon's blood picture and chemical structure [61].

As shown in Table 6.6, the main chemical composition of dragon blood includes flavonoid derivatives, volatile oil, phenols, and steroids. In addition, there are components such as dimethyl-erythroid, dimethyl-draconis, yellow Draconis resin hydrocarbon, and terpenoids [62].

Two-way regulation of hemostasis and activating blood circulation: it plays an antithrombotic role by inhibiting platelet aggregation and prolonging the anticoagulant activity.

Hypoglycemic effect: Shanshan and other studies [63] show that the supercritical fluid extract of dragon blood has an excellent hypoglycemic impact on diabetic mice induced by alloxan.

Antibacterial effect: its mechanism may be related to inhibiting the formation of bacterial biofilm and reducing the appearance of extracellular polymer.

Cardiovascular protection: studies have shown that dragon blood extract can effectively reduce the resting heart rate of experimental animals and has a specific antagonistic effect on arrhythmias induced by barium chloride injection and adrenaline injection.

They promote wound healing and antitumor activity.

6.4.2 Wound dressing

6.4.2.1 Wound healing theory

Skin function is performed by specialized cells found in two main skin layers (epidermis and dermis, as shown in Fig. 6.15). In addition to these two layers, subcutaneous tissue under the dermis supports the dermis. The epidermis is the outermost skin, about 120 μm thick, composed of many cells closely connected at different stages of differentiation, forming stratified squamous epithelium.

Table 6.6 Main chemical components of Dracaena Draconis.

Flavonoids	Chalcone	2,4-dihydroxy-5-methyl-6-methoxychalcone
		4,40-dihydroxy-2,6-dimethoxydihydrochalcone
	Flavane	7-hydroxy-6,8-dimethyl-2,5-dimethoxyflavane
	Flavone	7-hydroxy-8-methyl-2,5-dimethoxyflavane
Other	Triterpenoids	6-methyl-2,5-dimethoxyflavan-7-ol
	Steroids	–
	Saponins	–

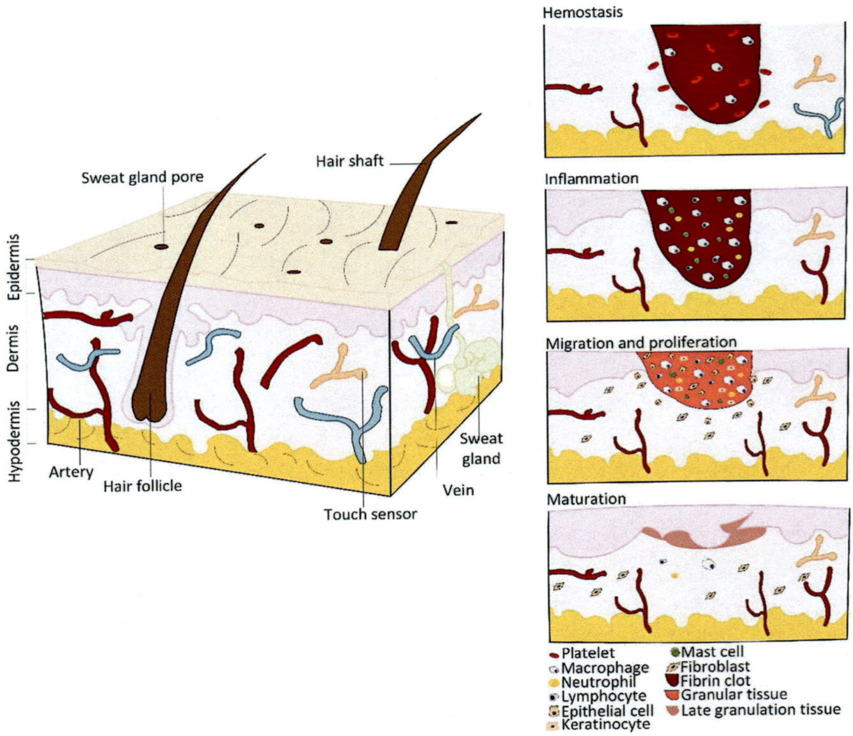

Figure 6.15 Skin structure and wound healing stage [64].

Wound healing is generally divided into five stages: hemostasis, inflammation, migration, survival, and maturation. During hemostasis, platelets gather to promote coagulation and hemostasis. The inflammatory stage cooccurs with hemostasis. At this stage, neutrophils and phagocytes successively enter and penetrate the injured area to destroy bacteria and eliminate debris in dying cells and the damaged matrix. The following steps, migration and diffusion, can also be collectively called the new organization formation stage because they are interdependent. The infiltration of new epithelial cells characterizes migration into the damaged area to replace dead cells and reduce inflammation during migration. The proliferation stage includes covering all damaged areas with epithelial cells and macrophages.

In contrast, fibroblasts and endothelial cells move to the damaged areas to form granular tissue composed of a new matrix and blood vessels. The final stage, maturation, includes a remodeling process in which fibroblasts cover all damaged surfaces to form a new skin layer, ideally without leaving scars. Through this complex wound-healing process, the skin can self-regenerate, but in the face of some severe and challenging wounds, it is necessary to use grafts or dressings to accelerate wound healing [65].

6.4.2.2 Classification of wound dressing

Wound dressings are divided into traditional and new, as shown in Fig. 6.16.

Traditional wound dressing has a limited protective effect on wounds and is not suitable for some serious and complex injuries. An ideal wound dressing requires a variety of properties, including appropriate swelling ratio, oxygen permeability, wound exudate absorption and preservation of water around the damage, and antiinflammatory and sterilization to promote wound healing. Nanofibrous membranes and foam wound dressings are widely attractive because of their potential to simulate the natural state of various tissues.

6.4.2.3 Nanofiber wound dressing

Compared with the gold standard for wound dressing, including membrane, foam, microfiber, and mesh fiber, nanofiber dressing usually has higher porosity so that it can permeate water and oxygen more effectively, better change nutrients, and discharge metabolic waste. [68] In addition, the small adhesion and high surface area of nanofiber dressing

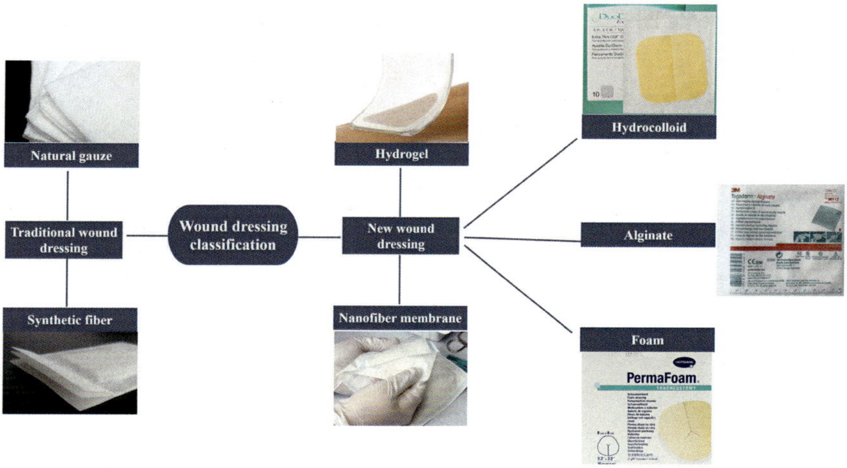

Figure 6.16 Classification of the wound dressing [66,67].

can enhance hemostasis. The small pore size of nanofiber dressing can protect the wound from bacterial infection and inward growth of cells/tissues and provide excellent compliance compared with microfiber and mesh commercial dressings to cover better and safeguard the damage from the disease. Most importantly, the expanded surface area of nanofiber dressing can effectively load/incorporate drugs. Therefore, they show great potential in developing advanced bioactive dressings.

Electrospinning technology is considered the most straightforward and economical way to prepare nanofiber membranes. Electrospinning fiber membrane has the following characteristics:

1. Morphological: the surface is smooth without a bead.
2. Porosity: it has a highly interconnected porous structure. Appropriate porosity provides a system easy for cell adhesion and value-added, realizes the exchange of gas, nutrients and liquid, and finally realizes wound healing. According to the research, the material applied to tissue engineering should have a porosity higher than 90%.
3. Water vapor transmittance: water vapor transmittance is used to create a suitable moist environment for wound healing.
4. In addition, proper mechanical properties and flexibility in dry and wet conditions can better carry out clinical operation and treatment and protect the wound area; during tissue regeneration, sufficient mechanical support should be provided for cell growth and proliferation.

6.4.3 Wound dressing containing dragon blood

Cao [69,70] of Beijing University of Chemical Technology dissolved dragon blood in PVP to prepare the solution and electrospinned the fiber. Finally, they processed it into a film design, a new wound excipient. Next, we introduce the preparation of the dragon blood wound dressing through a specific experiment.

As shown in Fig. 6.17, the positive pole of the high-voltage power supply is connected to the spinning needle, and the negative pole is grounded and connected to the collection plate, so a high-voltage electric field is formed between the hand and the collection plate. Using an orthogonal experimental design, three levels and three factors are shown in Table 6.7.

The spinning solution with a dragon blood content of 15%, 20%, and 30%, respectively, was prepared. Make an orthogonal experiment group,

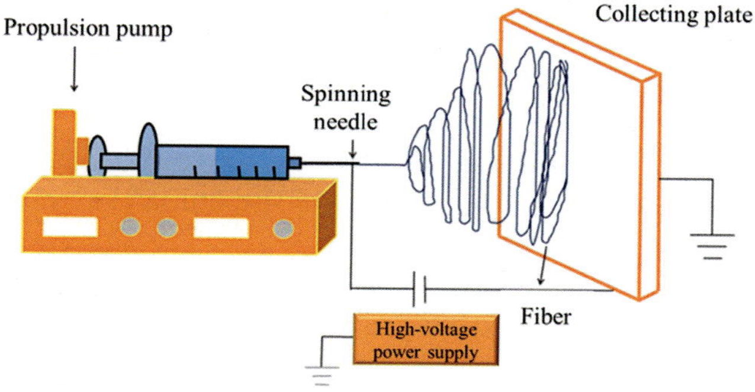

Figure 6.17 Solution electrospinning equipment [69].

Table 6.7 Levels and factors of experimental design [69].

Level	Factor		
	X voltage (kV)	*Y* distance (cm)	*Z* flow (mL/h)
1	18.00	13	0.30
2	20.00	15	0.50
3	22.00	17	0.70

and test and characterize. The results showed that the fiber diameter affected the dissolution of the dragon blood. The finer the fiber diameter, and the larger the surface area, the greater the dissolution. The optimum spinning conditions were found by comparing the SEM images of nine groups of data in the orthogonal experiment. Three groups of electrons spun fibers were prepared under this condition, which was recorded as A (15%), B (20%), and C (30%), and observed by SEM. Finally, the fibers with PVP content of 15% were smooth and uniform in diameter distribution. With the increase in PVP content, the fiber diameter became larger. The infrared test results showed no chemical reaction between dragon blood and PVP. XRD results showed that adding dragon blood had little effect on the crystallinity of the fiber. The thermal property test results show that the endothermic peak of the dragon blood is mainly about 95°C, and the endothermic peak of PVP is about 120°C. A sizeable endothermic peak is formed at 110°C after blending, which indicates that PVP and dragon blood include a low eutectic mixture in the fiber. After mixing, the crystal form distribution of the drug changes, resulting in a slight shift in the endothermic peak of the drug. This result corresponds to the difference in the XRD crystal form above. With the increase of Draconis content, this endothermic peak did not change significantly, which shows that the drug content has little effect on the endothermic peak. The drug dissolution curve showed that the dissolution decreased gradually with increased drug content.

The complete experimental results show the above-prepared dragon blood nanofiber wound dressing improves the utilization rate of dragon blood.

6.5 Fibrous plaster

6.5.1 Traditional plaster

6.5.1.1 Introduction to traditional plaster

Plaster [71], an external use of traditional Chinese medicine, was called thin paste in ancient times. Generally speaking, the plaster is divided into matrix and drug. The matrix is the same, made of vegetable and animal oil, and the medicine varies according to the symptoms. Traditional Chinese medicine paste can be roughly divided into stiff paste and ointment—the classification of plaster is shown in Fig. 6.18. Hard plaster is what we usually call plaster, among which black plaster is the most common. The theory of traditional Chinese medicine guides the ointment. The drug is dissolved in a suitable matrix, coated on the side of paper, cloth, or skin, pasted on the affected area, absorbed through the skin epidermis, and entered the blood circulation to treat diseases.

With the progress of science and technology, plaster preparation has significantly progressed from traditional black plaster and traditional Chinese medicine ointment to cataplasm and transdermal drug delivery systems imported from abroad and the application of nanotechnology.

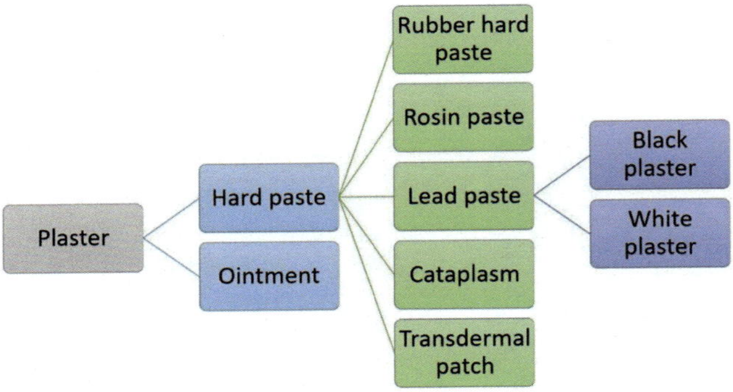

Figure 6.18 Plaster classification.

6.5.1.2 Traditional plaster preparation process

Black plaster is a paste made from medicinal materials, edible vegetable oil, and red lead (the main component is P_3O_4), which is applied to the back-mounting material of the skin. The traditional black plaster preparation [72] is shown in Fig. 6.19.

Sesame oil, cottonseed oil, soybean oil, and peanut oil are commonly used in the extraction stage of restorative materials, most of which is sesame oil. Drugs are usually extracted by frying, but most drugs are polar substances, and frying can only extract fat-soluble components and some nonpolar substances in drugs. Therefore, there are also relevant research reports on various water frying, alcohol immersion, or mixed extraction methods.

The disadvantage of refining at high temperatures is that oxidation or decomposition produces irritating low-level decomposition products. Therefore, going through the past "fire poison" step is necessary. The standard methods are water spray, water immersion, and water explosion. The water spraying method is simple, time-saving, and labor-saving, but the effect is not apparent.

The matrix of traditional Chinese medicine plaster is different from that of black dressing, which is mainly rubber. Currently, conventional Chinese medicine ointment production in China specifically includes solvent spreading and hot-pressing spreading methods [73]. The solvent spreading method evolved from the process of zinc oxide paste. The pulping process is relatively mature, and this method is widely used in domestic pharmaceutical factories. The solvent spreading method uses a lot of gasoline, and the content of filler zinc oxide is 35%–40%. The hot pressing spreading method is similar to rubber products such as adhesive tape in the production process. Domestic pharmaceutical companies rarely use it, and gasoline is not used. The filler is lithopone 45%–50%.

Traditional Chinese medicine ointment has strong adhesion and is easy to carry. Still, a large amount of gasoline is needed in the preparation process, which can easily cause environmental pollution.

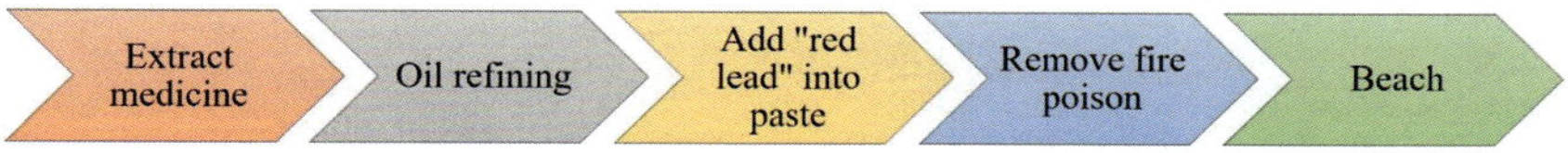

Figure 6.19 Process flowchart of traditional black plaster preparation.

The matrix used in cataplasm is hydrophilic. Compared with the first two, cataplasm has the advantages of controllable dosage, extensive drug loading, small absorption area, stable blood drug concentration, convenient use, good compatibility with skin, and aging resistance. It can be exposed and pasted repeatedly, and the administration can be terminated at any time. The preparation of cataplasm includes matrix molding and preparation molding, and the process flow is shown in Fig. 6.20 [72].

Although the preparation of traditional black plaster and traditional Chinese medicine ointment has been relatively perfect, there are still many problems. For example, the red pill often used in preparing black application pollutes the human body and the environment, gradually eliminating it. Most gasoline is used to prepare traditional Chinese medicine ointment, which also causes environmental pollution and high cost. Traditional plaster has many problems due to its limited matrix and preparation methods, such as poor drug absorption and matrix permeability. Application of advanced technologies such as solid dispersion and nanotechnology in plaster preparation has become a new direction.

6.5.2 Electrospun fibrous plaster

Electrospun nanofibers have a broad application prospect in biomedicine because of their unique structure. Aiming at the problems of traditional plaster, Xu [74] analyzed its existing problems, such as drug absorption, and took this as the starting point to prepare a black gypsum/polyethylene

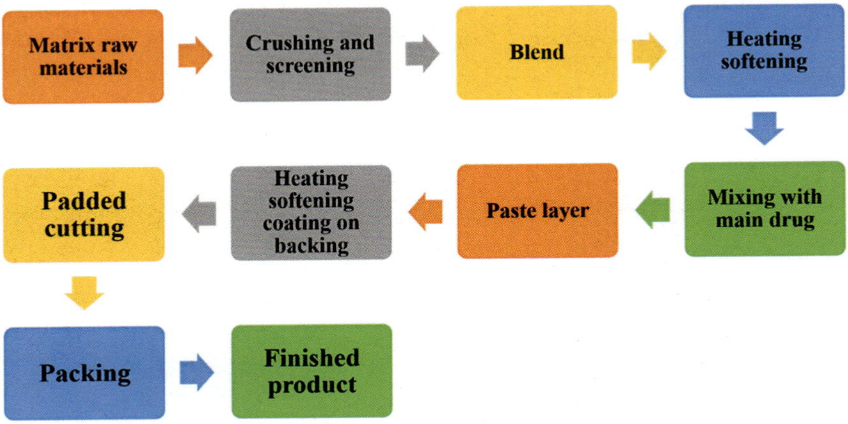

Figure 6.20 Cataplasm preparation process flow.

glycol (PEG) composite fiber by considering solid dispersion technology, drug release carrier, and electrospinning technology.

The raw material used for the black plaster/PEG composite fiber membrane is commercial PEG (P103730, Mn = 20000), as well as dried blood, borneol, red dan, musk, and sesame oil. The schematic diagram of the preparation process of black plaster is shown in Fig. 6.21:

Afterward, black plaster/polyvinyl alcohol composite fibers were prepared by electrospinning. The schematic diagram of the upward melt electrospinning device independently designed and assembled by Xu [74] is shown in Fig. 6.22. They consider that the solution electrospinning technology involves problems caused by solvent residue. In the electrospinning process, the toxic solvent in the fiber diffuses out, causing harm to the human body. Therefore melt electrospinning without solvent is used. Spinning parameters are shown in Table 6.8.

Subsequently, Xu carried out the performance characterization of the prepared composite fibers, including the observation of fiber morphology and the investigation of drug dissolution. [74]

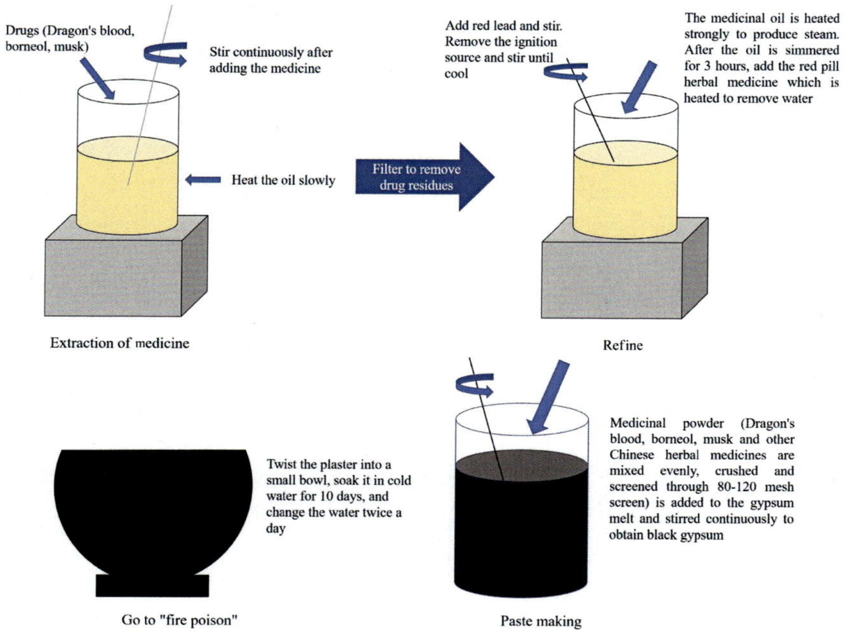

Figure 6.21 Preparation process of black plaster.

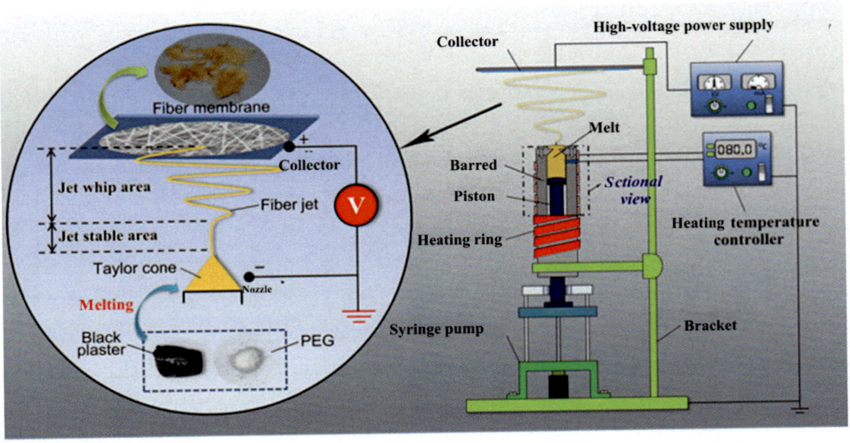

Figure 6.22 Schematic of upward melt electrospinning device and electrospinning principle [74].

6.5.2.1 Fiber morphology

The morphology and diameter distribution of electrospun fiber are shown in Fig. 6.23. Pure peg cannot be electrospun. As shown in Fig. 6.23, the pure black gypsum fiber has obvious adhesion. It is believed that the jet temperature at the collector is too high, which prevents the fiber from completely solidifying, so the contact positions between the fibers are combined. When a peg is added, the adhesion improves gradually. The analysis is that PEG solidifies rapidly under cold conditions in composite fiber stretching in the air to avoid bonding between fibers. Then, there is the fiber diameter. It can be seen that with the increase of PEG content, the fiber diameter becomes larger and larger, and the fiber morphology gradually deteriorates, indicating that when the spinning conditions are the same, the properties of the spinning solution affect the fiber diameter. According to the data, PEG has a higher viscosity than black plaster at the same heating temperature. In the two processes of Taylor cone formation and jet whipping, the influence of surface tension needs to be overcome. Under a specific spinning voltage, the higher the melt viscosity, the more excellent the resistance to fiber drawing. Therefore, larger diameter fibers were collected.

6.5.2.2 Drug dissolution

As an example, dragon blood, a poorly soluble drug, the dissolution data of traditional black plaster and black gypsum fiber film are compared, as

Table 6.8 Experimental parameters and results of black plaster and PEG composite fibers [74].

No.	Black plaster (wt. %)	PEG (wt. %)	Collection distance (cm)	Voltage (kV)	Temperature (°C)	Morphology	Average diameter (μm)	SD
1	100	0	8	40	80	Poor	2.71	1.119
2	70	30	8	40	80	Good	2.93	1.03
3	35	65	8	40	80	Good	15.32	5.76
4	15	85	8	40	80	Good	17.78	9.45
5	0	100	8	40	80	Poor	–	–

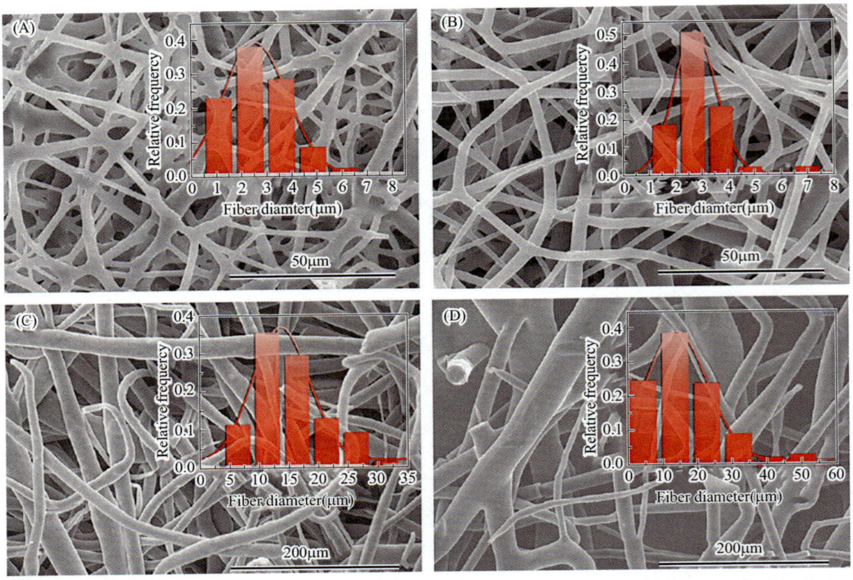

Figure 6.23 SEM images and diameter distribution of (A) pure blaster black plaster fiber, (B) 30 wt.% PEG, (C) 65 wt.% PEG, and (D) 85 wt.% PEG composite fiber. [74].

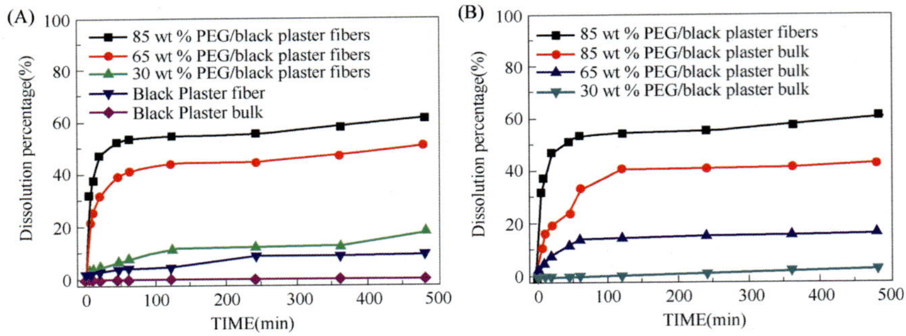

Figure 6.24 Dissolution rate of Draconis Sanguis in vitro of black plaster composite fibers (A) and with different PEG contents (B) [74].

shown in Fig. 6.24. Conventional block index black plaster dissolution is zero, while the composite fiber containing 85 wt.% PEG shows an improved dissolution of dragon blood drug of about 60%. The dissolution rate of composite fiber containing 65 wt.% PEG is 50.7%. The solubility of PEG containing 30% by weight is 18%. In addition, the dissolution rate of pure black plaster fiber is about 9%, which is greatly improved compared with the traditional block black plaster. The analysis shows that

PEG enhances the wettability of the drug while making the drug into fiber increases the dispersion of the drug, which also benefits from the large specific surface area and porosity of the fiber membrane itself.

6.5.3 Session summary

From the content of Section 6.4, it is not difficult to conclude that electrospinning parameters and spinning solution are important factors affecting the morphology and structure of fibers. In addition, we note the fiber membrane made above does not mention the harm of red lead, so it still needs further study. In scientific research, finding problems, countermeasures, and many data are essential steps before beginning research. After selecting the preparation scheme, we should also think hard and thoroughly understand the relationship between the preparation and product performance.

6.6 Preparation and properties of controlled release fiber

Controlled release fiber is the fiber with the function of controlling drug release. The control of drug dose and fiber morphology can improve the stability and utilization of drugs and reduce the toxic and side effects. The difference between the controlled-release system and the standard drug-release system can be seen in Fig. 6.25. This section introduces the drug-controlled-release system and the carrier materials used to prepare controlled-release fibers and lists the preparations of the two controlled-release fibers.

6.6.1 Drug-controlled release system

Drug-controlled release systems can be divided into the following five types according to different mechanisms.

6.6.1.1 Diffusion control system

Diffusion control systems are generally divided into matrix and container reservoirs. In the form of a container, the drug is embedded in the polymer membrane and released into the environment through diffusion in the polymer. The matrix form is the drug combined with the polymer in the form of dissolution. The two records of spherical diffusion drug-controlled release systems are shown in Fig. 6.26.

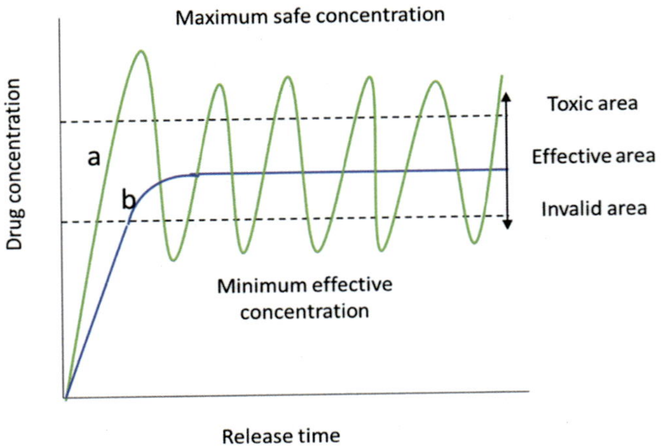

Figure 6.25 Relationship between release time and drug concentration of controlled and ordinary release systems. A joint release system B controlled release system **[75]**. *Adapted from the literature of [75].*

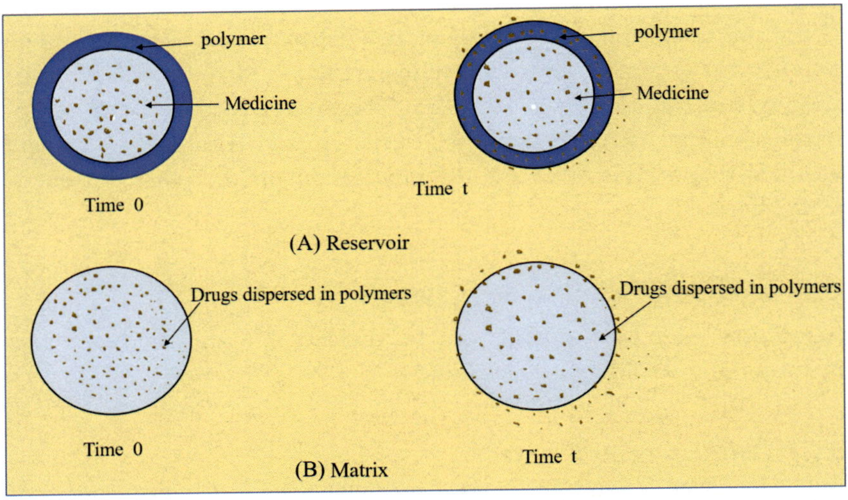

Figure 6.26 Cross section of two forms of the spherical diffusion-controlled drug release system. (A) Reservoir; (B) Matrix. *Adapted from the literature of [75].*

6.6.1.1.1 Reservoir system

The diffusion of the reservoir system conforms to Fickian's first law (Eq. 6.1). That is, the greater the concentration gradient, the greater the diffusion flux.

$$J = D\mathrm{d}C/\mathrm{d}Z \tag{6.1}$$

where "*J*" is the molar migration of the drug, and "*D*" is the diffusion coefficient through the polymer membrane, which is independent of the drug concentration. "*C*" is the drug concentration, and "*Z*" is the displacement of the drug through the polymer.

Most reservoir systems are plane, cylindrical, and spherical, ignoring the edge effect. Eq. (6.2) is derived from Eq. (6.1):

$$M = \frac{DKA\Delta C}{\delta} \cdot t \tag{6.2}$$

"*M*" represents the total amount of drug released when the time is "*t*," "*K*" is the distribution coefficient of the drug in the membrane and environment, "*A*" is the total area of the polymer membrane, "*δ*" is the film thickness. Where "Δ*C*" is constant,

"d*M*/d*t*" is consistent with Eq. (6.2). At this time, the release rate does not change with time, that is, zero-order release, which is the ideal result in drug-controlled release diffusion. In this system, when the drug amount is large enough to reach the saturated concentration in the design, if "Δ*C*" does not change within a certain period, it reaches zero-order release. The release was controlled by adjusting the film thickness according to the release law when designing the experiment. The applications of reservoir systems in the medical field include capsule and hollow fiber.

6.6.1.1.2 Matrix system

In the matrix system, drugs are combined with polymers in the form of dissolution or dispersion. It can also be divided into dissolved drug systems and dispersed drug systems.

For the dissolved drug system, the drug is dissolved in the polymer, and the diffusion rate of the drug can be expressed by Eq. (6.3):

$$\delta C_i/\delta_t = D_{ip}\delta^2 C/\delta Z^2 \tag{6.3}$$

When M_i/M <0.6, and M_i represents the drug release time of *T*, *M* Max is the maximum drug release rate for planar membranes.

$$M_i/M_\infty = 4\sqrt{(D_{ip}t/\Pi\delta)} \tag{6.4}$$

Eq. (6.4) shows that the dissolution system with simple geometry cannot achieve zero-order release.

In the dispersed drug system, the drug is distributed in the polymer in the form of a dispersion, and the dispersion concentration is greater than the solubility of the drug in the polymer. Higuchi uses the pseudo-steady-state solution model to analyze this system and obtains:

$$M_i = A\sqrt{[D_{ip}C_i(2C_0 - C_{is})t]} \tag{6.5}$$

It shows that the dispersion system with simple geometry is also challenging to achieve zero-order release.

6.6.1.2 Chemical control system

The chemical control system includes a degradation system and a side-chain system. There are two degradation systems: reservoir and matrix. They can be divided into homogeneous degradation (occurring inside the polymer) and heterogeneous degradation (only appearing on the system's surface). The side-chain system refers to drugs connected with polymer molecules through chemical bonds, and the chemical bonds break under the action of water or enzyme to release drugs—the schematic diagrams of the two systems are shown in Figs. 6.27 and 6.28.

6.6.1.3 Solvent activation system

This system is related to the permeability of solvent, which is usually composed of two mechanisms: permeability and swelling.

The permeation system can release the drug evenly, and the drug release rate is only related to the solubility. The earliest permeation system is a hollow semipermeable membrane, as shown in Fig. 6.29. When the

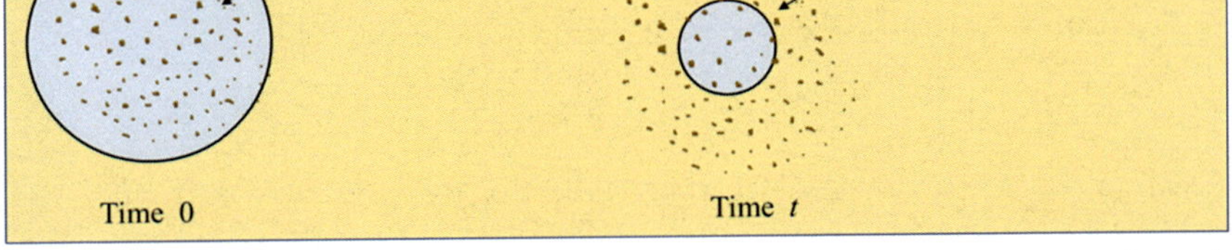

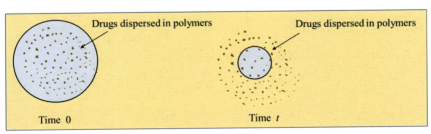

Figure 6.27 Cross-section diagram of surface degradation system. *Adapted from the literature of Li YX, Feng XD. Controlled drug release system and its mechanism. Polymer Bulletin 1991;(1): 19–27.*

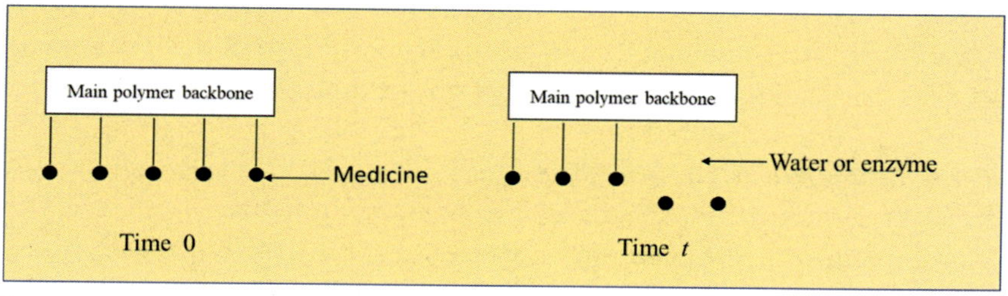

Figure 6.28 Schematic diagram of the side-chain system. *Adapted from the literature of Li YX, Feng XD. Controlled drug release system and its mechanism. Polymer Bulletin 1991;(1): 19–27.*

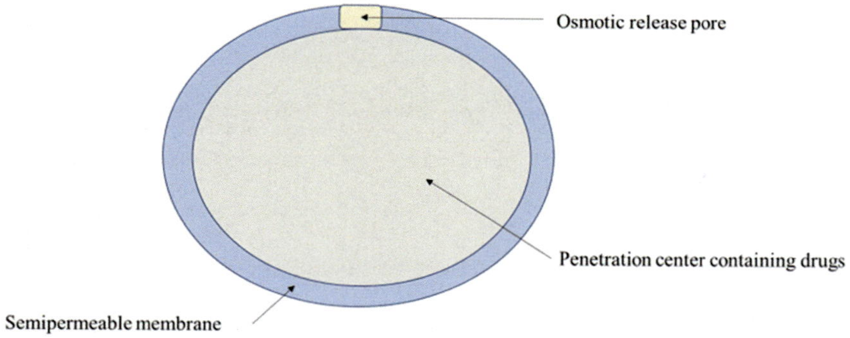

Figure 6.29 Hollow semipermeable membrane.

semipermeable membrane is in contact with water or biological liquid, water enters through the semipermeable membrane and drugs are released through tiny holes.

In the swelling system, glassy polymers that can swell are needed, such as (ethylene-vinyl acetate copolymer) EVA and PVA. The drug is dissolved or dispersed in the polymer. Generally, it is not released. When the solvent diffuses into the polymer, the polymer swells. At this time, the polymer chain relaxes because the temperature is lower than TG, and the drug is released. This is the principle of swelling permeation.

6.6.1.4 Magnetic control system

The drug and magnetic powder are mixed and evenly dispersed in the polymer in this system. Under normal circumstances, the drug release is the same as usual, but when the external magnetic field is increased, the drug release is accelerated. The mechanism of the magnetic controlled release system is not precise. Some researchers believe that the vibration of magnetic particles may drive the movement of drug particles.

6.6.1.5 *Intelligent drug delivery system*

The most prominent feature of an intelligent drug delivery system is that it can immediately terminate the drug release according to the body's perception or other control signals—the smart drug release system classification is shown in Table 6.9.

6.6.2 Polymer drug-controlled release carrier

The controlled release of macromolecular drugs has many advantages: stable drug release, effective use of drugs, reducing drug toxicity, and being closer to the source of disease. Whether the material can be degraded can be divided into degradable pharmaceutical controlled-release carrier and nondegradable drug carrier. The source of the materials can be divided into natural polymer carriers and synthetic polymer carriers [79].

6.6.2.1 *Natural polymer carrier*

Natural polymer carriers usually have good biocompatibility and tissue activity. Natural biodegradable polymers as drug carriers mainly include CS, alginate, agar, fibrin, and collagen. Yan [80] used soybean protein isolate (SPI) as raw material and compound with biocompatible polymers (PVA) and natural polymer (sodium alginate: SA), successfully prepared SPI-based polymer composite microgels and gel beads, and loaded with Schiff base metal complex antioxidants, and studied their slow-release properties. The experimental results showed that SPI/SA-SHMP had better-sustained release properties for 5-FU. The cumulative release rate is high, especially in the buffer solution with pH = 7.4; the cumulative release rate can reach 79.61%. SPI/SA-SHMP gel beads can be an ideal enteric drug carrier material. Han and others [81] prepared Glycidyl Methacrylate Modified CS/poly (N-isopropyl acrylamide) (CS-GMA/PNIPAAm) hybrid polymer network hydrogel by photopolymerization technology. The hydrogel has dual sensitivity to pH and temperature. Hybrid hydrogels can be used as drug-controlled release carriers.

6.6.2.2 *Synthetic polymer carrier*

Synthetic polymers can make up for the poor mechanical properties of natural materials. Polymers such as polyphosphate [82], PU [83], and polyanhydrides have good biocompatibility and physiological properties

Table 6.9 The classification of the intelligent drug release system [76–78].

	PH responsive	Temperature responsive	Bioactive molecular responsive	Field response type
Controlled release principle	Oral drugs go through different pH environments, such as saliva and digestive juice, and the pH change is acidic, neutral weak alkaline. The controlled release of the drug can be achieved by wrapping the drug with a material sensitive to pH change	The conformational energy of the macromolecule chains of thermosensitive gels varies with temperature	There are three main categories: Glucose response Antigen response Thrombin response	Force field response: A force field response release system is constructed according to the characteristics that the extracellular matrix of tissue in a dynamic mechanical environment can release growth factors under the stimulation of the force field
Use occasion	Oral drug delivery system	High temperature shrinkage: when the temperature rises, the gel is rapidly dehydrated, the hydrophobic groups attract each other, and the macromolecular chain contracts. The low-temperature shrinkage type is the opposite	–	–

and can be biodegradable. They can effectively control the drug release according to zero-order kinetics in the process of sustained release. Therefore, it has become the primary type of synthetic polymer carrier. Zhou [84] synthesized a novel temperature-responsive side-chain photo-sensitive Block Copolymer Poly[(*n*-isopropyl acrylamide-*co*-N,N-dimethyl acrylamide)-block propyl alkyl-4-azobenzoate] (P (NIPAM *co* DMAA)—*b*-pazohpa) loaded with paclitaxel (PTX) by atom transfer radical polymerization. The release of PTX in micelles can be regulated by changing temperature and light stimulation. The developed block copolymer can be used as a drug-controlled release carrier for cancer treatment.

6.6.3 Controlled release fiber

After the introduction in the previous sections, we have a preliminary understanding of the characteristics of nanofibers. Because of its small diameter and large specific surface area, it has great potential for preparing and applying controlled-release materials.

6.6.3.1 Core–shell nanofibers

Core–shell nanofibers have the characteristics of high porosity, large specific surface area, and good mechanical strength. Its unique core–shell structure can be used in many fields, such as controlling the release of drugs and bioactive factors. This section introduces a method to prepare HA/CS core-shell PEC nanofibers with high-molecular-weight hyaluronic acid (HA) polyanion and CS polycation as raw materials through built-in electric field induction (Fig. 6.30). PEC, a polyelectrolyte complex, is a composite fiber separated by mixing two polymers with opposite charges and electrostatic induction. CS is a polycation that migrates outward along the direction of the electric field line. HA is a polyanion that migrates inward in the opposite direction. This leads to phase separation before they become fibers. Under the action of the electric field, they gather in the inner layer and outer layer, respectively, to form a core–shell structure.

Combretastatin A-4 phosphate (CA4P) was encapsulated in HA/CS nanofiber membrane to study its drug release characteristics. Because CA4P is an anionic drug with negative ions in water, most CA4P moves with HA to form inner fibers. Fig. 6.31 shows the drug release kinetics curve of mixed and core–shell fiber. It can be seen from Fig. 6.31 that the hybrid HA/CS/CA4P nanofiber members containing CA4P have the phenomenon of sudden drug release. After 36 h, the drug stops releasing, and the core–shell HA/CS/CA4P nanofiber members show gradual and continuous disclaimer.

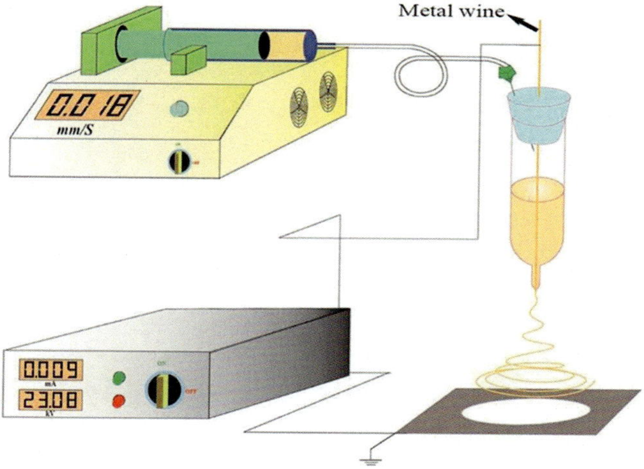

Figure 6.30 Schematic representation of the electrospinning setup [85].

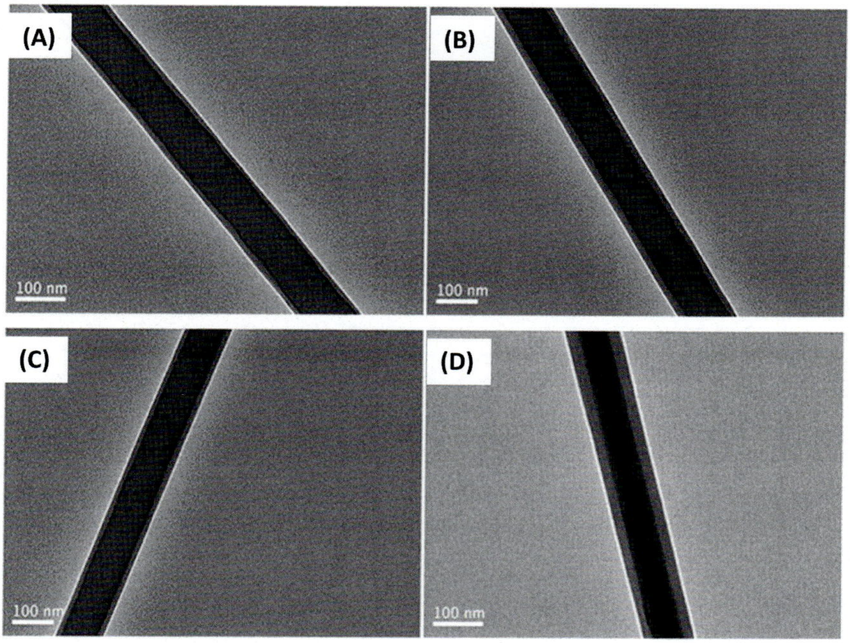

Figure 6.31 TEM images of HA/CS nanofibers electrospun from different ratios of HA and CS: (A) 9/1, (B) 8/2, (C) 7/3, and (D) 6/4 [85].

For the fiber membrane, the controlled release principle is that the core–shell HA/CS/CA4P nanofiber membranes prepared by the electric field induction method are negatively charged and easy to hydrolyze. After adding aqueous anionic drugs, they are wrapped in the core layer. When HA dissolves, the pills dissolve, but the CS in the outer layer hinders the release of drugs. Therefore, the drug release of core–shell HA/CS/CA4P nanofiber members is gradual, while the medications in HA/CS/CA4P hybrid fibers are randomly distributed in the threads. When HA dissolves, the drugs dissolve and release suddenly (Fig. 6.32).

Electrospun core–shell nanofibers prepared by electric field induction control the sudden release of drugs and have sustained release. They can be applied to some diseases requiring long-term drug treatment. The prepared natural core–shell polyelectrolyte nanofibers have excellent application potential as wound dressing and tissue engineering drug carriers.

6.6.3.2 Polylactic acid/ polyhydroxy butyrate fiber

Polyhydroxy butyrate [86] (PHB) is the first discovered polyhydroxy alkyl (PHA) and an essential organic compound. PHB was discovered by lemoigne in Bacillus megaterium (lemoigne, 1926). Its mechanical properties are similar to traditional plastics such as polypropylene or polyethylene. It can be extruded, molded, spun into fibers, and made into films, which can be used to prepare hybrid polymers with other synthetic polymers. PHB is a

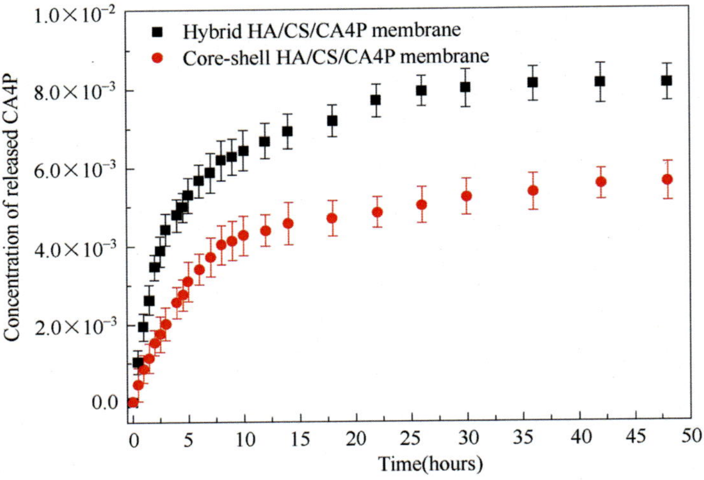

Figure 6.32 Drug-release kinetic curves of hybrid HA/CS and core–shell HA/CS nanofiber membranes [85].

biocompatible, nontoxic, biodegradable polymer suitable for biomedical applications [87]. PHB has high crystallinity because of its perfect stereoregularity and high purity. However, PHB's excessive brittleness, poor processability and thermal stability limit its potential application as raw materials.

Polylactic acid (PLA) is strictly different from PLLA in structure. Because PLA has a chiral carbon, there are L-lactic acid, that is, L-lactic acid and D-lactic acid, that is, D-lactic acid. Generally, L-lactic acid produced by corn fermentation is primarily l-lactic acid, which forms L-lactide. Then ring-opening polymerization obtains L-PLA (i.e., PLLA), but the default PLA is PLLA. PLLA has biodegradability, biocompatibility, and good mechanical properties.

Ultrathin PHB/PLA fiber [88] was prepared by solution electrospinning technology (Fig. 6.33). The solution electrospinning technology will not be repeated to explore how ultrathin fibers' structural and morphological characteristics determine fragment kinetics, drug diffusion rate, and corresponding drug release curve. Fibers with different ratios were prepared, and their various structures and morphology were tested and observed. Probe electronspin resonance technology can effectively study the chain segment dynamics of polyester molecules in PHB/PLA fibers and corresponding felts. This method is widely used in studying macroscale objects, especially PHB/PLA films with similar polyester ratios.

The SEM results showed that PHB/PLA fiber appeared alternately as elliptical and cylindrical fragments at low drug concentrations. When the

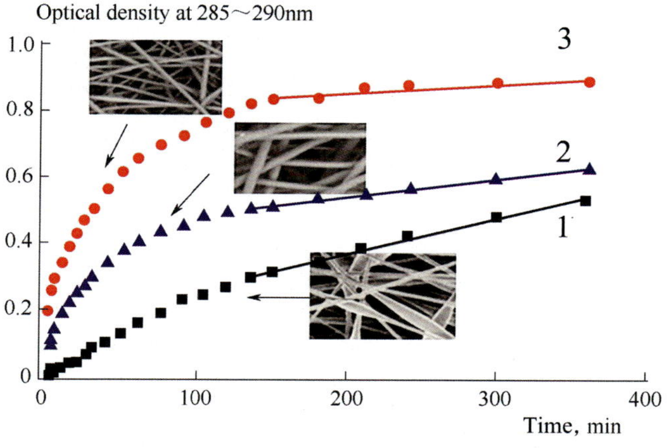

Figure 6.33 Typical kinetic profiles of DPD controlled release from PHB fibers. DPD concentrations are (1) 1 wt.%, (2) 2 wt.%, and (3) 5 wt.%. Micrographs illustrate the shape of fibers (× 1000) [88].

drug content was high (> 3%), the abnormal ellipsoidal structure disappeared, and all fibers were cylindrical. The kinetic release curves were different with different fiber morphology. The multifactor study of PLA/PHB ultrathin fibers with different geometries will help experts understand the operation prospect of submicron drug delivery carriers.

6.7 Bacteriostatic mask

In 2020, New Coronavirus infection broke out around the world. Among many means of epidemic prevention, wearing masks is undoubtedly the most effective and direct means to isolate the virus and prevent its spread. On December 8, 2020, Anhui Box Health Technology Co., Ltd., launched a global "inactivated protective mask" containing copper oxide, which caused a sensation worldwide. The inactivation characteristic of the copper oxide mask on COVID-19 is reported from Cu^{2+}. The antibacterial effect of metals has attracted extensive attention.

6.7.1 Introduction to the antibacterial agent

Antibacterial agents can be divided into organic, natural, and inorganic, according to their composition. Organic antibacterial agents include quaternary ammonium salts and organic metals. Low-molecular organic antibacterial agents have poor heat resistance and are volatile when used, which can no longer meet people's needs. Therefore high-molecular organic antibacterial agents have emerged, and organic and inorganic antibacterial agents have been combined to improve their application performance. For example, the Japanese mix of organo-polysiloxane, hydrolyzable silane, triazole-containing compounds and zeolite, and the resulting products have an excellent antibacterial effect and are not interfered by ultraviolet light but also have good compatibility with polymers. CS, a natural antibacterial agent, is a kind of high-molecular sugar in animals. CS has good antibacterial properties and is often used to prepare wound dressings. In addition, animal antibacterial agents also include natural peptides and amino acids. Plants include fruits and vegetables, Chinese herbal medicine and spices, such as licorice and ginger, which have antibacterial effects. Microorganisms are derived from bacteria, fungi, and microorganisms. Although there are many natural antibacterial agents, their thermal stability is poor, and it is difficult to produce on a large scale. Inorganic antibacterial agents are mainly metal antibacterial agents, common metals endowed with specific biological functions, such as copper, silver, zinc, manganese, and a group of nonbiogenic metals with less research, such as lithium, strontium, and gallium. Each metal has its unique antibacterial mechanism, which may be one or more [89–91].

6.7.2 Copper oxide mask

The mask containing copper oxide is shown in Fig. 6.34. This nonmedical mask is a five-layer structure, with two layers of copper oxide nonwoven fabric inside and outside, two layers of melt-blown fabric between the inner and outer layers, and the innermost layer is ordinary non-woven fabric. The inner and outer copper oxide nonwoven fabrics contain copper oxide masterbatches and PP masterbatches. The extreme copper oxide nonwoven fabrics can kill those exposed to COVID-19, while the inner nonwoven fabrics are designed for asymptomatic infected people. If asymptomatic infected people appear, these people will become effective transmitters of the novel coronavirus. In this mask, the inner copper oxide nonwoven fabrics can kill the virus exhaled by asymptomatic infected people, reducing the communication risk. The function of melt-blown cloth is to filter more than 95% of particles.

The copper oxide nonwoven fabric is imported from Medcu in Israel. The raw material ratio and manufacturing process are unknown. Therefore we found an antibacterial and antiviral nanofiber membrane and its preparation method and application.

Liu [92] invented an antiviral nanofiber membrane. The preparation process is to mix silver nanoparticles, copper oxide nanoparticles, and

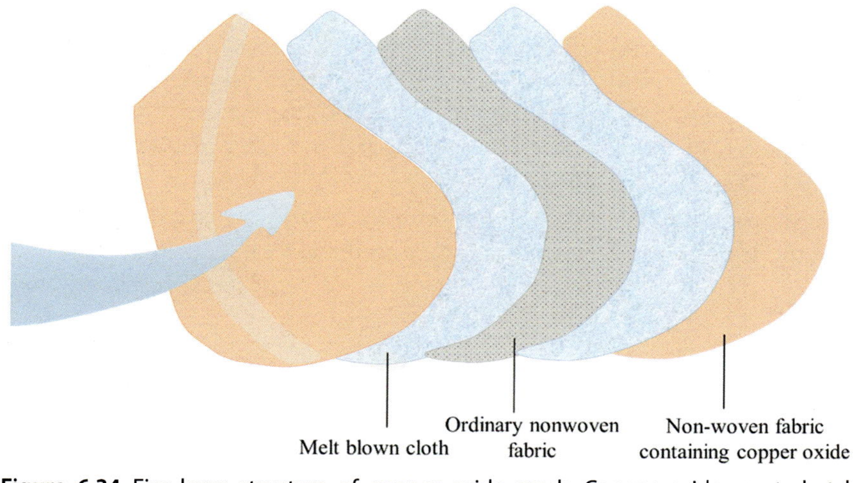

Figure 6.34 Five-layer structure of copper oxide mask. Copper oxide masterbatch and nonwoven PP particles are combined and injected into the nonwoven fiber.

hexafluoro isopropanol solution of PLA with ultrasonic stirring to prepare a spinning solution and electrospun to obtain a nanofiber membrane. It can not only block and filter but also contact and kill viruses. Nanostructures guarantee better breathability, and PLA is degradable and green.

6.7.3 High-molecular quaternary ammonium salt mask

High-molecular quaternary ammonium salt destroys bacterial cell membranes and viral capsid protein through side chain structure, resulting in bacterial death and virus inactivation. Quaternary ammonium salts are organic antibacterial agents that are ammonium ion compounds substituted with four hydrogen atoms from an alkyl group.

The cell structure of bacteria is divided into Gram-positive bacteria and Gram-negative bacteria. The outer membrane of Gram-negative bacteria is negatively charged. The R group is generally an organic group, with N forming a practical antibacterial component. X (Fig. 6.35) is a group of silver ions such as fluorine, chlorine, bromine, and iodine. When at least one of the R groups is a substituent with a carbon chain of 8–18, it is called a single-chain quaternary ammonium salt; when there are two, it is a double-chain quaternary ammonium salt; when N^+ is 2, it is a double quaternary ammonium salt, and when it is greater than or equal to 3, it is a poly quaternary ammonium salt.

Antibacterial agents include natural, organic, and inorganic antibacterial agents. Inorganic antibacterial agents are mainly metal antibacterial agents. According to different action mechanisms, metal antibacterial agents can be divided into metal ion antibacterial agents and photocatalytic metal oxide types. The general order of antibacterial properties of metal ions is $Ag^+ > Hg^{2+} > Cu^{2+} > Cd^{2+} > Cr^{3+} > Ni^{2+} > Pb^{2+} > CO^{2+} > Zn^{2+} > Fe^{3+}$. After consulting much literature, it is found that the order of antibacterial properties of metal ions is not unified, and the most frequently cited is adopted. The antibacterial property of silver ions is the best, but silver salt has strong photosensitivity and can easily change color when exposed to light. Mercury, lead, and cadmium have high toxicity and can cause cancer. They

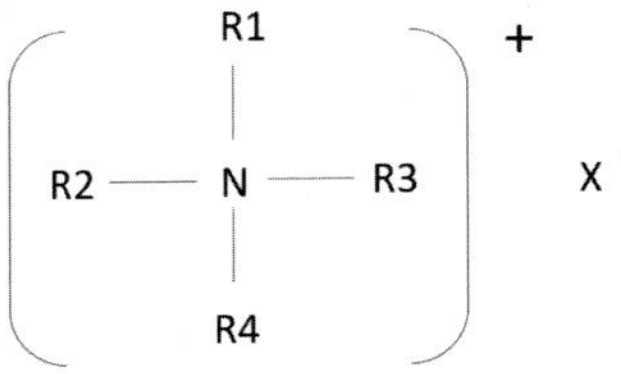

Figure 6.35 General formula of quaternary ammonium salt structure.

are rarely used. The colors of nickel, copper, and cobalt limit their application. Zinc ion has poor antibacterial activity. The representative of photocatalytic metal oxide is titanium dioxide. Its antibacterial effect can only be realized under the action of ultraviolet light, water, or oxygen. The antibacterial applicability of several commonly used medical metal materials is compared and summarized in Table 6.10.

Because of the increase of complex diseases and the disadvantages of poor mechanical strength of polymer materials, modern biomedical materials are developing toward composites. At the same time, the content introduced in this chapter is not rich enough. The primary purpose is to give students a deeper understanding of the various applications of polymer materials.

We introduce the contents of this chapter from principle, material, and preparation to application: artificial blood vessels, antituberculosis bacillus composite drug carrying fiber, wound dressing containing dragon blood, fiber plaster, controlled-release fiber, and bacteriostatic mask. It can help readers establish a closed-loop thinking mode of the performance application structure. However, the application of polymer materials in biomedicine is more than that.

The basic units of all living substances are organic molecules, and human beings are complex organic combinations composed of various functional polymers. Therefore, from the perspective of molecular design theory, different functional polymers used in biomedicine may be synthesized artificially. We believe more and more biomedical polymers will be applied in reality to improve the level of people's living.

Table 6.10 Common metal antibacterial materials [89,91,93–100].

Group	Antibacterial effect
Silver antibacterial material	Gram-negative bacteria > Gram-positive bacteria The inhibitory effect on yeast is weak The inhibitory effect on mold is the weakest
Copper antibacterial material	Gram-negative bacteria > Gram-positive bacteria The sterilization efficiency is more than 99%.
Zinc antibacterial material	*Escherichia coli* *Streptococcus* *Staphylococcus aureus* Broad spectrum sterilization
Titanium antibacterial material	*E. coli* *Pseudomonas aeruginosa* *S. aureus* *Aspergillus* Dental fungus *Salmonella*

Appendix A (Fig. A1 and Table A1)

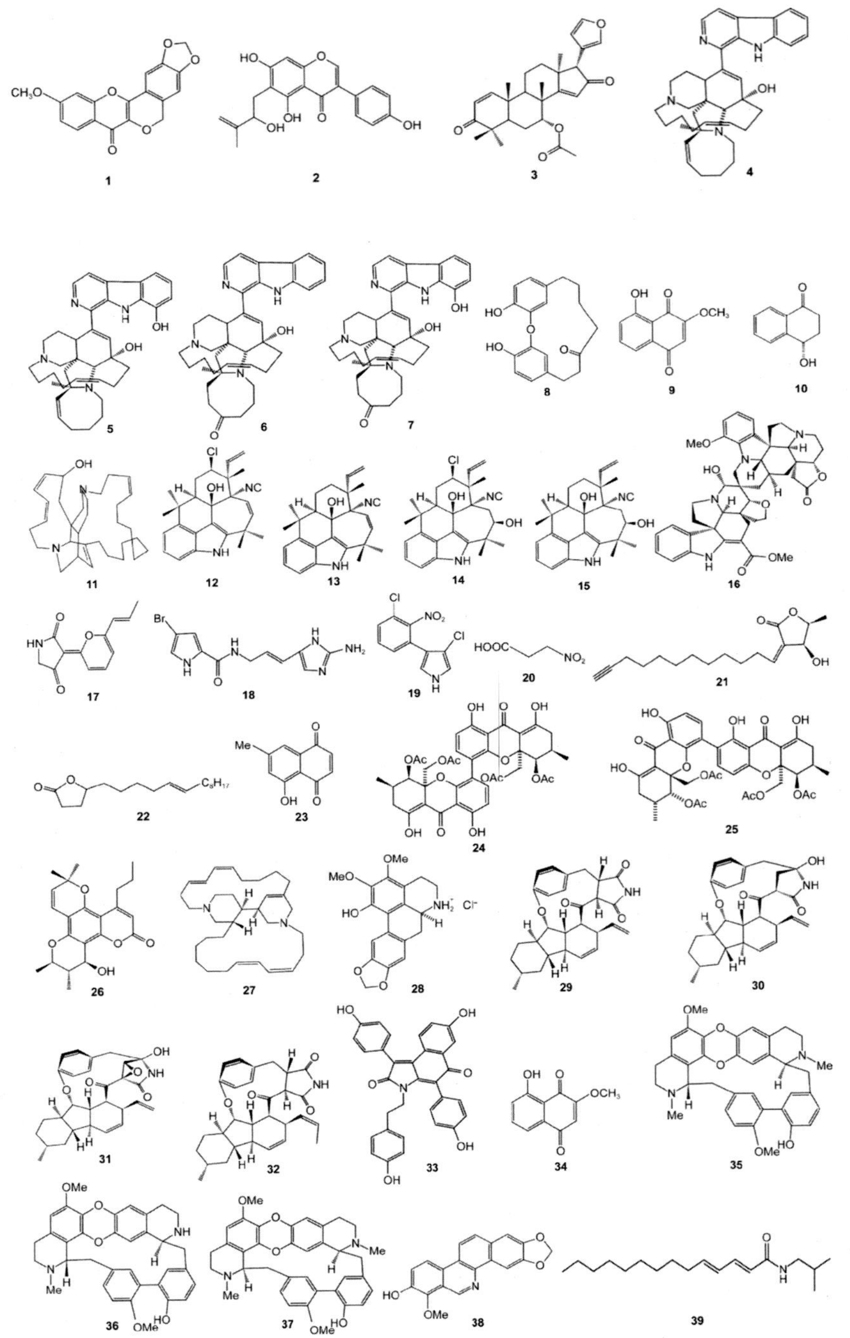

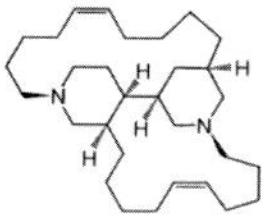
40

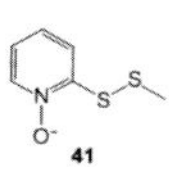
41

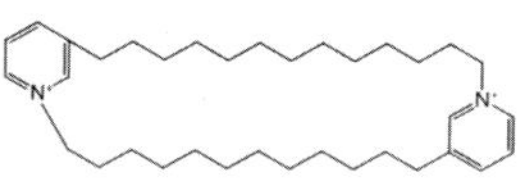
42

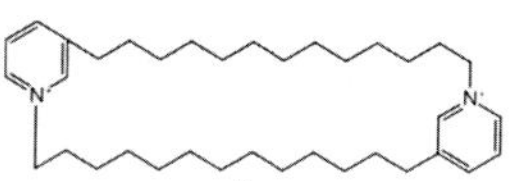
43

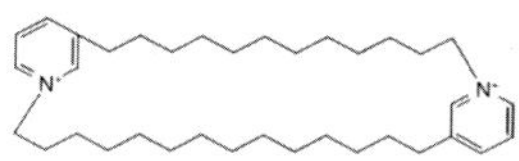
44

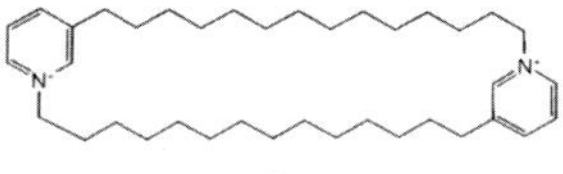
45

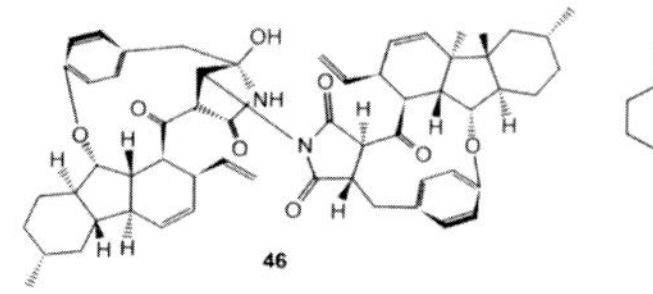
46

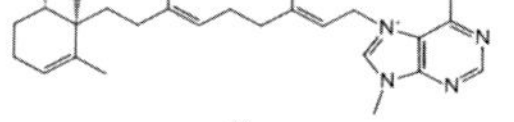
47

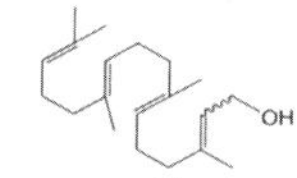
48

49

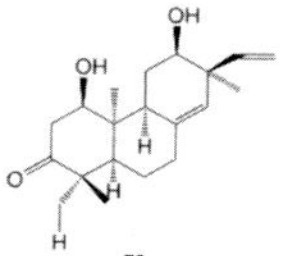
50

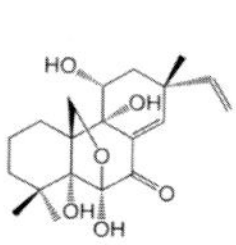
51

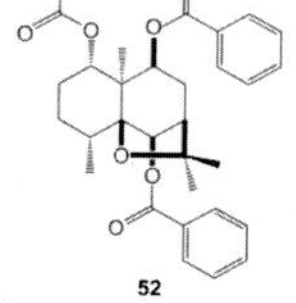
52

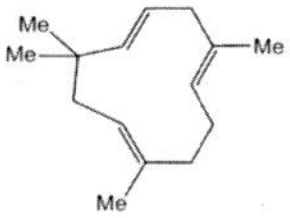
53

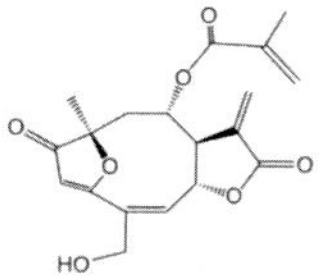
54

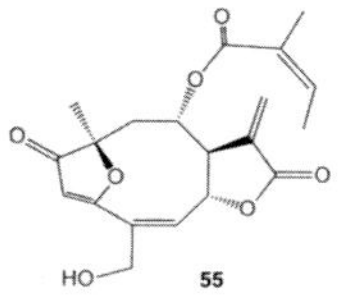
55

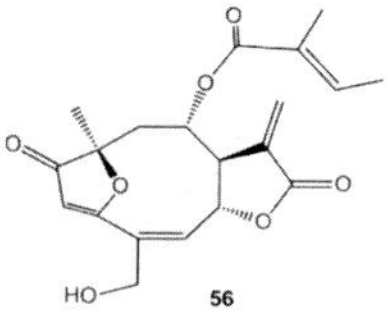
56

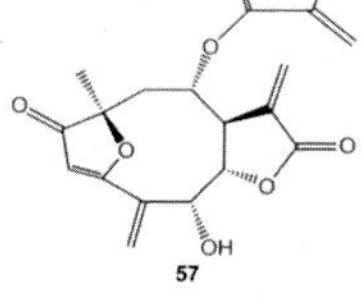
57

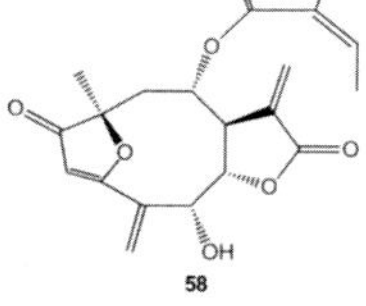
58

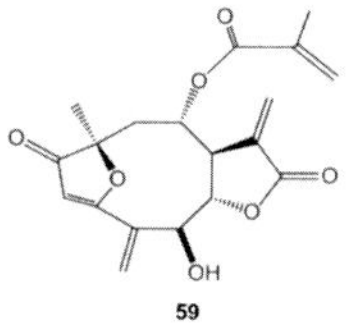
59

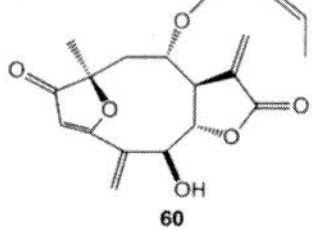
60

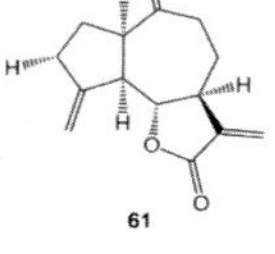
61

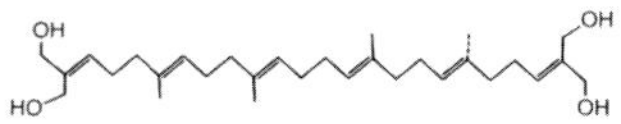
62

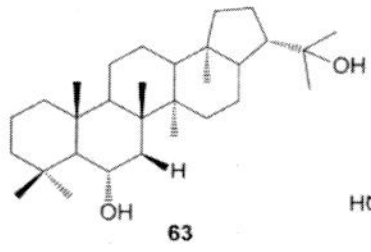
63

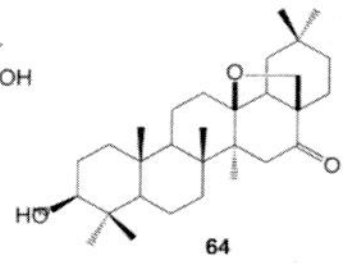
64

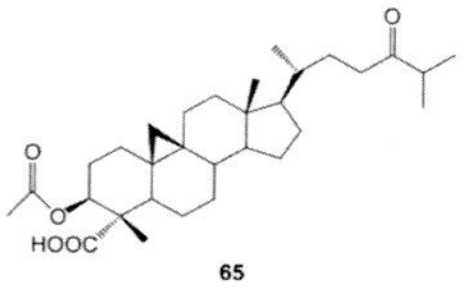
65

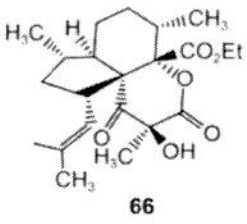
66

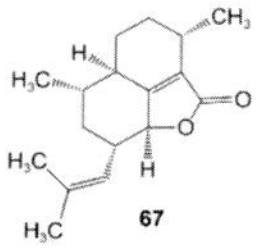
67

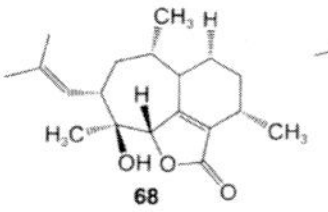
68

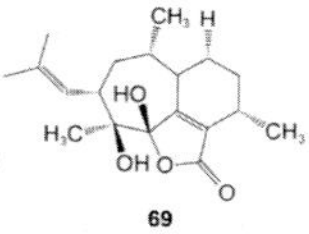
69

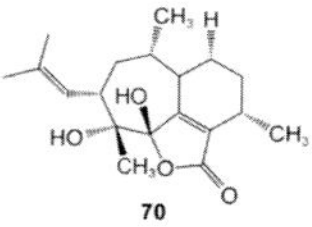
70

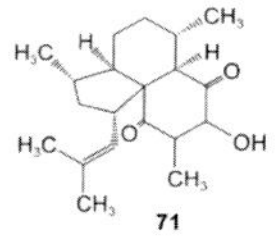
71

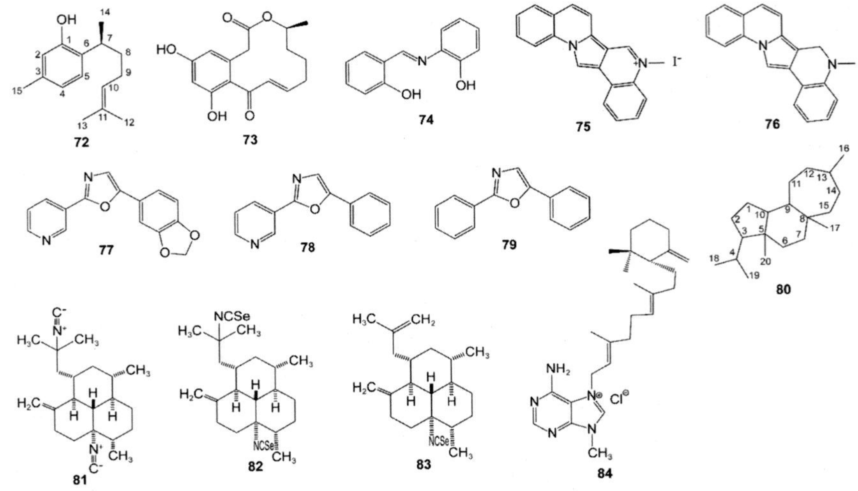

Figure A1 The structural formula of small molecules which MIC are below in 10 μg·mL^{-1}.

Table A1 MIC lower than 10 μ Small molecule name and structural formula of Gxm/L [37].

Compound no	Compound name	Source	Mic value/μg/mL	MBC value/μg/mL
1	3,4-methylenedi oxy-10-methoxy-7-oxo [3] benzopyrano[4,3-b] benzopyran	*Derris indica*	6.25	–
2	Laburnetin	*Ficus chlamydocarpa*	4.88	19.53
3	Azadiradione	*Chisocheton siamensis* (root)	6.25	–
4	manzamine A	*Indonesian sponge*	1.53	–
5	(+)-8-hydroxymanzamine A	*Indonesian sponge*	0.91	–
6	manzamine E	*Indonesian sponge*	3.76	–
7	manzamine F	*Indonesian sponge*	2.56	–
8	Engelhardione	*Engelhardia roxburghiana* (root)	0.2	–
9	3-methoxyjuglone	*E. roxburghiana*(root)	0.2	–
10	(-)-4-hydroxy-1-tetralone	*E. roxburghiana*(root)	4	–
11	ingenamine G	*Pachychalina* sp.	8	–
12	ambiguine K	*Fischerella ambigua*	2.8	–
13	ambiguine L	*F. ambigua*	4.5	–
14	ambiguine M	*F. ambigua*	3.3	–
15	ambiguine N	*F. ambigua*	10.9	–
16	Globospiramine	*Voacanga globosa*	4	–
17	Vermelhotin	(Unknown marine fungi)CRI247–01	3.1	–
18	Hymenidin	*Prosuberites laughlini*	6.1	–

(Continued)

Table A1 (Continued)

Compound no	Compound name	Source	Mic value/μg/mL	MBC value/μg/mL
19	Pyrrolnitrin	*Streptomyces pyrrocinia*	4	–
20	3-nitropropionic acid	*Phomopsis*	0.4	–
21	lincomolide B	*Cinnamomum kotoense*	2.8	–
22	Micromolide	*Micromelum hirsutum* (Stem bark)	1.5	–
23	7-methyljuglone	*Euclea natalensis*	0.5	–
24	phomoxanthone A	*Phomopsis*	0.5	–
25	phomoxanthone B	*Phomopsis*	6.25	–
26	(+)-calanolide A	*Calophyllum lanigerum*	3.13	–
27	halicyclamine A	*Haliclona* sp.	6.25	–
28	3-methoxynordomesticine hydrochloride	*Ourisia macrophylla*	4	–
29	hirsutellone A	*Hirsutella nivea* BCC 2594	0.78	–
30	hirsutellone B	*H. nivea* BCC 2594	0.78	–
31	hirsutellone C	*H. nivea* BCC 2594	0.78	–
32	hirsutellone D	*H. nivea* BCC 2594	3.125	–
33	denigrin C	*Dendrilla nigra*	4	–
34	3-methoxyjuglone	*Juglans cathayensis*(Root bark)	5	–
35	Tiliacorinine	*Tiliacora triandra*	6.2	–
36	2'-nortiliacorinine	*T. triandra*	3.1	–
37	Tiliacorine	*T. triandra*	3.1	–

38	Decarine	*Zanthoxylum capense*	1.6	–
39	N-isobutyl-(2E,4E)-2,4-tetradecadienamide	*Z. capense*	6.2	–
40	neopetrosiamine A	*Neopetrosia proxima*	7.5	–
41	2-(methyldithio)pyridine-N-oxide	*Allium stipitatum*	1.25	–
42	cyclostellettamine B	*Pachychalina* sp.	4	–
43	cyclostellettamine C	*Pachychalina* sp.	4	–
44	cyclostellettamine D	*Pachychalina* sp.	8	–
45	cyclostellettamine F	*Pachychalina* sp.	8	–
46	hirsutellone F	*H. nivea* BCC 2594	3.12	–
47	agelasine F	*Agelas* SP.	3.13	–
48	Geranylgeraniol	*Pedilanthus* tithymaloides	1.56	–
49	(E)-phytol	*Lucas volkensii Pourthiaea lucida*	2	–
50	lecheronol A	*Sapium haematospermum*	4	–
51	diaporthein B	*Diaporthe* sp.	3.1	–
52	1α-acetoxy-6β,9β-dibenzoyloxy-dihydro-β-agarofuran	*Celastrus vulcanicola*	6.2	–
53	α-humulene	*Polyalthia cerasoides*	6.25	–
54	Goyazensolide	*Camchaya calcarea*	3.1	–
55	Centratherin	*C. calcarea*	3.1	–
56	Lychnophorolide	*C. calcarea*	6.2	–
57	Isogoyazensolide	*C. calcarea*	1.5	–
58	Isocentratherin	*C. calcarea*	3.1	–
59	5-epi-isogoyazensolide	*C. calcarea*	3.1	–
60	5-epi-isocentratherin	*C. calcarea*	3.1	–

(*Continued*)

Table A1 (Continued)

Compound no	Compound name	Source	Mic value/μg/mL	MBC value/μg/mL
61	Dehydrocostuslactone	*Saussurea loppa*	2	–
62	Tetrahydroxysqualene	*Rhus taitensis*	10	–
63	Zeorin	*Sarmienta scandens*	8	–
64	Aegicerin	*Clavija procera*	3.1	–
65	bonianic acids B	*Radermachera boniana*	9.9	–
66	Elisabetholide	*Pseudopterogorgia elisabethae*	6.25	–
67	Amphilectolide	*P. elisabethae*	6.25	–
68	sandresolide A	*P. elisabethae*	6.25	–
69	sandresolide B	*P. elisabethae*	6.25	–
70	sandresolide C	*P. elisabethae*	6.25	–
71	elisabethin H	*P. elisabethae*	6.25	–

References

[1] Ma XX. Academician chenxuesi: preparation and application of biomedical polymer materials. High Tech and Industrialization 2020;294(11):23–5.
[2] Samand P, Sheila M, Frederik C. The tissue-engineered vascular graft-past, present, and future. Tissue Engineering. Part B, Reviews 2015;22(1):68–100.
[3] Wei Ao, Chen NL. Development of artificial blood vessels abroad. Industrial Textiles 2000;(12):1–4 16.
[4] Jia LX, Wang L, Ling K. Development course and direction of the artificial blood vessel. Shanghai Textile Technology 2003;3:52–3.
[5] Zhang FX, Zhou JJ, Zhang H. Development and the prospect of the vascular graft. Chinese Journal of Practical Surgery 2012;32(12):1049–51.
[6] Song FL, Hu DX. Research progress of artificial blood vessels. Chinese Journal of Modern Medicine 2013;23(2):65–70.
[7] Sun H, Yan YS. Chen, et al. Application of silk fibroin in small caliber artificial blood vessel. China Tissue Engineering Research 2014;18(16):2576–81.
[8] Huang FH, Sun LZ, Zheng J. Thermal crosslinking collagen coating of the woven polyester artificial blood vessel and its experimental study. Chinese Journal of Molecular Cardiology 2008;8(6):357–61.
[9] Colla G, Porto L. Development of artificial blood vessels through tissue engineering. BMC Proceedings 2014;8(4):45.
[10] Parodi JC, Palmaz JC, Barone HD. Transfemoral intraluminal graft implantation for abdominal aortic aneurysms. Annals of Vascular Surgery 1991;5(6):491–9.
[11] Chen XM, Wang ZH, Yu Jun, et al. Endovascular treatment of abdominal aortic aneurysm (report of 2 cases). Chinese Journal of General Surgery 1999;2:46–9.
[12] Liu K, Wang N, Wang W, et al. A bio-inspired high strength three-layer nanofiber vascular graft with structure guided cell growth. Journal of Materials Chemistry B 2017;20(5):3758–64.
[13] Famaey N, Verhoeven J, Jacobs S, et al. In situ evolution of the mechanical properties of stretchable and non-stretchable ePTFE vascular grafts and adjacent native vessels. The International Journal of Artificial Organs 2014;37(12):900–10.
[14] Shakelsetal, li CJ, Zou T, et al. Development status and prospect of artificial blood vessel. Industrial Textiles 2019;37(3):1–5.
[15] Ou Y, Chen X. Development and application of polymer artificial blood vessel. China tissue Engineering Research and Clinical Rehabilitation 2009;13(28):5401.
[16] Mercante LA, Scagion VP, Migliorini FL, et al. Electrospinning-based (bio)sensors for food and agricultural applications: a review. TrACTrends in Analytical Chemistry 2017;91:91–103.
[17] Miguel SP, Figueira DR, Simões D, et al. Electrospun polymeric nanofibres as wound dressings: a Review. Colloids and Surfaces B: Biointerfaces 2018;169:60–71.
[18] Ying SZ, Qi D, Hong JY, et al. Photocrosslinked maleilated chitosan/methacrylated poly (vinyl Alcohol) bicomponent nanofibrous scaffolds for use as potential wound dressings. Carbohydrate Polymers 2017;168:220–6.
[19] Li YM. Potential application of electrospun polycaprolactone composite fiber in vascular tissue engineering. Doctoral thesis, Jilin: Jilin University, 2019.
[20] Wang H, Feng Y, Behl M, et al. Hemocompatible polyurethane/gelatin-heparin nanofibrous scaffolds formed by a bi-layer electrospinning technique as potential artificial blood vessels. Frontiers of Chemical Science and Engineering 2011;5 (3):392–400.
[21] Wang HY. Construction and performance of electrospun polyurethane based small caliber artificial blood vessel. Doctoral thesis, Tianjin: Tianjin University, 2012.

[22] Zhang Y. Regulation of surface micro nano structure of corn gluten film based on electrospinning technology and its effect on cells. Master's thesis, Shanghai: Shanghai Jiaotong University, 2017.

[23] Yang J, Wang J, Fu L, et al. Electrospun Core−shell fibers for high-efficient composite cathode-based solid oxide fuel cells. Energy and Fuels 2021;35(2):1768−78.

[24] Zeng XW, Feng Q, Zhang YY, et al. Research progress of puerarin on inflammation related diseases. Chinese Pharmacological Bulletin 2018;34(1):8−11.

[25] Wei SY. Research progress of puerarin on cardiovascular protection and its mechanism. Chinese Journal of traditional Chinese medicine 2015;40(12):2278−84.

[26] Williamson MR, Black R, Kielty C. PCL−PU composite vascular scaffold production for vascular tissue engineering: attachment, proliferation and bioactivity of human vascular endothelial cells. Biomaterials 2006;27(19):3608−16.

[27] Li W. Double layer artificial blood vessel promotes vascular regeneration. Master's thesis, Tianjin: Nankai University, 2018.

[28] D KJ, D LC, Christopher B, et al. Artificial blood vessel: the holy grail of peripheral vascular surgery. Journal of Vascular Surgery 2005;41(2):349−54.

[29] Aper T, Wilhelmi M, Gebhardt C, et al. Novel method for the generation of tissue-engineered vascular grafts based on a highly compacted fibrin matrix. Acta Biomaterialia 2016;29:21−32.

[30] liu FZ, Liu MX, Wang YH, et al. Research progress on the application of 3D printing technology in the medical field. Progress in Materials in China 2016;35 (5):381−5.

[31] Han TL, Sun YD, Zhou J, et al. Research progress of 3D printing technology in vascular surgery. Chinese Journal of Vascular Surgery (Electronic Edition 2017;9 (4):301−5.

[32] Han LJ, sun Y, Yu XL, et al. Application status of 3D printing technology in cardiovascular diseases. Southwest National Defense Medicine 2019;29(12):1267−9.

[33] Liu BW, Chen Z, Wang S, et al. Application status and prospect of 3D printing technology in vascular surgery. Chinese Journal of Practical Surgery 2016;36(3):341−3.

[34] Liu Y, Ke X, Yu L, et al. Composite bioabsorbable vascular stents via 3d bioprinting and electrospinning for treating stenotic vessels. Journal of Southeast University (English Edition) 2015;31(2):254−8.

[35] Koch A, Mizrahi V. Mycobacterium tuberculosis. Trends in Microbiology 2018;26 (6):555−6.

[36] Yin CY, Leng DL, He ZG. Research progress of anti tuberculosis drugs. Journal of Shenyang Pharmaceutical University 2015;32(1):77−84.

[37] Wang RF, An TT, Jiang CH, et al. Research progress of natural small molecules against *Mycobacterium tuberculosis*. Chinese Journal of New Drugs 2020;29 (20):2331−40.

[38] Hua JC, Li Z, Gong JX, et al. Progress in preparation, properties and applications of hydrogel fibers. New Chemical Materials 2015;43(10):16−18 21.

[39] Li MF, Zhang T, Sun W, et al. Study on the forming process of tissue engineering hydrogel fiber based on sodium alginate. Aerospace Medicine and Medical Engineering 2012;25(4):291−5.

[40] Milena I, Nevena M, Nadya M, et al. Electrospun non-woven nanofibrous hybrid mats based on chitosan and pla for wound-dressing applications. Macromolecular Bioscience 2009;9(1):102−11.

[41] Shen XY. Preparation, structure and properties of uhmw-pan based pH responsive porous hollow gel fibers. Doctoral thesis, Shanghai: Donghua University, 2001.

[42] High performance fibers and other synthetic fibers. Chemical Fiber Abstract 2014;43 (6):21−4.

[43] He ZQ. Preparation of PVA/Sericin hybrid fiber by gel spinning and its structure and physical properties. Foreign Silk 2001;37(5):14–18.
[44] Yuan YS, Chen L, Zhao YP, et al. pva-g-pnipa temperature sensitive gel fiber. Journal of Tianjin University of Technology 2012;31(4):6–9.
[45] Lin H, Peng C, Wu W. Fibrous hydrogel scaffolds with cells embedded in the fibers as a potential tissue scaffold for skin repair. Journal of Materials Science: Materials in Medicine 2014;25(1):259–69.
[46] Zhou RJ. Research on hydrogel fiber manufacturing and biomedical application based on electrospinning process. Doctoral thesis, Zhejiang: Zhejiang University, 2019.
[47] Pedde R. Emerging biofabrication strategies for engineering complex tissue constructs. Advanced Materials 2017;29(19). Available from: https://doi.org/10.1002/adma.201606061.
[48] Onoe H, Okitsu T, Itou A, et al. Metre-long cell-laden microfibres exhibit tissue morphologies and functions. Nature Materials 2013;12(6):584–90.
[49] Bi L. Experimental study on tuberculosis of bone and joint. Master's thesis, Xian: The Fourth Military Medical University, 2007.
[50] Fei ZQ, Hu YY. Anti bone tuberculosis sustained release drugs: selection of carrier materials. China Tissue Engineering Research 2015;19(21):3387–91.
[51] Li DW. Study on local long-acting drug sustained-release composites for the treatment of bone tuberculosis. Master's thesis, Guangzhou: Southern Medical University, 2016.
[52] Chen HM. Study on slow and controlled release characteristics of coaxial drug loaded tissue engineering bone scaffold. Doctoral thesis, Xinjiang: Xinjiang University, 2020.
[53] Li K. Construction and characteristics of drug loaded anti tuberculosis bone scaffold based on three-dimensional printing technology. Master's thesis, Shanghai: The Second Military Medical University, 2016.
[54] Miao GH. Preparation and properties of mesoporous bioactive glass for drug and gene carrier research. Master's thesis, Guangzhou: South China University of technology, 2014.
[55] Zhang ZY, Li YB, Qi TN, et al. 3D printed polylactic acid resin humerus combined with bioactive coating promotes osteoblast adhesion and antibacterial ability. China Tissue Engineering Research 2020;24(16):2485–92.
[56] Wang F, Zhou H, Guo YC, et al. Biocompatibility and osteogenic activity of nano hydroxyapatite/chitosan/polylactide scaffold in vitro. China Tissue Engineering Research 2014;18(8):1198–204.
[57] Joseph C, Danielle M. Essential role of wnt3a-mediated activation of mitogen-activated protein kinase P38 for the stimulation of alkaline phosphatase activity and matrix mineralization in $C_3H_{10}t^{1/2}$ mesenchymal cells. Endocrinology 2007;148(11):5323–30.
[58] Liu YQ, Li CX, Chen J, et al. Electrospun high bioavailable rifampicin–isoniazid-polyvinylpyrrolidone fiber membranes. Applied Nanoscience 2021;11(8):2271–80.
[59] Zhen D, Panfei H, Bi L. Gatifloxacin-1,2,3-triazole-isatin hybrids and their antimycobacterial activities. Archiv Der Pharmazie Chemistry in Life Science 2019;352(10):1900135.
[60] Chen MH, Chen JH, Jia K, et al. Comparative study on three methods of preparing rifampicin and isoniazid containing antituberculosis tetralogy tablets. Journal of Practical Medicine 2017;34(7):641–4.
[61] Lang GZ, Li CJ, Gaohu TY, et al. Bioactive flavonoid dimers from Chinese dragon blood, the red resin of *Dracaena cochinchinensis*. Bioorganic Chemistry 2020;97:103659.
[62] Zhang L, Wang XP, Huang XW, et al. Research progress on chemical constituents and pharmacological effects of dragon blood and dragon blood. China Modern Applied Pharmacy: Journal of Pharmacy Education and Practice 2019;36(20):2605–11.

[63] Xiu SS, Yong KL, Chen LL, et al. Hypoglycemic effect and mechanism of dracaena draconis supercritical extract. Natural Products Research and Development 2005;6:766–8.
[64] Dias J, Granja P, Bártolo P. Advances in electrospun skin substitutes. Progress in Materials Science 2016;84:314–34.
[65] Lima TDPDL, Passos MF. Skin wounds, the healing process, and hydrogel-based wound dressings: a short review. Journal of Biomaterials Science 2021;32(14):1910–25.
[66] Medical gauze dressing. <http://www.jlfulang.com/product-view-4-27.html>; Aug 6, 2022 (accessed 3.6.20).
[67] Biosynthesis of a carboxymethylated cellulose complex for wound dressing, <https://zhuanlan.zhihu.com/p/37695918>; June 5, 2018 (accessed 3.6.20).
[68] Homaeigohar S, Boccaccini AR. Antibacterial biohybrid nanofibers for wound dressings. Acta Biomaterialia 2020;107:25–49.
[69] Cao K. Preparation of drug loaded fibers by electrospinning and their drug release properties. Master's thesis, Beijing: Beijing University of chemical technology, 2018.
[70] Cao K, Liu Y. Preparation of dragon blood wound dressing by electrospinning. In: Summary of the Fourth China Electrospinning Conference (CICE 2016); 2016, p. 17.
[71] Zhong BX, Liu WZ, Qin Ana, et al. Historical evolution of external plaster. Journal of Changchun University of Traditional Chinese Medicine 2011;27(1):134–7.
[72] Liu WZ, Ana Qin, Zhong BX, et al. Research progress of traditional plaster preparation technology. Journal of Central University for Nationalities (Natural Science Edition) 2010;19(4):73–6.
[73] Tang LP, Zhu CX, Zhang Y, et al. Development prospect of Chinese traditional plaster in the 21st century. Shi Zhen, Guoyi, Guoyao 2001;2:155–6.
[74] Xu Y, Li K, Liu Y, et al. Black plaster composite fiber prepared by upward electrospinning. Journal of Applied Polymer Science 2019;136(24):47662.
[75] Li YX, Feng XD. Controlled drug release system and its mechanism. Polymer bulletin 1991;1:19–27.
[76] Leganés J, Rodríguez A, Arranz M, et al. magnetically responsive hydrophobic pockets for on–off drug release. Materials Today Chemistry 2022;23:100702.
[77] Werzer O, Tumphart S, Keimel R, et al. Drug release from thin films encapsulated by a temperature-responsive hydrogel. Soft Matter 2019;15(8):1853–9.
[78] Jia J, Wang C, Chen K, et al. Drug release of yolk/shell microcapsule controlled by Ph-responsive yolk swelling. Chemical Engineering Journal 2017;327:953–61.
[79] Tian W, Fan XD, Chen WX, et al. Research progress of polymer carriers for controlled drug release. Polymer Materials Science and Engineering 2006;4:19–23.
[80] Yan GF. Preparation and application of soybean protein isolate composite polymer gel material. Master's thesis, Lanzhou: Northwest Normal University, 2017.
[81] Han J. Studies on chitosan and its derivatives as drug controlled release carriers and scaffolds. Doctoral thesis, Beijing: Beijing University of Chemical Technology, 2011.
[82] Zuwala K, Riber CF, Løvschall KB, et al. Macromolecular prodrugs of ribavirin: polymer backbone defines blood safety, drug release, and efficacy of anti-inflammatory effects. Journal of Controlled Release 2018;275:53–66.
[83] Polo fonseca L, Trinca RB, Felisberti MI. Amphiphilic polyurethane hydrogels as smart carriers for acidic hydrophobic drugs. International Journal of Pharmaceutics 2018;546(1):106–14.
[84] Zihao Z, Jianguo Z, Qing C, et al. Temperature and photo dual-stimuli responsive block copolymer self-assembly micelles for cellular controlled drug release. Macromolecular Bioscience 2020;21(3):2000291.
[85] Ma H, Chen G, Zhang J, et al. Facile fabrication of core-shell polyelectrolyte complexes nanofibers based on electric field induced phase separation. Polymer 2016;110:80–6.

[86] Ali AQ, Kannan TP, Ahmad A, et al. In vitro genotoxicity tests for polyhydroxy butyrate – a synthetic biomaterial. Toxicology In Vitro 2007;22(1):57–67.
[87] K C, Y L, A OA, et al. Plla-phb fiber membranes obtained by solvent-free electrospinning for short-time drug delivery. Drug Delivery and Translational Research 2018;8(1):291–302.
[88] Iordanskii A, Karpova S, Olkhov A, et al. Structure-morphology impact upon segmental dynamics and diffusion in the biodegradable ultrafine fibers of polyhydroxy butyrate-polylactide blends. European Polymer Journal 2019;117:208–16.
[89] Le ZW, Ling XL, Yue XX. Research status and development trend of antibacterial materials. Journal of Chengdu Textile College 2016;33(2):58–66.
[90] Zhang KH, Lin SB, Tan SZ. Research status and development trend of organic antibacterial agents. Paint Industry 2005;63(5):45–9.
[91] Guo Y. Application and development of antibacterial materials. Journal of Chengdu Textile College 2017;34(1):206–9.
[92] Liu W, Liu Y, Yang JH, et al. An antibacterial and antiviral nanofiber membrane and its preparation method and application. China Patent, CN-112709013-A; 2020.
[93] Lv Z, R KF, Yang H, et al. Synthesis and antibacterial activity of nano ZnO with different morphologies. Journal of Wuhan University (Science Edition) 2013;59 (1):47–50.
[94] Nan L, Liu YQ, Yang WC, et al. Study on antibacterial properties of antibacterial stainless steel containing copper. Journal of Metals 2007;10:1065–70.
[95] Song MM, Zhang HY, Zhang W, et al. Study on the inhibitory effect of silver antibacterial materials on food contaminating bacteria. Food Research and Development 2009;30(6):43–7.
[96] Yu YN, Yao TM, Meng SY, et al. Research progress of medical metal inorganic antibacterial materials. Journal of Clinical Military Medicine 2017;45(7):768–70.
[97] Lei L, Shang J, Dong JH, et al. Antibacterial properties of silver nanoparticles and application of antibacterial modification on implant surface. Chinese Journal of Bone and Joint Surgery 2013;6(5):465–9.
[98] Liu F, Zhang Z, Chen FY, et al. Preparation and antibacterial properties of nano silver antibacterial materials. Chinese Journal of Disinfection 2012;29(12):1063–5.
[99] Xu RK. Research progress of nano antibacterial materials. Liaoning Chemical Industry 2013;42(4):371–3 388.
[100] Wang J, Shui H, Ji ZJ, et al. Research progress of silver inorganic antibacterial materials. Material Guide 2013;27(17):59–64 78.

Index

Note: Page numbers followed by "*f*" and "*t*" refer to figures and tables, respectively.

R

V

W

Y

Z

Printed in the United States
by Baker & Taylor Publisher Services